Gauging What's Real

Gauge theories have provided our most successful representations of the fundamental forces of nature. How, though, do such representations work? Interpretations of gauge theory aim to answer this question. Through understanding how a gauge theory's representations work, we are able to say what kind of world our gauge theories reveal to us.

A gauge theory's representations are mathematical structures. These may be transformed among themselves while certain features remain the same. Do the representations related by such a gauge transformation merely offer alternative ways of representing the very same situation? If so, then gauge symmetry is a purely formal property since it reflects no corresponding symmetry in nature.

Gauging What's Real describes the representations provided by gauge theories in both classical and quantum physics. Richard Healey defends the thesis that gauge' transformations are purely formal symmetries of almost all the classes of representations provided by each of our theories of fundamental forces. He argues that evidence for classical gauge theories of forces (other than gravity) gives us reason to believe that loops rather than points are the locations of fundamental properties. In addition to exploring the prospects of extending this conclusion to the quantum gauge theories of the Standard Model of elementary particle physics, Healey assesses the difficulties faced by attempts to base such ontological conclusions on the success of these theories.

Richard Healey is Professor of Philosophy at the University of Arizona.

Gauging What's Real

The Conceptual Foundations of Contemporary Gauge Theories

Richard Healey

OXFORD
UNIVERSITY PRESS

Great Clarendon Street, Oxford OX2 6DP

Oxford University Press is a department of the University of Oxford.
It furthers the University's objective of excellence in research, scholarship,
and education by publishing worldwide in

Oxford New York

Auckland Cape Town Dar es Salaam Hong Kong Karachi
Kuala Lumpur Madrid Melbourne Mexico City Nairobi
New Delhi Shanghai Taipei Toronto

With offices in

Argentina Austria Brazil Chile Czech Republic France Greece
Guatemala Hungary Italy Japan Poland Portugal Singapore
South Korea Switzerland Thailand Turkey Ukraine Vietnam

Oxford is a registered trade mark of Oxford University Press
in the UK and in certain other countries

Published in the United States
by Oxford University Press Inc., New York

© Richard Healey 2007

The moral rights of the author have been asserted
Database right Oxford University Press (maker)

First published 2007
First published in paperback 2009

All rights reserved. No part of this publication may be reproduced,
stored in a retrieval system, or transmitted, in any form or by any means,
without the prior permission in writing of Oxford University Press,
or as expressly permitted by law, or under terms agreed with the appropriate
reprographics rights organization. Enquiries concerning reproduction
outside the scope of the above should be sent to the Rights Department,
Oxford University Press, at the address above

You must not circulate this book in any other binding or cover
and you must impose the same condition on any acquirer

British Library Cataloguing in Publication Data

Data available

Library of Congress Cataloging in Publication Data

Data available

Typeset by Laserwords Private Limited, Chennai, India
Printed in the UK
by
MPG Books Group

ISBN 978–0–19–928796–3 (Hbk.)
   978–0–19–957693–7 (Pbk.)

10 9 8 7 6 5 4 3 2 1

Contents

Preface	ix
Acknowledgements	xiii
Introduction	xv
1. What is a gauge theory?	1
1.1 Classical electromagnetism: a paradigm gauge theory	3
1.2 A fiber bundle formulation	7
1.2.1 Electromagnetic interactions of quantum particles	14
1.2.2 Electromagnetic interactions of matter fields	18
2. The Aharonov–Bohm effect	21
2.1 Fiber bundles	26
2.2 A gauge-invariant, local explanation?	31
2.3 Geometry and topology in the Aharonov–Bohm effect	40
2.4 Locality in the Aharonov–Bohm effect	44
2.5 Lessons for classical electromagnetism	54
3. Classical gauge theories	58
3.1 Non-Abelian Yang–Mills theories	58
3.1.1 The fiber bundle formulation	64
3.1.2 Loops, groups, and hoops	70
3.1.3 Topological issues	74
3.2 A fiber bundle formulation of general relativity	77
3.2.1 A gravitational analog to the Aharonov–Bohm effect	78
4. Interpreting classical gauge theories	82
4.1 The no gauge potential properties view	83
4.2 The localized gauge potential properties view	85
4.2.1 Problems defining theoretical terms	91
4.2.2 Leeds's view	99
4.2.3 Maudlin's interpretation	102
4.3 The non-localized gauge potential properties view	104
4.4 A holonomy interpretation	111
4.4.1 Epistemological considerations	112
4.4.2 Objections considered	119
4.4.3 Semantic considerations	122
4.5 Metaphysical implications: non-separability and holism	123

vi CONTENTS

5. Quantized Yang–Mills gauge theories   129
   - 5.1 How to quantize a classical field   131
   - 5.2 Coulomb gauge quantization   133
   - 5.3 Lorenz gauge quantization   135
   - 5.4 Classical electromagnetism as a constrained Hamiltonian system   136
   - 5.5 The free Maxwell field as a Hamiltonian system   139
   - 5.6 Path-integral quantization   141
   - 5.7 Canonical quantization of non-Abelian fields   143
   - 5.8 Path-integral quantization of non-Abelian fields   145
   - 5.9 Interacting fields in the Lagrangian formulation   146

6. The empirical import of gauge symmetry   149
   - 6.1 Two kinds of symmetry   150
   - 6.2 Observing gauge symmetry?   155
   - 6.3 The gauge argument   159
   - 6.4 Ghost fields   167
   - 6.5 Spontaneous symmetry-breaking   169
   - 6.6 The θ-vacuum   175
   - 6.7 Anomalies   182

7. Loop representations   184
   - 7.1 The significance of loop representations   185
   - 7.2 Loop representations of the free Maxwell field   186
   - 7.3 Loop representations of other free Yang–Mills fields   192
   - 7.4 Interacting fields in loop representations   195
   - 7.5 The θ-vacuum in a loop representation   197
   - 7.6 Conclusion   198

8. Interpreting quantized Yang–Mills gauge theories   200
   - 8.1 Auyang's event ontology   200
   - 8.2 Problems of interpreting a quantum field theory   203
     - 8.2.1 Particle interpretations   205
     - 8.2.2 Bohmian interpretations   209
     - 8.2.3 Copenhagen interpretations   212
     - 8.2.4 Everettian interpretations   215
     - 8.2.5 Modal interpretations   218

9. Conclusions   220

A. Electromagnetism and its generalizations   229

B. Fiber bundles   233

C. The constrained Hamiltonian formalism   248

D.	Alternative quantum representations	257
E.	Algebraic quantum field theory	265
F.	Interpretations of quantum mechanics	272
	F.1 The Copenhagen interpretation	272
	F.2 Bohmian mechanics	274
	F.3 Everettian interpretations	276
	F.4 Modal interpretations	278

Bibliography   280
Index   287

Preface

The quest for the deep structure of the world, begun by the pre-Socratic philosophers some two and a half thousand years ago, was pursued with great success by physicists during the 20th century. The revolutions wrought by relativity and quantum theory in the first quarter century created the scaffolding around which theories of the fundamental forces of nature were then constructed. After some mid-century struggles, by the end of the century these theories had been consolidated into two "packages." Gravity has now been successfully described up to cosmological distances by Einstein's general theory of relativity, while what has come to be known as the Standard Model has been successfully confirmed in its predictions of the behavior of the so-called elementary particles and their interactions down to the tiny subnuclear distance scales explored by high energy accelerators. While physicists keep trying to combine these into a single overarching theory, now is an appropriate time to pause and take stock of what has already been achieved.

By "taking stock," I don't mean offering a detailed historical account lauding scientists and their achievements, though the history is fascinating and there is plenty of praise to spread around. Nor do I mean popularizing contemporary theories of fundamental forces to engage the interest of the casual reader by sub-stituting metaphor for faithful exposition and anecdote for critical analysis. What is needed is a careful logical and philosophical examination of the conceptual structure and broader implications of contemporary theories of fundamental forces like that which followed the rise of relativity and quantum theory.

Logical positivists like Rudolf Carnap and Moritz Schlick, along with many other philosophers including Bertrand Russell and (especially) Hans Reichenbach, eagerly undertook such an examination of the physical theories of their day. Karl Popper acknowledged Einstein's relativity as a major stimulus to the development of his own critical realist view of scientific investigation, while rejecting what he saw as the instrumentalism infecting Bohr and Heisenberg's understanding of quantum theory. The philosophy of physics has since become a flourishing branch of the philosophy of science, raising the level of foundational discussion of relativity and quantum theory within this small interdisciplinary community of philosophers, physicists, and mathematicians. But science has moved on, and there are as yet few (if any) sustained investigations into the conceptual foundations of the theories of fundamental forces that emerged as an important culmination of twentieth century physics.

I offer this book as both a contribution and a stimulus to the study of the conceptual foundations of gauge theories of fundamental interactions. I have changed my mind and corrected my own misunderstandings often enough while working toward it to realize quite clearly that the book does not represent the last word—or even my last word—on this topic. If I have conveyed a sense of its intellectual interest and importance, while provoking some reader to do a better job, then I will be satisfied.

To whom is this book addressed? To my fellow philosophers of physics, certainly—but I set out to reach a far wider class of readers. There is much food for thought here for other philosophers; not only for philosophers of science, but also for general epistemologists and metaphysicians. We have here a paradigm of reliable, objective knowledge of matters very distantly related to human sensory experience. But what are the scope and limits of this knowledge, and by what processes and what reasoning have we come by it? Indeed, how exactly should we understand the content of theoretically mediated knowledge claims in this domain? Do fundamental forces act at a distance, and if not how (and on what) do they act? Do traditional distinctions between matter and force even make sense in these theories, in the absence of any clear ontology of localized particles or fields?

I write also for students and practitioners of physics and other sciences. The distinction between philosophy and science may be convenient for librarians and educational administrators, but it marks no true intellectual boundary. Physics is a source of both novel concepts and intractable conceptual problems. Understanding these concepts and dealing with these problems is the common concern of the physicist and the philosopher. Each can learn from the other, and both can learn from the mathematician. Professional commitments are only a potential hindrance to what is best viewed as an essentially collaborative enterprise.

Students of physics typically have little time to reflect on the foundations of their discipline, and even find themselves discouraged from such reflections by their teachers. I hope some who, like me, were initially attracted to physics by the revolutionary insights it promised into the nature of reality will find a way to sneak a peek at this book. Experts in this field of theoretical physics will doubtless find things in this book with which to disagree. I encourage such disagreement on foundational questions as both inevitable and desirable among scholars. We can all learn from it. I hope other intellectually curious scientists and mathematicians will find in this book a useful introduction to some basic concepts of gauge theories, even if they decide to skip over some of its more philosophical parts.

Finally, this book is for anyone who is attracted to relatively serious popularizations of physics but repelled by their superficiality and fuzzy thinking. It is addressed to a reader who finds textbooks narrowly technical, reluctant to address basic questions about how our theories portray our world, or unconvincing in their analyses and arguments as to why we should understand them one way rather than another.

What distinguishes the class of gauge theories is a certain kind of abstract symmetry in their mathematical formulation. Since the correct interpretation of this gauge symmetry is key to understanding gauge theories, there is a fair amount of mathematics in this book. In writing it, I have had in mind a reader who is familiar with the basic concepts of calculus and linear algebra, and willing to be introduced to some novel mathematical ideas relevant to this field. I have tried to introduce such ideas gradually, with the aid of diagrams, starting in chapter 1. A more systematic development appears in a number of appendices (especially A, B, C), which the reader is advised to consult if and when (s)he feels the need to do so.[1] While the mathematics is necessary for a full comprehension, I encourage the reader not to get bogged down in technical details, since the main ideas and conclusions of the book are accessible without them. A reader who is impatient to learn what these are may wish to skip chapters 3, 5, and 7 and the later sections of chapter 6, at least on a first reading.

The book divides naturally into two acts. The first four chapters are concerned with classical gauge theories, while the last four focus on quantum gauge theories. While quantum mechanics makes only a cameo appearance in the first act, in the second it plays a starring role. I have added three brief appendices (D, E, F) for the benefit of readers unfamiliar with quantum mechanics and its conceptual problems, and chapter 5 is included, with a rapid introduction to quantum field theory, even though it contains material that may be very familiar to readers with a physics or mathematics background. The main philosophical arguments and conclusions of the book will be found in even-numbered chapters—itself a fact with no philosophical significance.

I have been thinking about gauge theories for a long time now—over ten years—and during that time I have incurred a variety of intellectual, personal, and financial debts, at least some of which I would like to acknowledge here and repay (metaphorically, for the financial ones!) elsewhere. I would like to thank the participants in a workshop on gauge theories in Tucson in the spring of 1998, and specifically Gordon Belot, Steven Leeds, Paul Teller, and Chuang Liu also for their stimulating publications and conversations; Harvey Brown for his friendship and support in Oxford and elsewhere, as well as for introducing me to the important work of Jeeva Anandan; my present and former colleagues at the University of Arizona, especially Jenann Ismael, Shaughan Lavine, and David Chalmers, for many helpful conversations; to Jeff Barrett and David Malament at UC Irvine, where I talked on "Gauge Potentials: Physical Reality or Mathematical Fiction?" in October 2000, after giving a talk with the same title the previous June, at the quadrennial meeting of Visiting Fellows of the Center for Philosophy of Science at the University

---

[1] To supplement appendix B, which presupposes some familiarity with differential geometry, the reader may wish to consult a text such as (Nakahara, 1990).

xii PREFACE

of Pittsburgh, in San Carlos de Bariloche, Argentina; Guido Bacciagaluppi for inviting me to participate in a workshop in Berkeley in May 2002, and again for the invitation to speak at the University of Freiburg in July, 2003; and Huw Price, both for an invitation to talk at a workshop he organized in Sydney, Australia in March 2004, and also for arranging invitations on my behalf to talk at the Australian National University and the University of Queensland; the Centre for Philosophy of Natural and Social Science at the London School of Economics, where I was a Visiting Fellow in the first half of 2003; Michael Redhead and other Sigma Club members for inviting me to talk to them in London in May 2003; the Oxford reading group in philosophy of physics, who first heard some of my early ideas on loop representations in February 2003; the organizers of the New Directions in the Foundations of Physics conference in Maryland in May 2005, for an invitation to talk on loop representations and the significance of gauge symmetry; Shufang Su for her lectures that woke me from my dogmatic slumbers to the mysteries of large gauge transformations and the θ-vacuum, Ian Aitchison for listening to my half-baked ideas on how to solve them, and Domenico Giulini, for patiently explaining why these ideas needed more oven time; J. B. Kennedy and James Mattingly for the stimulus of their different views on how to understand the Aharonov–Bohm effect; Tim Maudlin for his tenacious but still unsuccessful attempts to set me straight on the localized reality behind gauge potentials and its metaphysical significance; various anonymous referees of my earlier papers on the topics of this book for the constructive criticisms and the suggestions pointing me to important work including that by Gambini and Pullin (1996) and Gribov (1977); readers of earlier versions of the present work, including my colleagues William Faris and Kurt Just, for catching errors and offering helpful advice (only some of which I have taken, at my peril!); the members of two seminars I taught at the University of Arizona, who not only suffered through my early attempts to get my ideas straight but also contributed greatly to that process by their spirited challenges; and to Janet Berge for turning my primitive electronic diagrams into works of art. Much of the research that led up to this book was funded by the National Science Foundation, including my participation in the conference on the Conceptual Foundations of Quantum Field Theory held at Boston University in March 1996. In particular, this material is based upon work supported by the National Science Foundation under Grant No. 0216918: I gratefully acknowledge their support. Last but not least, thanks to Julie for putting up with me while I worked on the manuscript.

Acknowledgements

Several chapters of this book include material I have adapted from my previous publications.
Passages in chapter 2, as well as a version of figure 2.1, first appeared in "Nonlocality and the Aharonov–Bohm Effect," ©1997 by *Philosophy of Science*.
Material from "On the Reality of Gauge Potentials," ©2001 by *Philosophy of Science*, appears in several places. These include a version of figure 1.5 as well as passages from chapters 2 and 4.
I thank Elsevier for permission to reprint passages from two of my publications, as follows: "Nonseparable Processes and Causal Explanation", *Studies in History and Philosophy of Science 25*, pages 337–74, ©1994, with permission from Elsevier; and "Gauge Theories and Holisms," *Studies in History and Philosophy of Science Part B: Studies in History and Philosophy of Modern Physics 35*, pages 643–66, ©2004, with permission from Elsevier. I draw on the latter especially in chapter 3, section 3.2.
Passages in chapter 4, section 4.2.1 first appeared in "Symmetry and the Scope of Scientific Realism," chapter 7 of W. Demopoulos and I. Pitowsky (eds.) *Physical Theory and its Interpretation: Essays in Honor of Jeffrey Bub*, pages 143–60 ©2006 Springer, with kind permission of Springer Science and Business Media.

Introduction

What are gauge theories, and why care about their conceptual foundations? For now, think of the term 'gauge' as simply labeling, rather than describing, a prominent class of physical theories. Gauge theories are at the heart of contemporary physics, arguably our surest guide to the basic structure of the world. Anyone interested in learning about what our world is like at the deepest level cannot afford to ignore this guide. But what kind of world do our gauge theories reveal to us? That is the key question for any analysis of the conceptual foundations of contemporary gauge theories. Vague as it is, the attempt to clarify and answer it is the main aim of this book, though many other fascinating issues arise along the way. The question would be answered by an interpretation of each of our gauge theories—an account of what the world is like if that theory is true. A slightly more liberal understanding of the interpretative task has been recommended by one prominent contemporary philosopher, Bas Van Fraassen, when introducing his 1991 book on quantum mechanics (p. 4):

When we come to a specific theory, the question: *how could the world possibly be the way this theory says it is?* concerns the content alone. This is the foundational question *par excellence*, and it makes equal sense to realist and empiricist alike.

Van Fraassen writes as a constructive empiricist—one who takes the goals of science to be met by construction of theories that correctly describe all observable matters, whether or not they are right about what lies behind them. Accordingly, he is a pluralist about interpretations, maintaining that though our understanding of a theory is enhanced by providing multiple interpretations, it is neither possible nor necessary effectively to address the question as to what the theory is really telling us about the underlying structure of the world. He takes that to be not a scientific but a metaphysical question, best avoided by philosophy as well as science. My approach will also be guided by empiricist scruples intended to guard against acceptance of unwarranted, or even meaningless, claims allegedly implied by scientific theories, and by gauge theories in particular. But, unlike van Fraassen, I take empiricist principles to be sufficiently powerful to discriminate among alternative interpretations of a physical theory. Accordingly, the key interpretative question addressed here is the following: *What beliefs about the world are (or would be) warranted by the empirical success of this (gauge) theory?*

A convincing answer to this question must proceed in several stages. It will be based on a detailed study of the mathematical structure of particular gauge theories, laying bare the key elements whose representative status needs to be ascertained. This clears the way for the presentation of alternative interpretations of various gauge theories, differing primarily in their accounts of which elements of mathematical structure should be accorded physical significance, and what that implies for a world truly described by the theory. The implications are remarkable: prominent gauge theories turn out to portray a world with surprising and counterintuitive features. Some interpretations describe it as radically indeterministic, others as grossly non-local, yet others as significantly holistic. In each case it will be important to assess the metaphysical implications, and to consider whether they show we could never be warranted in believing our world is like that. But even if metaphysical objections can be overcome, there remains the basic epistemological question as to which, if any, interpretation is (or would be) best supported by the empirical success of the theory in question.

A gauge theory addresses its subject matter via a representation with redundant elements: in its standard formulation, there are more variables than the independent physical magnitudes whose behavior it seeks to describe. But this representation is not unique—the theory offers (an infinite set of) distinct representations of ostensibly the same physical situation. These representations are related by *gauge symmetries*—mathematical transformations from one representation to another. Structures that are preserved by a gauge symmetry are said to be *gauge invariant*. The significance of gauge invariance and the interpretation of gauge symmetry are key issues in the conceptual foundations of gauge theories. On the one hand, any structure that fails to be gauge invariant is often dismissed as "mere gauge"—an artefact of the representation—by contrast with gauge-invariant structures, which the theory uses to represent elements of physical reality. On this view, gauge symmetry is a purely formal feature of a gauge theory's representational framework with no physical significance. But some have thought of a gauge symmetry as a transformation with physical content, analogous to moving an entire system three feet to the north, or setting it in uniform motion. This view lies behind the so-called gauge argument, which seeks to derive, or at least suggest, the existence of interactions of specific kinds from a requirement of gauge symmetry. Moreover, the usual explanation of many of the more exotic phenomena described by contemporary gauge theories (such as instantons and the θ-vacuum, ghost fields, and spontaneous symmetry breaking) apparently proceeds in a particular gauge, which raises the question as to whether the theory portrays this phenomenon as real, or as a mere artefact of that choice of gauge with no physical import.

It is a unifying principle of this book that since gauge symmetry is indeed a purely formal requirement, no physical consequence of a gauge theory can depend on a choice of gauge. A single-minded adherence to this principle will unmask a variety of mathematical elements as mere "surplus structure," while motivating a gauge-invariant account of any genuine physical phenomena. But gauge invariance is not yet gauge independence: if the phenomena themselves do not depend on an arbitrary choice of gauge, why do our theories even need to mention gauge in accounting for them? Here there is an analogy with contemporary textbook formulations of the general theory of relativity and other space-time theories in the language of modern differential geometry. Such formulations are often said to be "coordinate free," while earlier formulations were merely coordinate invariant. Since a choice of space-time coordinates is purely arbitrary, no physical consequence of general relativity can depend on such a choice. Older formulations of the theory were always in terms of some arbitrary coordinate system, while contemporary formulations in terms of geometric objects, including tensor fields, on differentiable manifolds make no explicit mention of coordinates.

If gauge symmetry is a purely formal requirement, then choice of gauge in a gauge theory is analogous to choice of a space-time coordinate system. Is it possible to formulate a gauge theory without mentioning gauges? For many gauge theories, this turns out to be possible: such theories are not merely gauge symmetric, but demonstrably gauge free. Gauges may be introduced to simplify calculations, but they need not appear even implicitly in the foundational principles of the theory. This contrasts with "coordinate free" formulations of space-time theories like general relativity. For there is still an implicit reference to coordinate systems in such modern treatments, in as much as a differentiable manifold is defined in terms of an atlas of coordinate charts. It is true that a principle of gauge symmetry has undoubtedly played an important heuristic role in the development of contemporary gauge theories, while such a principle cannot even be stated in a "gauge-free" formulation of such a theory. But the elimination of gauge turns out to have important implications for the interpretation of theories arrived at using this heuristic. Gauge symmetry may be a ladder that can be kicked away after it has been climbed.

This proves to be true for some gauge theories but not for others, or so I shall argue. The argument depends on whether the theory can be reformulated without loss of predictive power so as to eliminate any localized structure corresponding to a gauge potential associated with a given field. But predictive power is not the only theoretical virtue at issue here. Even where such an empirically equivalent reformulation is available, it does not follow that it is preferable. For one may still claim that the original formulation offers a more satisfactory explanation of the common empirical content. In that case, inference to the best explanation might warrant the extension of belief to a localized structure corresponding to a gauge potential after all. Gauge symmetry would then be reinstated as a significant requirement—not on this structure, but rather on how the theory represents it. So it is that further pursuit of the argument for the elimination of gauge leads into a broader arena of philosophical discussion.

Those claiming that locally defined gauge potentials ground a superior explanation of the predictive success of a gauge theory emphasize what they consider the unpalatable brute facts that must be swallowed to account for the phenomena without them. A gauge-free account appeals to structures that are not only intricately inter-related but also metaphysically problematic. In the simplest case (classical electromagnetism interacting with quantum particles) such an account ascribes properties to (or on) a loop of empty space that are not fixed by properties of anything located at points around the loop, while the properties ascribed to distinct loops are not independent, but satisfy a set of simple relations. There is nothing like this elsewhere in physics. Moreover, the account violates a key metaphysical tenet of at least one prominent philosopher, David Lewis.[1]

An argument for a gauge-free interpretation of a gauge theory therefore needs philosophical work. It must defend an unconventional metaphysics against both bad arguments and inchoate distaste. And it must outline and advocate a modestly realist epistemology that endorses some inferences to theoretical structure behind the phenomena while rejecting others as unwarranted by the evidence.

The first gauge theory acknowledged by physics describes a classical field—the electromagnetic field. For physicists of the early 20th century, prior to the rise of quantum mechanics, there was little temptation to take gauge seriously by inferring the existence of any localized structure corresponding to the potentials of electromagnetism. But things changed when quantum mechanics accounted for the behavior of charged particles directly in terms of these potentials. This predicted surprising phenomena like the Aharonov–Bohm effect, to be described in chapter 2. Their subsequent experimental observation seemed to warrant an inference to the existence of such localized structure as offering the best explanation of the observed phenomena.

The main goal of the first part of this book (chapters 1–4) is to argue that this is a bad inference: in fact the observed phenomena provide some support for a contrary conclusion—that the gauge potentials of a classical theory like electromagnetism are best understood as representing *non*-localized structure. The philosophical interest here derives not only from the metaphysical implications of the conclusion, but also from the semantic as well as epistemological lessons to be learnt by analyzing the structure of the argument for that conclusion.

But subsequent advances in physics have weakened the argument in an interesting way. On the one hand, quantum mechanical predictions of Aharonov–Bohm-type phenomena continue to be verified in increasingly sophisticated and convincing experiments. But on the other hand, the classical gauge theories that account for these phenomena (specifically, classical electromagnetism) are no longer believed to be empirically adequate because they cannot account for phenomena in other domains. Experimental studies

---

[1] Specifically, it violates Lewis's (1986) principle of Humean supervenience, while conforming to his (1983) patchwork principle.

of interactions at very high energies have provided a wealth of data in support of theories comprising what is now known as the Standard Model. These theories are all gauge theories, but, unlike classical electromagnetism, they are *quantum*, not classical, field theories. Classical electromagnetism is inadequate to account for all experimental observations, including those at high energies. It has therefore been superseded, first by quantum electrodynamics, and then again by the unified electroweak theory incorporated into the Standard Model.

The second part of the book explores the status of gauge symmetry and gauge potentials in quantized gauge theories such as those of the Standard Model. The main point here is that gauge symmetry remains a purely formal symmetry of the models of a conventional formulation of a quantized gauge field theory. After introducing these theories in chapter 5, in chapter 6 I first clarify what I mean by a purely formal symmetry, and then go on to defend this point against possible objections. If it is correct, then it becomes interesting to explore the possibility of reformulating quantized gauge field theories by eliminating gauge from their models. Loop representations of quantized gauge field theories hold out just that prospect: they are introduced in chapter 7. Chapter 8 then begins the quest for an interpretation of quantized gauge field theories like those of the Standard Model. Imposing barriers block that quest. After more than 80 years, there is still no agreement on how to understand *any* quantum theory, let alone quantum field theories, which pose formidable additional problems of interpretation. Any conclusions must therefore remain tentative: I give mine in the final chapter. We have much still to learn about our world from contemporary gauge theories. This book is presented as a rallying cry for others to join in and further the quest.

1
What is a gauge theory?

Fundamental physics lays claim to describe the basic structure of our world; support for this claim rests on the empirical evidence for its theories. According to contemporary physics, every interaction is of one or more of four basic types, corresponding to the strong, weak, electromagnetic, and gravitational forces. The strong interaction is described by a successful theory, quantum chromodynamics. Another successful theory portrays the weak and electromagnetic interactions as different aspects of a unified electroweak interaction. Many predictions of these theories have been severely tested and verified in sophisticated laboratory experiments. Together, they constitute the theoretical underpinnings of the so-called Standard Model of elementary particles—the culmination of the 20th century's attempts to understand the nature and behavior of the basic constituents of the world, and a foundation for current attempts to deepen that understanding. The gravitational interaction is currently best described by Einstein's general theory of relativity. In recent years an increasing number of its predictions have been confirmed to high accuracy by astronomical observations.

While all these theories have rather similar structures, the theories incorporated in the Standard Model resemble each other more closely than each resembles general relativity. Most obviously, the former are quantum theories, while the latter is classical. But a second point of contrast will figure importantly in what follows. One way of saying what all these theories have in common is to call them all gauge theories, and this is correct on one understanding of what constitutes a gauge theory. But on a narrower understanding, the term 'gauge theory' is correctly applied only to quantum chromodynamics and unified electroweak theory, while general relativity does not count as a gauge theory. Although 'gauge theory' is a term of art that may be used in whatever way is most appropriate in a given context, reasons may be offered for using it one way rather than another. Those favoring a particular usage of the term have generally done so as part of their advocacy of a preferred formal framework for the presentation of theories of a kind they found particularly interesting.

Two such frameworks figure prominently: fiber bundles, and constrained Hamiltonian systems. Each is flexible enough to permit the formulation of all the theories with which we are concerned; and the two formulations complement one another so that a full understanding requires both. But it is not just a historical accident that a particular theory is generally formulated one way rather than another.

The gauge theories of the Standard Model are known as Yang–Mills theories to honor the contributions of two physicists, Yang and Mills (1954) to our understanding of theories of this type. Such theories have a natural and elegant formulation within the fiber bundle framework. The present chapter introduces fiber bundles in a way that presupposes no prior acquaintance with them. A fuller mathematical exposition is provided in appendix B. The exercise of formulating general relativity within the fiber bundle framework[1] makes it clear just how it differs from Yang–Mills theories. These differences bear significantly on how the theories should be interpreted, or so I shall argue in chapter 4.

General relativity is a classical field theory, while the Yang–Mills theories of the Standard Model are quantum field theories. The usual approach to formulating a quantum theory is to "quantize" a prior classical theory, and one important technique for quantizing a classical theory requires that the classical theory first be taken to describe a class of what are called Hamiltonian systems. When the second half of the book focuses on quantized Yang–Mills theories, it will become important to know what a Hamiltonian system is, and what makes such a system constrained. So an introduction to constrained Hamiltonian systems will be postponed until chapter 5. Appendix C provides a fuller mathematical exposition.

It is a historical accident that the term 'gauge' is applied to our theories. To my knowledge, the term originated as a translation of the German word 'eich', which first appeared in this context in a paper by Herman Weyl (1918). In that paper, published immediately after Einstein's general theory of relativity, Weyl proposed a unified theory of gravity and electromagnetism that proved to be empirically inadequate. The fundamental magnitudes of the theory were the space-time metric and the electromagnetic potential. The theory was invariant under a joint transformation of these quantities involving a linear transformation of the metric at each space-time location that could be viewed as a (location-dependent) change of scale for measuring lengths and durations. Such a transformation was naturally called a "local" gauge transformation in so far as a change of length scale at each location is equivalent to a change in the gauge with respect to which lengths are compared there. Since this was a *joint* transformation, its effect on the electromagnetic potential also came to be known as a gauge transformation, even though this had no natural interpretation in terms of changes of length scales.

In his theory, Weyl attempted to extend Einstein's idea that tiny spatial and temporal distances may vary with location in space and time. He allowed them also to depend on the path by which such a location is reached: technically,

---

[1] See chapter 3, section 3.2.

this meant replacing a (semi-)Riemannian metric by a metric that was non-integrable. Weyl subsequently abandoned this idea of a non-Riemannian metric, but, after the development of quantum mechanics, Weyl (1929) instead paired the *same* transformation of the electromagnetic potential with a location-dependent (and again non-integrable) phase transformation of the wave-function so as to arrive at a joint transformation of just these magnitudes. Noting that Maxwell's equations for electromagnetism and Dirac's quantum-mechanical wave equation were invariant under this new joint transformation, he persisted in referring to that property of the equations as gauge invariance, even though it was no longer associated with any change of length scale.

Others conformed to Weyl's usage, and generalized it in two stages. In the first stage, any (possibly) location-dependent transformation of a theory involving the potential for some field which leaves that field and the rest of the theory invariant came to be called a gauge transformation. More generally still, an *arbitrary* smooth transformation of variables at each point in (space-)time that preserves the physical content of a theory came to be known as a gauge transformation, whether or not it involves the potential for some field that is left invariant. A gauge theory in the most general sense is then any theory whose physical content is preserved by such a (possibly) location-dependent transformation of variables. Of course, application of the term then crucially depends on determining the physical content of the theory, which may or may not be identified with its empirical or observational content.

When one inquires into the formalism of gauge theories, it is helpful to bear in mind a simple example of such a theory. This can be used both to motivate and to illustrate the general mathematical structures involved. Classical electromagnetic theory provides perhaps the simplest example of a successful gauge theory: it is described in the next section.

1.1 Classical electromagnetism: a paradigm gauge theory

Consider a battery of the sort you find at the local hardware store. The markings on a typical AA battery state that it supplies electricity at 1.5 volts. This means that there is a potential difference of 1.5 volts between the terminals of the battery—at least when the battery is in standard conditions. In such conditions, if each terminal is connected by a copper wire to a conducting plate and the plates are placed 1 centimeter apart, then the battery will give rise to an electric field of 1.5 volts per centimeter between these plates. But what will the potential then be at each plate? Will the positive terminal be at a potential of $+1.5$ volts, and the negative at zero; the positive be at zero and the negative at -1.5 volts; or what? If these are silly questions, as they seem to be, it is because they have a false presupposition: that there is such a thing as the absolute value of the electric potential at a place, over and above the various potential differences between that place and others. Rejecting that presupposition, one is free to assign any real number to a location as its electric potential in volts, provided one then respects the actual potential differences between that place and others when going on to assign them electric potentials also. Different such assignments would be related by a simple transformation in the electric potential, consisting of the addition of the same real number to the potential at each point. Following Weyl's lead, it has become customary to call this a gauge transformation. But it will prove useful to adopt the more specific term *potential transformation* to refer to a transformation that adds a value of some magnitude (such as a real number) to a potential at each spatial (or, more generally, spatiotemporal) location. A change from one description of the battery to an entirely equivalent description is made by performing a potential transformation that adds the *same* real number to the electric potential everywhere.

It is important to note that it is an empirical question whether or not electric potentials have any absolute significance. Faraday noted this in the days prior to Maxwell's seminal formulation of his equations of classical electromagnetic theory. He considered it an important enough question to warrant an elaborate and potentially dangerous experimental investigation. Accordingly, Faraday constructed a hollow cube with sides 12 feet long, covered it with good conducting materials but insulated it carefully from the ground, and electrified it so that it was at a large potential difference from the rest of his laboratory. As he himself put it (Maxwell 1881, p. 53),

I went into this cube and lived in it, but though I used lighted candles, electrometers, and all other tests of electrical states, I could not find the least influence on them.

However careful and extensive the experimental investigations of Faraday and others, these could not, of course, definitively establish that electric potentials have no absolute significance. But this was a consequence of the highly successful classical electromagnetic theory devised by Maxwell and his successors. According to Maxwell's theory, the electric field and all other measurable magnitudes are invariant under a transformation that adds the same real number to the value of the electric potential everywhere. This theory displayed a similar, and even more extensive, indeterminateness in a second potential associated with magnetic fields. Unlike the electric potential, this has a direction as well as a magnitude, whose value at a point is therefore represented by a spatial vector known as its magnetic vector potential. The magnetic field $\mathbf{B}(\mathbf{x})$ at point $\mathbf{x}$ is related to this vector potential $\mathbf{A}(\mathbf{x})$ by the equation

$$\mathbf{B} = \nabla \times \mathbf{A} \qquad (1.1)$$

Here $\boldsymbol{\nabla}\times$ is the differential operator also known as *curl*, such that the components of $\boldsymbol{\nabla}\times\mathbf{A}$ are $(\partial A_z/\partial y - \partial A_y/\partial z, \partial A_x/\partial z - \partial A_z/\partial x, \partial A_y/\partial x - \partial A_x/\partial y)$ in a rectangular Cartesian coordinate system, where the coordinates of $\mathbf{x}$ are (x, y, z). Thus while the electric potential φ at $\mathbf{x}$ is represented by a real number—the value of a scalar field $\varphi(\mathbf{x})$—the magnetic potential is represented by a vector field $\mathbf{A}(\mathbf{x})$.

Not only are all measurable magnitudes of Maxwell's theory invariant under a transformation that adds the same vector to the vector representing the magnetic vector potential everywhere, but the following transformation in the magnetic vector potential also leaves the magnetic field given by 1.1 unchanged:

$$\mathbf{A} \to \mathbf{A} - \boldsymbol{\nabla}\Lambda \qquad (1.2)$$

where $\Lambda(\mathbf{x})$ is any (suitably differentiable) scalar field, and $\boldsymbol{\nabla}\Lambda$ is a vector representing its gradient, with components $(\partial\Lambda/\partial x, \partial\Lambda/\partial y, \partial\Lambda/\partial z)$. I shall refer to a transformation that adds an appropriate magnitude (real number or vector) to a potential at each spatial (or, more generally, spatiotemporal) location in such a way that the value of this magnitude varies smoothly from point to point as a *variable potential transformation*. All measurable magnitudes of Maxwell's theory are invariant not only when one adds the same vector to the magnetic vector potential everywhere, but also under a variable potential transformation in the magnetic potential of the form 1.2.

To allow for changing electric and magnetic fields, the potentials should be written instead as $\varphi(x)$, $\mathbf{A}(x)$ respectively, where x is short for (t, x, y, z)—the temporal and spatial coordinates of the instant and point at which the potential is evaluated. As shown in Appendix A, Maxwell's equations governing the behavior of electric and magnetic fields are invariant under the simultaneous potential transformations 1.2 and

$$\varphi \to \varphi + \partial\Lambda/\partial t \qquad (1.3)$$

where $\Lambda(x)$ is also now regarded as a function of time as well as spatial position, and the electric field $\mathbf{E}$ is related to the potentials by

$$\mathbf{E} = -\boldsymbol{\nabla}\varphi - \partial\mathbf{A}/\partial t \qquad (1.4)$$

As it is usually understood, classical electromagnetic theory implies that the electromagnetic history of a world is fully specified by the values of electric and magnetic fields everywhere at each moment, along with the associated charge and current densities. (After the advent of the theory of relativity, it was recognized that these values would not be invariant, varying instead from one reference frame to another in a specific manner.) The fields manifest themselves by inducing forces on electrically charged bodies which affect their accelerations. The force on a particle of charge e traveling with velocity $\mathbf{v}$ is given by the Lorentz force law

$$\mathbf{F} = e(\mathbf{E} + \mathbf{v} \times \mathbf{B}) \qquad (1.5)$$

in which the vector product $\mathbf{v} \times \mathbf{B}$ has components $(v_y B_z - v_z B_y, v_z B_x - v_x B_z, v_x B_y - v_y B_x)$. It follows that the content of classical electromagnetic theory is invariant under the potential transformations 1.2, 1.3.

These may be reexpressed by combining them into the variable potential transformations

$$A_\mu(x) \rightarrow A_\mu(x) + \partial_\mu \Lambda(x) \qquad (1.6)$$

As the index μ ranges over $0, 1, 2, 3$, the left-hand side of this equation yields the time and three spatial components of the *four-vector electromagnetic potential* $A_\mu = (\varphi, -\mathbf{A})$; while on the right-hand side, ∂_μ varies over its components $(\partial/\partial t, \partial/\partial x, \partial/\partial y, \partial/\partial z)$. The advantage of expressing variable potential transformations this way is that it makes manifest the way the potentials transform from one inertial reference frame to another, in accordance with the Lorentz transformations of the special theory of relativity. A_μ is called a four-vector because it transforms the same way as the four space and time coordinates (t, x, y, z) of a point event under Lorentz transformations.

A *gauge transformation* defined by $\Lambda(x)$ incorporates the potential transformation 1.6 (or the equivalent pair 1.2, 1.3); in classical physics this is all it includes. In this notation, the relations 1.1, 1.4 between the electromagnetic field and its potential may also be reexpressed as

$$F_{\mu\nu} = \partial_\mu A_\nu - \partial_\nu A_\mu \qquad (1.7)$$

Here $F_{\mu\nu}$ is a tensor whose components are related to the electric and magnetic fields in a particular frame as follows:

$$F_{\mu\nu} = \begin{pmatrix} 0 & E_x & E_y & E_z \\ -E_x & 0 & -B_z & B_y \\ -E_y & B_z & 0 & -B_x \\ -E_z & -B_y & B_x & 0 \end{pmatrix} \qquad (1.8)$$

Again, reexpressing the electric and magnetic fields in this way makes manifest their transformation properties under Lorentz transformations: they transform precisely so as to make 1.7 true in every inertial frame. An equation like 1.7 that takes the same form in every such frame is said to be *relativistically covariant*. (Note that units for distance and time have been chosen so that c, the speed of light, is equal to 1. This convention simplifies many equations, and it is followed throughout this book.)

1.2 A fiber bundle formulation

Despite its great empirical success, the theory described in the previous section is not empirically adequate. It cannot describe the structure, or account for the stability, of atoms; and it is incapable of correctly predicting many aspects of the observed behavior of charged particles like electrons, especially at high energies.

The stability of atoms, as well as many of their quantitative features, became clear after the mechanics governing their constituents was taken to be not classical but quantum. Quantum mechanics accounted for a great many aspects of the structure and behavior of atoms on the assumption that the electrons in an atom are negatively charged particles, subject to each other's electromagnetic field as well as that due to the much heavier, positively charged nucleus. At first, this field was itself still described classically by Maxwell's theory. But its action on the electrons could no longer be understood on the basis of the Lorentz force law 1.5. Instead, the effects of electromagnetism were represented by terms in the Hamiltonian operator that figures in the Schrödinger equation—the fundamental dynamic law the new quantum mechanics put in place of Newton's second law of motion. This equation possessed so-called stationary solutions—wave-functions for atomic electrons representing stable states of the atom (see appendix A). Transitions among these states induced by an external electromagnetic field could also be accounted for within this framework, permitting a detailed understanding of many observed features of atomic spectra and many other things.

But ever since Einstein's analysis of the photoelectric effect in terms of the exchange of discrete quanta of electromagnetic energy (subsequently called photons), it had been clear that classical electromagnetic theory could not account for all the properties of light or other kinds of electromagnetic radiation. And the exact electromagnetic properties of electrons, the finer details of atomic structure, and the high-energy behavior of charged particles were also inexplicable on the basis of classical electromagnetism, even after classical mechanics was replaced by quantum mechanics. Further progress required that electromagnetism itself be described by a quantum, rather than a classical, theory—a theory in which Maxwell's equations still figured, but now interpreted as governing operator-valued quantum fields rather than number-valued classical fields. Electrons themselves came to be regarded as the quanta of a distinct quantum field, so that their electromagnetic interactions were now treated by a quantum theory of interacting fields—quantum electrodynamics (QED). This theory established its empirical credentials by making some spectacularly accurate predictions, and served as a paradigm for the construction of the gauge field theories that currently underlie the Standard Model.

Certain features of the structure of electromagnetic theory were preserved throughout the course of this evolution. In particular, in each reincarnation, that theory displayed gauge invariance. Hence an examination of the nature and representation of the gauge invariance of the simple theory of classical electromagnetism laid out in the previous section will remain directly relevant to an analysis of the conceptual foundations of other versions of electromagnetic theory as well as other theories such as the gauge field theories underlying the Standard Model.

So far, the gauge invariance of classical electromagnetic theory has been taken to follow from the fact that the variable potential transformations 1.2, 1.3, 1.6 are *symmetries* of the theory, in the sense that the result of applying such a transformation to a model of the theory is itself a model of the theory satisfying the same fundamental equations—Maxwell's equations A.2 and the Lorentz force law 1.5. If this symmetry is a purely formal feature of classical electromagnetism's representative framework, then models related by such a transformation represent the very same physical situation; indeed, this is how the gauge symmetry of classical electromagnetism has generally been understood. On this understanding, even though electric, magnetic, or electromagnetic potentials appear in models as mathematical fields on a space-time manifold M, these are mere "surplus structure" and do not purport to represent any distribution of physical magnitudes over space-time: models with the same electromagnetic fields but different potentials represent the same state of affairs.

The theory of classical electromagnetism may be generalized in various ways. The space-time manifold within which electromagnetic phenomena occur may be allowed to have a geometry and even topology different from that of Newtonian or Minkowski space-time (the space-time of Einstein's special relativity). Maxwell's equations may be modified to allow for the possibility of magnetic monopoles—magnetic analogs to isolated electric charges; or they may be more radically changed so as to describe non-electromagnetic interactions, as first suggested by Yang and Mills (1954). It turns out that many such generalizations are facilitated if classical electromagnetism is first reformulated in a rather different mathematical framework in which an electromagnetic potential appears, not as a field on a space-time manifold M, but rather as what is called a *connection* on a *fiber bundle* with M as *base space*. In this framework, a gauge symmetry is a transformation of fiber bundle structures rather than fields on M. The fiber bundle formulation has led to new results in mathematics as well as making possible a deeper insight into the structure of gauge theories. Hence, following the work of Wu and Yang (1975) as well as Trautman (1980), it has become increasingly common to formulate electromagnetism and other gauge theories using the mathematics of fiber bundles. Indeed, Trautman considers the availability of such a formulation to be definitive of a gauge theory. He writes (p. 306)

For me, a gauge theory is any physical theory of a dynamic variable which, at the classical level, may be identified with a connection on a principal bundle. The structure group G of the bundle P is the group of gauge transformations of the first kind; the group $\mathcal{G}$ of gauge transformations of the second kind may be identified with a subgroup of the group $AutP$ of all automorphisms of P.

While this is an admirably concise statement of one important general understanding of what a gauge theory is, its application to particular theories will take some spelling out. The first step is to apply Trautman's understanding to classical electromagnetic theory. In this theory, it will turn out that an (electro)magnetic potential may be represented by the connection on a principal fiber bundle whose base space is a manifold representing space (-time). The bundle curvature represents the (electro)magnetic field. Gauge transformations turn out to be represented by various group operations on elements of the fiber bundle structure. Fiber bundles are perhaps best thought of in geometric terms, and that is how they will be described here. Appendix B complements this with a more complete algebraic treatment in the language of differential geometry.

The term 'fiber bundle' makes one think of a bundle of fibers, and this is not a bad image to start with. So think of a piece of fabric to which threads are attached at each point, as depicted in figure 1.1.

The threads illustrate the fibers, and the fabric to which they are attached illustrates the base space, of a fiber bundle. All the fibers resemble each other, and each has the same shape as a typical fiber that is not attached to the fabric at all. If the fabric in figure 1.1 is extended into a long strip, its ends may be joined together so that it forms a band that may or may not have a twist in it. An abstract fiber bundle has the structure of a (possibly) twisted product of two spaces—a base space and a typical fiber. It consists of a space (a manifold), each patch (neighborhood) of which has the structure of a (direct) product of two other spaces (also manifolds). These "patches" are smoothly joined up to form

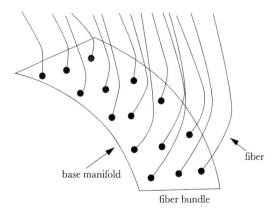

Figure 1.1. **A fiber bundle**

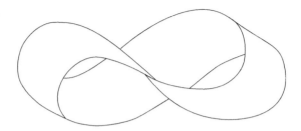

Figure 1.2. A Möbius strip

the *total space* E of the bundle: if this is itself a product, there is no "twist," and the bundle is said to be trivial. To get an idea of the difference between trivial and non-trivial bundles, it is helpful to bear in mind the contrast between a cylinder and a Möbius strip—the surface formed by rotating one end of a rectangle through two right angles out of the plane of the rectangle, and then gluing it to the other end to form a closed, twisted ribbon. Each is *locally* like the product of two spaces: a circle S and a line segment L with labeled end-points. But while one can smoothly map images of L onto "vertical" lines in the side of a cylinder all the way around, any map of images of L onto "vertical" lines in the side of a Möbius strip must incorporate discontinuities where arbitrarily close points on its edge are images of distinct end-points of L. For while the entire cylinder has two edges, the Möbius strip has only one—its "base" coincides with its "top."

A *fiber bundle* has a manifold M as *base space*, and a *projection map* $\pi : E \rightarrow M$ that associates each point m of M with points in E constituting the *fiber above* m, F_m: each F_m has the structure of another manifold, the *typical fiber* F. If U_m is a suitably small patch (open set) of M containing m, then each smooth map $\chi_m : \pi^{-1}(U_m) \rightarrow U_m \times F$ exhibits the local product structure of the neighborhood of a point $u \in F_m$: such a map is called a *local trivialization* of the bundle. The bundle is *trivial* if and only if some local trivialization χ_m can be extended into a smooth map $\chi: E \rightarrow M \times F$ on the whole manifold M. For $U \subseteq M$, a smooth map $\sigma: U \rightarrow E$ satisfying $\pi \circ \sigma = I$ (the identity map on U) is called a *section* of the bundle: this is *global* if $U = M$, it is *local* if $U \subset M$. A section picks out a privileged element of the fiber above each point and so may be thought to define a common origin or "baseline" for all these fibers.

By a simple extension, the cylinder and Möbius strip can be turned into examples of two sorts of fiber bundle that figure prominently in the fiber bundle formulation of electromagnetism—vector bundles and principal bundles. This is performed by extending the line segment L so that the typical fiber F becomes the infinite line of real numbers $\mathbb{R}$, and giving this the structure of a real vector space: one can think of a vector $\vec{x}$ in this space as an arrow starting from 0 and with tip at $x \in \mathbb{R}$. Any linear mapping of $\mathbb{R}$ onto itself will preserve this structure, and the set of all such mappings forms the group GL(1) of all linear transformations of a one-dimensional vector space. One can construct a fiber bundle E over S whose typical fiber is the vector space $\mathbb{R}$ by first placing a copy $\mathbb{R}_p$ of $\mathbb{R}$ "over" each point p of S,[2] and then requiring that p lie on an arc (open set) $U_p \subset S$, the patch of E "over" which has the same structure as the product of U_p with $\mathbb{R}$.[3] One way to do this results in an infinite cylinder standing on S, and a different way yields an infinitely "tall" Möbius strip.

Since the typical fiber here is a vector space, each of these is an example of a *vector bundle* with total space E, projection map π onto base space S, typical fiber $\mathbb{R}$, and structure group GL(1). Each element g of GL(1) acts directly on the typical fiber, but also indirectly on elements of E itself. The latter action is indirect since each point of $\mathbb{R}_p$ is associated with an element of $\mathbb{R}$ only through an arbitrary choice of local trivialization. But g's action on elements of E can be defined in a way that can be shown not to depend on this arbitrary choice.[4]

In general, the *structure group* G of a fiber bundle is a continuous group of transformations that map every F_m onto itself. In a *principal fiber bundle* $P(M, G)$, the typical fiber F is just the structure group G itself. In this case, we can take each element g of G to act (indirectly, via some arbitrary local trivialization) on an arbitrary element u of the total space P to give another element in the fiber

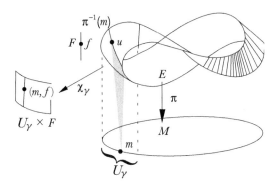

Figure 1.3. A fiber bundle (E, π, M, F, G). The total space E is represented by the Möbius strip and the fibers by lines isomorphic to the typical fiber F (with element f). The projection map π sends the fiber $\pi^{-1}(m)$ into $m \in M$. The base space M is covered by coordinate patches. The local trivialization χ_γ untwists the fibers and reveals the total space as a local product $U_\gamma \times F$.

---

[2] In other words, defining a projection map $\pi : E \to S$ with $\pi(u) = p$ for every $u \in \mathbb{R}_p \subset E$.

[3] That is to say, there is a smooth map h from $\pi^{-1}(U_p)$ onto $U_p \times \mathbb{R}$ that takes every point $u \in \pi^{-1}(U_p)$ into a point (q, x), where $\pi(u) = q \in U_p$: this h then constitutes a local trivialization of the bundle.

[4] This is shown in appendix B, where it is illustrated by figure B.1.

above the same point as u. A principal fiber bundle is trivial just in case it has a global section.[5] To each of the cylinder and Möbius strip there corresponds a different principal fiber bundle based on the circle S. In both cases the typical fiber is the group $GL(1)$, but while the principal bundle corresponding to the cylinder is trivial, that corresponding to the Möbius strip is not—it has no global section.

In a fiber bundle formulation of classical electromagnetism, the base space M is generally taken to be a manifold representing some connected region (possibly the whole) of space or space-time. The structure group is taken to be the group $U(1)$—a multiplicative group whose elements are of the form $g_\theta = e^{i\theta}$, where $i = \sqrt{-1}$ and $0 \leq \theta < 2\pi$. $U(1)$ has the same structure as the group of rotations about a point in a plane. As we shall see, its elements are used to represent changes in the phase of a quantum mechanical wave-function or matter field at a point. This group is *Abelian*, since its elements g_1, g_2 satisfy the commutative law $g_1 \circ g_2 = g_2 \circ g_1$. The principal fiber bundle $P(M, U(1))$ appropriate for the formulation of classical electromagnetism is trivial, provided the theory is taken to rule out magnetic monopoles.

The (electro)magnetic potential is represented by means of what is known as a *connection* on $P(M, U(1))$. Basically, a connection on a fiber bundle E is a rule that, for every point m in the base space M, pairs each smooth curve through m with a class of corresponding smooth curves in E, one through each point in the fiber above m, known as its *horizontal lifts*. It does this by specifying what is to count as the "horizontal" component of every vector at each point $u \in E$, since every such vector indicates which way some smooth curve through u is headed. This defines what it is for any point in the fiber F_m above each point m in M to be "lined up with" a corresponding point in each of the fibers $F_{m'}$ above neighboring points m' of M. The "vertical" component of any vector in E at each point u in E is readily specified without any such rule: it is defined by the tangents at u to curves in F_m, where $m = \pi(u)$. But this is not enough to fix its "horizontal" component, since the fiber bundle structure defines no notion of orthogonality in E. The connection adds the further geometric structure required to do this smoothly at each point u. Appendix B shows that in the case of a principal fiber bundle $P(M, G)$, this is provided by a geometric object called a Lie-algebra-valued one-form ω on P. Moreover, each section maps ω onto a Lie-algebra-valued one-form iA on (an open set of) M, where A is a covector field that may be identified with the (electro)magnetic (four-)vector potential! But while ω is a geometric object that is quite independent of sections, it corresponds to a particular A only relative to a given section. Consequently, a change of section induces a

---

[5] A vector bundle always has a global section, whether or not it is trivial. One global section is defined by the points of E that are mapped into the zero vector by (every) local trivialization. Note that this is a global section of both the cylinder and the Möbius strip vector bundles, even though the latter is not trivial.

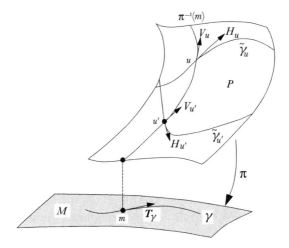

Figure 1.4. Connection on a principal fiber bundle

transformation $A \to A'$ which has all the mathematical properties of a gauge transformation 1.2 or 1.6.

But note that this transformation is not a bundle automorphism from Trautman's group $\mathcal{G}$, nor is it generated by an element of the structure group G. Indeed, it leaves the intrinsic geometric structure of the bundle itself unchanged, affecting only the way its connection is taken to be represented on the base space M.

In the fiber bundle formulation, the (electro)magnetic field is represented by the *curvature* of the principal fiber bundle $P(M, U(1))$. A precise definition of this notion is given in appendix B, but the basic idea of curvature may be simply explained geometrically. Let γ be a closed curve in M beginning and ending at m. For each point $u \in F_m$, the bundle connection defines a horizontal lift $\tilde{\gamma}$ of γ in the total space P that begins at u and ends at $v \in F_m$.[6] The bundle is curved just in case $v \neq u$ for some m and γ: the magnitude of the curvature at a point u is a measure of how far each infinitesimal curve $\tilde{\gamma}$ is from closing. The curvature may be represented by a geometric object Ω on P called a Lie-algebra-valued two-form. This is simply related to the Lie-algebra-valued one-form ω on P (it is the covariant derivative of ω—see appendix B).

A bundle section also maps Ω onto a Lie-algebra-valued two-form iF on (a subset of) M, where F is a tensor field that (for the purely magnetic case in which $A = \mathbf{A}$) may be identified with the magnetic field $\mathbf{B}$ or (for the full electromagnetic case $A = A_\mu$) the electromagnetic field strength $F_{\mu\nu}$. F is

---

[6] $\tilde{\gamma}$ is called a horizontal lift of γ since γ is the image of $\tilde{\gamma}$ under π and the tangent vector to $\tilde{\gamma}$ always lies in the horizontal subspace H of the tangent space at each point along $\tilde{\gamma}$.

14 I WHAT IS A GAUGE THEORY?

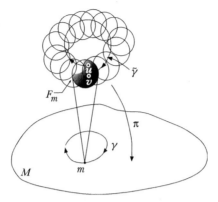

Figure 1.5. Holonomy and curvature in a fiber bundle

independent of the choice of section: this is a consequence of the fact that the structure group U(1) is Abelian.

The discussion up to this point has used classical electromagnetic theory by itself as an example to show how to unpack Trautman's definition of a gauge theory. But to illustrate the use of vector bundles and to explain his distinction between gauge transformations of the first and second kind, it is necessary to consider how a theory like electromagnetism is applied in a quantum mechanical context. It is easiest to continue the discussion by looking at how classical electromagnetism is applied in describing the behavior of quantum mechanical particles.

1.2.1 Electromagnetic interactions of quantum particles

The state of a spinless quantum particle may be represented by a wave-function $\Psi(\mathbf{x}, t)$ whose values are complex numbers. In the non-relativistic theory this is assumed to satisfy the Schrödinger equation

$$\hat{H}\Psi = i\hbar \partial \Psi/\partial t \qquad (1.9)$$

Here $\hat{H}$ is the Hamiltonian operator (which maps Ψ into another complex-valued function), $i = \sqrt{-1}$, and $\hbar = h/2\pi$, where h is Planck's constant. For particles of electric charge e and mass m subject only to an electromagnetic interaction whose potential $A_\mu = (\varphi, -\mathbf{A})$, $\hat{H}$ may be expressed by the equation

$$\hat{H} = \frac{(\hat{\mathbf{p}} - e\mathbf{A})^2}{2m} + e\varphi \qquad (1.10)$$

in which the momentum operator $\hat{\mathbf{p}} = -i\hbar\boldsymbol{\nabla}$. $\hat{H}$ acts on Ψ in the following way. First determine $-i\hbar\boldsymbol{\nabla}\Psi$ and subtract $e\mathbf{A}\Psi$, calling the result Ψ'. Now form $-i\hbar\boldsymbol{\nabla}\Psi'$ and subtract $e\mathbf{A}\Psi'$, calling the result Ψ''. Finally divide Ψ'' by $2m$ and add e times φ times Ψ.

Suppose Ψ is a solution to the Schrödinger equation with this Hamiltonian. Then

$$\Psi' = \exp[-(ie/\hbar)\Lambda(\mathbf{x},t)]\Psi \qquad (1.11)$$

is a solution to the Schrödinger equation with the transformed Hamiltonian

$$\hat{H}' = \frac{(\hat{\mathbf{p}} - e\mathbf{A}')^2}{2m} + e\varphi' \qquad (1.12)$$

where $\mathbf{A}', \varphi'$ are related to $\mathbf{A}, \varphi$ by the variable potential transformations 1.2, 1.3. Equation 1.11 is customarily called a gauge transformation of the wavefunction. If Λ is a constant, then the gauge transformation is called *"global,"* or said to be of the *first kind*: if Λ varies with $\mathbf{x}, t$, then the gauge transformation is called *"local,"* or said to be of the *second kind*: consistent with the terminology adopted for transformations of potentials, I shall refer to such wave-function transformations as *constant* (respectively *variable*) *phase transformations*.[7]

The value of the wave-function at a space-time point is a complex number—an element of the vector space $\mathbb{C}$ of such numbers. The transformation 1.11 multiplies the value $\Psi(\mathbf{x},t)$ by an element of the group U(1) of rotations in the complex plane. If Λ varies with $\mathbf{x}, t$, then the operation varies from point to point. This may readily be represented within the fiber bundle formalism by means of the *vector bundle* $\langle E, M, G, \pi_E, \mathbb{C}, P \rangle$ *associated to* the principal fiber bundle $P(M, \mathrm{U}(1))$. These bundles have a common base space—the space-time manifold (non-relativistic in this case), and a common structure group U(1). Each element in the fiber of E above m pairs some element of the vector space $\mathbb{C}$ of complex numbers with an element in the fiber of P above m, in such a way as to respect the action of the group G on the vector space $\mathbb{C}$.[8]

A particular state of a charged, non-relativistic quantum particle subject only to an electromagnetic interaction may now be represented by a global section s of the bundle $\langle E, M, G, \pi_E, \mathbb{C}, P \rangle$. How fast this state changes from point to point of M is specified by what is known as the covariant derivative $\boldsymbol{\nabla}s$ of s: this is uniquely determined by the connection on P. For every section σ of P,

---

[7] Each of the terms 'local' and 'global' has several distinct senses in this book. To prevent confusion, I will place each term within double quotes whenever it is used in the present sense.

[8] More explicitly, each point p in the total space E is an equivalence class of points $[(u,c)]$, with $u \in P, c \in \mathbb{C}$, and where $(u,c), (ug, g^{-1}c)$ are taken to be equivalent: the projection map π_E takes point p with representative (u,c) onto the same point m that u is projected onto by the projection map π of $P(M, \mathrm{U}(1))$. This construction is further explained and motivated in appendix B.

1 WHAT IS A GAUGE THEORY?

s determines a corresponding wave-function Ψ that assigns a value $\Psi(\mathbf{x},t)$ at every space-time point. A transformation of the wave-function in accordance with equation 1.11 may be taken to correspond to a change of section $\sigma \to \sigma'$ on P, since this alters the correspondence between s and Ψ. In the absence of magnetic monopoles, the principal bundle also has global sections. Each such section gives a different way of representing the principal bundle connection by a four-vector field A_μ on M. Associated with each different A_μ is a different way of representing the action of the covariant derivative ∇ by a covariant derivative operator D_μ acting on Ψ, which specifies how fast its values $\Psi(\mathbf{x},t)$ vary from point to point of M. A change of section $\sigma \to \sigma'$ alters A_μ in accordance with the transformation 1.6.

If Ψ is a solution to 1.9, then Ψ' is a solution to 1.9 with $\hat{H}$ replaced by $\hat{H}'$. Moreover, Ψ, Ψ' each represent exactly the same physical behavior of the quantum particle, subject to an electromagnetic interaction represented equally

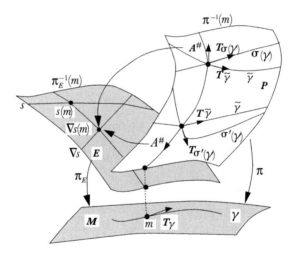

Figure 1.6. Quantum particles interacting with electromagnetism.

The electromagnetic potential is represented by the connection on the principal fiber bundle P, while the particles' wave-function is represented on an associated vector bundle E by a section s. The connection on P determines the covariant derivative ∇s of s. A choice of section σ for P determines how s represents the wave-function Ψ, and how ∇s represents $D_\mu \Psi$, as well as how the connection is represented by A_μ on M, the manifold representing space (-time). $\tilde{\gamma}$ represents a horizontal lift of curve γ on M. As explained in appendix B, $A^{\#}$ is the fundamental vector on P: its value in a section is directly related to A_μ.

by $A_\mu = (\varphi, -\mathbf{A})$ or $A'_\mu = (\varphi', -\mathbf{A}')$ respectively. In this way, a choice of potential for electromagnetism goes along with a choice of phase at each point for the particle's wave-function, and this simultaneous choice corresponds to a choice of section for the principal bundle P. These choices are depicted in Figure 1.6, which also illustrates how both principal fiber bundle and associated vector bundle share a common base space representing space-time.

A change from one section of the principal bundle P to another may be taken to represent a gauge transformation. Such a transformation does not affect the underlying bundle structure, including the connection on the principal fiber bundle that we have been taking to represent the electromagnetic potential, and its curvature, representing the electromagnetic field. It is tempting to conclude that the effects of electromagnetism on quantum particles are represented by a unique, invariant connection on the principal fiber bundle, and that a gauge transformation corresponds merely to a change from one "coordinatization" of this connection to another. Indeed, this way of reading the fiber bundle formulation of electromagnetism motivates a common strategy for interpreting this and other gauge theories whose adequacy will be a major concern of this book. But there is another way of representing a gauge transformation within the fiber bundle formalism, as the quote from Trautman makes clear.

Consider an automorphism of the principal fiber bundle $P(M, U(1))$ that acts only on the total space, mapping it smoothly onto itself so that the fiber above each point in the base space is unchanged, including the action on it of the structure group G: call this a *vertical* automorphism—a precise definition is given in appendix B. If one thinks of the fiber above each point as like a single ring of a combination lock with a continuous set of rings, then a vertical automorphism is like a smooth resetting of all the rings in the lock. Now each vertical automorphism of $P(M, U(1))$ will effect a transformation $\omega \to \bar\omega$ in the connection on $P(M, U(1))$ that represents the potential, thereby inducing a variable potential transformation $A_\mu \to \bar A_\mu$ in a representation of the connection on M given by a particular section σ of $P(M, U(1))$, and a corresponding variable phase transformation $\Psi \to \bar\Psi$. This explains why Trautman identifies such bundle automorphisms with gauge transformations of the second kind.

Clearly, a typical vertical automorphism will not preserve all the invariant structure of the bundle: as we have seen, it will change the bundle connection. It is tempting to call such a transformation an *active* gauge transformation, in contrast to a *passive* gauge transformation, consisting of a mere change of bundle section that preserves the bundle's intrinsic structure and leaves the connection unchanged. But this terminology should be resisted at this point, since to employ it is tantamount to endorsing a particular interpretation of the fiber bundle formalism.

1.2.2 Electromagnetic interactions of matter fields

The previous section showed how to represent the action of a classical electromagnetic field on quantum particles using fiber bundles. This representation may be used in a very successful hybrid theory, some of whose consequences will be studied in detail in the next chapter. But despite its empirical success, this is not how electromagnetism acts on matter according to our best contemporary theories. According to those theories, electrons and other elementary particles of matter arise as quanta of an associated quantum field. Such a matter field interacts with quantum "force" fields, such as a quantized electromagnetic field. According to contemporary theories, the form of electromagnetic and other interactions remains invariant when the interacting fields are jointly transformed according to quantum analogs to the variable potential transformations 1.6 and the constant or variable phase transformations 1.11. Such transformations are therefore said to constitute gauge symmetries of these interactions.

In a fiber bundle formulation, a quantized "force" field may be represented by a fiber bundle connection, and a quantized matter field with which it interacts may be represented by a section of an associated vector bundle. The interactions between these fields are then invariant under just the same bundle transformations that the previous section identified as gauge transformations, following Trautman.

Consider now one such theory describing electromagnetic interactions of charged matter. This will exhibit an analog to the transformation 1.11 for the case of a quantized matter field, as well as illustrating the distinction between gauge transformations of the first and second kinds in this new context. According to this theory, the interacting fields obey Euler–Lagrange equations that arise as equations of motion when Hamilton's principle is applied to the action $S = \int_{t_1}^{t_2} L dt$ for an interaction Lagrangian L. Perhaps the simplest field that could be used to describe charged particles is not the Dirac field that describes spinning electrons, but a complex Klein–Gordon field. Presenting this example of a Lagrangian formulation of a theory of charged spinless particles interacting electromagnetically will also prepare the way for the discussion of the so-called gauge argument in chapter 6. Fortunately, the example will serve its purpose if both matter and electromagnetism are represented as if they were classical fields. So the value $\phi(x)$ of the matter field at space-time point x will be assumed for now to be simply a complex number, while electromagnetism is represented by $A_\mu(x)$, a field whose value at each point x is a four-vector—a quadruple of real numbers in some inertial coordinate system.

Consider the following Lagrangian density for a complex-valued field $\phi(x)$:

$$\mathcal{L}_0 = (\partial_\mu \phi)(\partial^\mu \phi^*) - m^2 \phi^* \phi \tag{1.13}$$

whose quantized counterpart has free, charged, spinless quanta of mass m. Variation of the action $S = \int_{t_1}^{t_2} L dt$ for the Lagrangian $L_0 = \int \mathcal{L}_0 d^3 x$ with respect to independent variations in ϕ, ϕ^* and application of Hamilton's principle results in the following Euler–Lagrange equations: the Klein–Gordon equations for the free fields

$$\partial_\mu \partial^\mu \phi + m^2 \phi = 0 \tag{1.14}$$
$$\partial_\mu \partial^\mu \phi^* + m^2 \phi^* = 0 \tag{1.15}$$

Now the Lagrangian density $\mathcal{L}_0$ and resulting action S are invariant under the following transformation:

$$\phi \to \exp(i\Lambda)\phi \tag{1.16}$$

where Λ is a *constant*. This transformation is an element of the Lie group U(1). Noether's first theorem then implies the existence of a conserved Noether current

$$J^\mu = i(\phi^* \partial^\mu \phi - \phi \partial^\mu \phi^*) : \partial_\mu J^\mu = 0 \tag{1.17}$$

which in turn implies the conservation of electric charge, defined as

$$Q = e \int J^0 d^3 x \tag{1.18}$$

where e may be identified with the charge of the quanta of the field's quantized counterpart. Within this approach to gauge theories, it is because of the connection with Noether's first theorem that the transformation 1.16 is taken to be a gauge transformation of the first kind, a so-called "global" gauge transformation. I shall call it a constant phase transformation, in line with earlier terminology.

The Lagrangian density $\mathcal{L}_{EM}$ for source-free electromagnetism is invariant under both constant and variable phase transformations (see appendix A, where $\mathcal{L}_{EM}$ is given by equation A.11). If one adds $\mathcal{L}_{EM}$ to $\mathcal{L}_0$, the resulting action is therefore still invariant under the joint transformations 1.6 and 1.16, so the addition of electromagnetism preserves constant phase invariance of the combined theory. But if $\Lambda(x)$ in 1.16 is instead allowed to vary with x, then the action for this theory *fails* to be invariant under the resulting joint transformations. However, invariance under 1.6 and 1.16 with *variable* Λ is restored by adding a further interaction term to the total Lagrangian density, so that it becomes

$$\mathcal{L}_{tot} = \mathcal{L}_0 + \mathcal{L}_{em} + \mathcal{L}_{int} \tag{1.19}$$

where

$$\mathcal{L}_{int} = J^\mu A_\mu \qquad (1.20)$$

Now $\mathcal{L}_{tot}$ is invariant under the joint transformations 1.6 and

$$\phi \to \exp(i\Lambda(x))\phi \qquad (1.21)$$

The gauge symmetries that play such an important role in the gauge field theories underlying the Standard Model of contemporary high-energy physics are generalizations of the joint variable potential and phase transformations illustrated by 1.6, 1.21. The fiber bundle formulation helps one to understand why adding an interaction term like $\mathcal{L}_{int}$ to the Lagrangian density of a matter field such as $\phi(x)$ restores symmetry of the theory under gauge transformations of the second kind. Such a term is added automatically if ordinary derivatives like ∂_μ are replaced in the Lagrangian (and the resulting equations of motion) by so-called covariant derivatives D_μ, the need for which becomes apparent when the matter field is represented by a section of a vector bundle. The derivative of a bundle section specifies how fast the section is changing, as one moves from a point u on the section in the fiber above m to a neighboring point on the section that lies in the fiber above a neighboring point to m. This idea makes sense just in case there is a well-defined notion of "horizontal" motion from fiber to fiber in the bundle, deviations from which count as *changes* in a section through these fibers. Such a notion is precisely specified by a bundle connection. As explained in appendix B, the connection on a principal fiber bundle uniquely determines the covariant derivative on an associated vector bundle. The action of the covariant derivative on the vector bundle for electromagnetism may be represented on a coordinate patch $U \subseteq M$ by

$$D_\mu \phi = (\partial_\mu + ieA_\mu)\phi : D_\mu \phi^* = (\partial_\mu - ieA_\mu)\phi^* \qquad (1.22)$$

where the connection on the principal bundle associated to this vector bundle is represented on U by A_μ. It is easy to check that replacement in $\mathcal{L}_0$ of ordinary derivatives by these covariant derivatives results in $\mathcal{L}_0 + \mathcal{L}_{int}$.

2

The Aharonov–Bohm effect

As the previous chapter explained, in a wholly classical context, electromagnetism acts on charged particles only through the electromagnetic field that gives rise to the Lorentz force: the electromagnetic potential has no independent manifestations, and seems best regarded as an element of "surplus mathematical structure," in itself representing nothing physical. But the situation is different in the quantum domain. Phenomena such as the Aharonov–Bohm (AB) effect there provide vivid illustrations of the fact that there is more to classical electromagnetism than just the field. The effects of electromagnetism on the phase (and subsequent observable behavior) of charged particles that pass through a region of space are not always wholly determined by the electromagnetic field there while they pass. As Aharonov and Bohm pointed out in their seminal paper (1959), in the quantum domain it is not the field but the electromagnetic potential itself that appears to give rise to these effects.

Quantum mechanics predicts that when a beam of charged particles has passed through a region of space in which there is no electromagnetic field, an interference pattern can be produced or altered by the presence of a static magnetic field elsewhere. This was first experimentally confirmed by Chambers (1960), and since then has been repeatedly and more convincingly demonstrated in a series of experiments including the elegant experiments of Tonomura *et al.* (1986)—Peshkin and Tonomura (1989) provide a useful review. In accordance with Lorentz covariance, there is also an electric Aharonov–Bohm effect. The interference pattern produced by electrons that have passed through a region in which there may be no electromagnetic field would be different if the electric field *outside* that region were suitably varied without affecting the electromagnetic field experienced by the electrons. But this effect is harder to demonstrate experimentally. A simple example of the magnetic Aharonov–Bohm effect is depicted in figure 2.1, in which it is assumed that the solenoid is closely wound and very long.

If no current flows through the solenoid behind the two slits, then the familiar two-slit interference pattern pictured in figure 2.2 will be detected on the screen. But if a current passes through the solenoid, generating a constant magnetic field **B** *confined to its interior* in the z-direction parallel to the two

2 THE AHARONOV–BOHM EFFECT

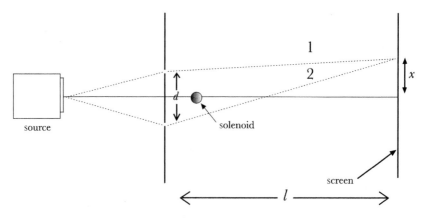

Figure 2.1. The Aharonov–Bohm effect

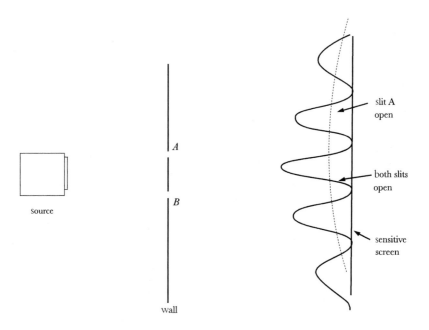

Figure 2.2. Two-slit interference pattern

slits, the maxima and minima of the interference pattern will all be shifted, as in figure 2.3, by an amount

$$\Delta x = \frac{l\lambda}{2\pi d} \frac{|e|}{\hbar} \Phi \tag{2.1}$$

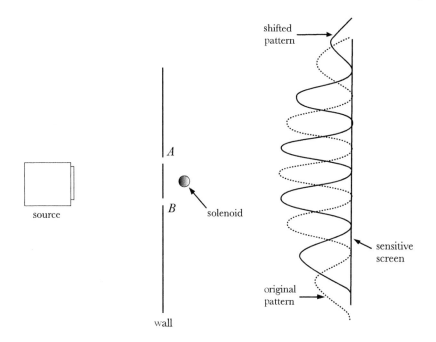

Figure 2.3. Fringe shift in the Aharonov–Bohm effect

where d is the slit separation, l is the distance to the screen, $\lambda = h/p$ is the de Broglie wavelength of the electrons in the beam (with negative charge e and momentum p), and Φ is the magnetic flux through the solenoid.

How is this phenomenon to be explained? At first sight, it appears that the magnetic field inside the solenoid must have some kind of non-local effect on the electrons, since **B** is zero everywhere outside the solenoid, in the region through which they must pass on their way from the slits to the screen. But Aharonov and Bohm denied that the effect was non-local, claiming instead that it arose from a purely local interaction with the magnetic vector potential **A** (or more generally the electromagnetic potential A_μ). They concluded that while in classical mechanics this potential could be regarded as just a mathematical device for conveniently calculating the physically real (electro)magnetic field, quantum mechanics shows that it is itself a physically real field. This view was endorsed and widely promulgated by Feynman in his famous *Lectures on Physics* (Feynman, Leighton, and Sands 1965a, volume 2).

Aharonov and Bohm (1959) first presented the effect as a theoretical consequence of quantum mechanics prior to any experimental demonstration. They derived this consequence by solving the Schrödinger equation 1.9 for scattering of an electron beam by an infinitely long and infinitely thin solenoid. In the Hamiltonian 1.10, they set the electric potential φ but not the magnetic vector potential **A** equal to zero for $r > 0$, even though the magnetic field $\mathbf{B} = \nabla \times \mathbf{A} = 0$ outside the solenoid: rather, they took **A** to be a vector potential arising from the magnetic flux Φ inside the solenoid, with cylindrical components $A_z = A_r = 0$, $A_\theta = \Phi/2\pi r$. A simplified quantum mechancal derivation for the setup pictured in figure 2.1 follows.

Consider two paths by which an electron might be thought to arrive at the same point on the screen: one passing through the upper slit, the other through the lower slit. If the difference in path lengths is a, then there will be a corresponding phase difference δ given by

$$\delta = \frac{2\pi a}{\lambda} \qquad (2.2)$$

For x much less than l,

$$a \approx \frac{xd}{l}, \text{ and so } \delta \approx \frac{2\pi xd}{\lambda l} \qquad (2.3)$$

If no current passes through the solenoid, then we have the ordinary two-slit interference experiment. The condition for constructive interference between these paths is $\delta = 2n\pi$, and so an interference maximum will appear on the screen at a distance $x \approx n\lambda l/d$ from the axis of symmetry, for each integer n.

Passing a current through the solenoid produces a magnetic field inside it (directed towards you) and a magnetic vector potential **A** both inside and outside. This produces an additional phase difference of $(e/\hbar)\mathbf{A}.\mathbf{dr}$ in the electrons' wave-function between point **r** and point $\mathbf{r} + \mathbf{dr}$. The total additional phase change over a path is then

$$\delta = \frac{e}{\hbar} \int \mathbf{A}.\mathbf{dr} \qquad (2.4)$$

This will induce an additional phase difference between two paths from source to screen of

$$\Delta(\delta) \equiv \delta_2 - \delta_1 = \left(\frac{e}{\hbar} \int_2 \mathbf{A}.\mathbf{dr}\right) - \left(\frac{e}{\hbar} \int_1 \mathbf{A}.\mathbf{dr}\right) \qquad (2.5)$$

Now if the solenoid is close to the slits and very small, then the direct path from source to screen through the top slit will go around the top of the solenoid, and the direct path from source to screen through the bottom slit will go around the bottom of the solenoid. Hence the additional phase difference between such paths will be given by

$$\Delta(\delta) = \frac{e}{\hbar} \oint \mathbf{A}.\mathbf{dr} \qquad (2.6)$$

where the integral is now taken around the closed curve formed by tracing a path from source to screen via the lower slit, and then returning from screen to source via the upper slit—a path that encloses the solenoid. It follows that

$$\Delta(\delta) = \frac{e}{\hbar} \oint \mathbf{A} \cdot d\mathbf{r} = \frac{e}{\hbar} \int \nabla \times \mathbf{A} \cdot d\mathbf{S} = \frac{e}{\hbar} \Phi \qquad (2.7)$$

This additional phase difference is independent of x, and so the locations of interference maxima and minima are all shifted upward by the same amount, namely

$$\Delta x = -\frac{l\lambda}{2\pi d} \Delta(\delta) = \frac{l\lambda}{2\pi d} \frac{|e|}{\hbar} \Phi \qquad (2.8)$$

Note that the vector potential $\mathbf{A}$ appears explicitly and ineliminably in this and other standard textbook explanations of the Aharonov–Bohm effect in terms of the action of classical electromagnetism on quantum particles. At first sight, this may appear to violate gauge symmetry, since $\mathbf{A}$ is not invariant but transforms according to 1.2 under a gauge transformation. Moreover, Aharonov and Bohm's analysis was based on the particular choice of gauge noted. But they themselves stress in their conclusion that their discussion does *not* call into question the gauge invariance of the theory, in so far as all empirical consequences of the potentials depend only on the gauge invariant quantity $\oint \mathbf{A} \cdot d\mathbf{r} = \int \mathbf{B} \cdot d\mathbf{S}$, which in turn depends only on the magnetic field $\mathbf{B}$.

Still, we are left with a puzzle. Even though all empirical consequences of classical electromagnetism in the quantum domain are a function of the (electro)magnetic field alone, this dependence is non-local: according to the theory, the size of the magnetic Aharonov–Bohm effect in region R depends on the magnitude of the magnetic field in a non-overlapping region S, even if the electromagnetic field and any other possible non-electromagnetic influences in R are held constant. But the attempt to restore locality by attributing the effect to the action of the magnetic vector potential $\mathbf{A}$ in region R leads to other difficulties.

Because $\mathbf{A}$ is gauge dependent, equations governing its evolution are neither gauge covariant nor deterministic. This is brought out most clearly by the Hamiltonian formulation (see section 5.4, and appendix C), in which $\mathbf{A}$ develops in accordance with the equation

$$\dot{\mathbf{A}} = -\mathbf{E} - \nabla \varphi \qquad (2.9)$$

Here, and in subsequent equations, a dot above a symbol indicates the rate of change of the quantity symbolized with respect to time. In 2.9, φ is an arbitrary smooth function, and so equation 2.9 does not retain its form under the gauge transformation 1.2. Moreover, specifying the values of $\mathbf{A}$ and $\mathbf{E}$ everywhere at an initial time determines the value of $\mathbf{A}$ at a later or earlier time only up to an arbitrary function $\int \nabla \varphi \cdot dt$. If the vector $\mathbf{A}$ does represent a genuine physical field associated with classical electromagnetism, over and above **E** and **B**, then this theory radically underdetermines how that field changes with time. Moreover, to treat **A** as a physical field violates the gauge symmetry of classical electromagnetism, even though the empirical content of that theory is still gauge invariant in the quantum domain. This raises an epistemological difficulty, for while the lack of gauge invariance theoretically privileges a particular gauge for **A**, the theory itself entails that no observation or experiment is capable of revealing that gauge. These difficulties will be further explored in chapter 4.

2.1 Fiber bundles

The discussion of the Aharonov–Bohm effect up to this point has not mentioned the fiber bundle formulation of classical electromagnetism sketched in the previous chapter (section 1.2). But many mathematicians, and some physicists and philosophers, have taken this formulation to offer a more natural or intrinsic geometric account of the effect that may provide insight into locality and the role of gauge symmetry. For example, in the conclusion of their seminal paper, Wu and Yang say (1975, p. 3856)

It is a widely held view among mathematicians that the fiber bundle is a natural geometric concept. Since gauge fields, including in particular the electromagnetic field, are fiber bundles, *all gauge fields are thus based on geometry*. To us it is remarkable that a geometrical concept formulated without reference to physics should turn out to be exactly the basis of one, and indeed maybe all, of the fundamental interactions of the physical world.

All the gauge theories considered in this book admit of a fiber bundle formulation, and this offers a perspective which can deepen one's understanding of them. But the beauty of the mathematics can also prove a confusing distraction from attempts to resolve important interpretative issues. Worse still, it can mislead one into thinking that these are either rendered unimportant by, or admit of a straightforward technical resolution in, this new framework. This is true in particular of the fiber bundle formulation of classical electromagnetism, or so I shall argue later in this chapter. But it is helpful to introduce some concepts of the fiber bundle formulation of electromagnetism into the discussion at this point in order to facilitate the exposition of an interpretative strategy which I shall generalize in chapter 3 and argue for in chapter 4.

Recall that all empirical consequences of classical electromagnetism in the region occupied by the magnetic Aharonov–Bohm effect depend only on the gauge-invariant quantity $\oint_C \mathbf{A}.d\mathbf{r}$ around closed curves C in that region. Note that these quantities already determine the component of the magnetic field in any direction $\hat{n}$ at each point by fixing the flux $\mathbf{B}.d\mathbf{S}$ through each infinitesimal element of area $d\mathbf{S} = \hat{n}|d\mathbf{S}|$ at that point. Wu and Yang (1975) noted further that these quantities actually contain redundant information about the effects of electromagnetism here on particles of charge e, since those effects are already determined by what they call the *phase factor*

$$\exp\left(-\frac{ie}{\hbar}\oint_C \mathbf{A}.\mathbf{dr}\right) \qquad (2.10)$$

Their point was that different magnetic fluxes through the solenoid $\Phi, \bar{\Phi}$ will lead to identical Aharanov–Bohm effects if $\bar{\Phi} = \Phi + \frac{2n\pi\hbar}{e}$. The relativistic generalization of this result is that, in any region, the effects of electromagnetism on quantum particles of charge e are wholly determined by the *Dirac phase factor* for each closed curve C in space-time in that region

$$\exp\left(\frac{ie}{\hbar}\oint_C A_\mu dx^\mu\right) \qquad (2.11)$$

As Wu and Yang put it (1975, p. 3846) (in the passage that follows I have changed units to set the velocity of light equal to 1, and modified their notation to conform to the present conventions),

The field strength $F_{\mu\nu}$ underdescribes electromagnetism, i.e. different physical situations in a region may have the same $F_{\mu\nu}$. The phase $\frac{e}{\hbar}\oint A_\mu dx^\mu$ overdescribes electromagnetism, i.e. different phases in a region may describe the same physical situation. What provides a complete description that is neither too much nor too little is the phase factor 2.11.

They also point out that the phase factor for a curve C arises naturally within the fiber bundle formulation—it is just the holonomy of C. Let me explain. The electromagnetic potential is represented by the connection on a principal bundle with structure group U(1)—the multiplicative group of complex numbers of modulus 1. A connection on a principal fiber bundle determines a corresponding set of holonomies as follows. Consider a horizontal lift of a closed curve C beginning and ending at m. For each point u in the fiber above m, the horizontal lift through u will trace out a corresponding curve in the bundle that returns to a point v in the fiber above m. In this way the connection maps the fiber above m onto itself (see figure 1.5). In the case of electromagnetism one can show (see appendix B) that this map is induced by the action of an element g_C of the structure group called the *holonomy* of C and that this is independent of m.

The wave-function is represented by a section of a vector bundle associated to the principal U(1) bundle—a smooth selection of a single element of the fiber above each point in the base space. This assigns a vector (a complex number) as the value of the wave-function at each point of space(-time). The connection on the principal bundle specifies parallel transport of vectors in the associated vector bundle and hence the result of parallel transporting a vector around C. This gives rise to a linear map of the space of vectors at each point onto itself. Again, in the case of electromagnetism this map is generated by (a matrix representation of) an element of the bundle's structure group, also naturally referred to as the *holonomy* of C, that is independent of C's starting point and also independent of the section. Since the wave-function is complex-valued, the representation here is just the fundamental representation of U(1). Consequently the holonomy of a closed curve C through m multiplies the wave-function at m by a complex number of modulus 1: this is just (the complex conjugate of) the Dirac phase factor for C! The generalization of these ideas to non-Abelian gauge theories involves certain complexities that are best postponed to the next chapter.

What does it mean to say that the electromagnetic potential is represented by the connection on a principal fiber bundle with structure group U(1), and how is this related to its representation by a vector field A_μ? As explained in appendix B, the connection may be identified with a certain mathematical object ω—a Lie-algebra-valued one form on the total space of the bundle. A section of the bundle is a smooth map from each point in the base space onto the fiber above that point. To each section σ there corresponds a map σ^* (the pull-back of σ) from one-forms on the total space onto one-forms on the base space such that $\sigma^*\omega = iA_\mu$ (i appears here as the generator of the Lie algebra of U(1) to convert the real-valued one-form A_μ into a Lie-algebra-valued one-form). A change of section from σ to $\bar\sigma$ gives rise to a different one-form, or covector field, $\bar A_\mu$ that corresponds to the same connection in a different gauge, where $\bar A_\mu$ is related to A_μ by the gauge transformation 1.6.

One way to understand the notion of gauge symmetry in the fiber bundle formulation is as symmetry under changes of section of the principal fiber bundle representing electromagnetism: each such change results in a corresponding change in what wave-function of charged particles is represented by a given section of an associated vector bundle. For the effect of such a change of section is just to induce the transformations 1.6 and 1.11, the standard form for gauge transformations of classical electromagnetism interacting with charged quantum particles. Such a change of section leaves the bundle connection invariant. It is tempting, therefore, to take the connection ω to offer an intrinsic, gauge-invariant, representation of electromagnetism in the Aharonov–Bohm effect, while any particular A_μ from a class of gauge-equivalent vector fields represents electromagnetism only in a particular section, associated with a particular choice of representation for the particles' wave-function. This temptation is strengthened by the realization that the electromagnetic field strength $F_{\mu\nu}$ may also be thought to derive from a geometric object defined on the same principal fiber bundle as ω, namely the the *curvature* of the bundle. The curvature is represented by the covariant derivative of ω—a two-form Ω (see appendix B). Ω offers a measure of the amount by which the horizontal

2.1 FIBER BUNDLES

lift of any closed curve fails to close. It is closely analogous to the Riemannian curvature of a spatial or space-time manifold. $F_{\mu\nu}$ derives from Ω in just the same way that A_μ derives from ω, namely as $\sigma^*\Omega = iF_{\mu\nu}$, where σ^* is the pull-back of Ω corresponding to the section σ. Because U(1) is an Abelian group, this pull-back is independent of the section σ, reflecting the gauge invariance 1.7 of the electromagnetic field strength.

But this is not the only way to understand gauge transformations and gauge symmetry in the fiber bundle formulation. It is not the way Trautman understands these notions, and it does not mesh well with the emphasis placed by Wu and Yang on the Dirac phase factor as offering just the right description of electromagnetism. The alternative is to understand gauge symmetry as symmetry under a class of fiber bundle automorphisms—structure-preserving maps of principal and associated vector bundles onto themselves. Appendix B spells out the details, but the basic idea is that a vertical bundle automorphism maps each of the total space and base space of a bundle smoothly onto itself in such a way as to preserve both the projection map and the action of the structure group on the fiber above each point. A non-trivial vertical automorphism maps the connection ω on the principal U(1) bundle onto a different connection $\tilde{\omega}$ with the same holonomies for all closed curves, while producing corresponding changes in parallel transport of vectors in the associated vector bundle that also leave all its holonomies fixed. If one takes electromagnetic gauge transformations to be represented in the fiber bundle formulation by vertical bundle automorphisms, as Trautman would recommend, then while the holonomies of all closed curves are invariant under gauge transformations, the connection of which these are holonomies is not. The standard transformation equations 1.6 and 1.11 may than be reinterpreted as showing how the connection and wave-function *as represented in a given section* change under a gauge transformation. They make vivid the fact that there is a continuous infinity of distinct connections with the same holonomies for all closed curves. Just like the holonomies, both the electromagnetic field two-form Ω and its representation in a given section remain unchanged under such a gauge transformation.

There is a close analogy between electromagnetic gauge transformations and the space-time transformations associated with the general covariance of general relativity and other space-time theories. If one thinks of choice of a bundle section as analogous to choice of a coordinatization of space-time, then the first way of understanding gauge transformations takes them to be like changes from one coordinate system to another, while the second way of understanding gauge transformations takes them to be like smooth mappings of the space-time manifold onto itself. In both cases, there is a substantial interpretative issue as to whether the manifold mappings involved in the second way of understanding transformations should be understood "actively," so that its effect is to represent a distinct physical scenario, or "passively," so that its effect is to switch between two different representations of the same physical scenario. I take up this issue in chapter 4, where I argue against an "active" interpretation of gauge transformations, even when these correspond to bundle automorphisms rather than mere changes of section.

Does the fiber bundle formalism shed any light on the difficulties with locality and radical indeterminism in phenomena like the Aharonov–Bohm effect? At most it offers another perspective on these difficulties with a natural generalization to the non-Abelian gauge theories discussed in the next chapter.

Suppose that one takes the bundle connection to represent an electromagnetic influence distinct from, though related to, that of the electromagnetic field. The relation may be expressed by saying that Ω is the covariant derivative of ω or, less abstractly, by the usual relation 1.7 between the tensor and vector fields $F_{\mu\nu}$ and A_μ that correspond to the pull-backs of these objects in a particular section. One can think of this relation either as a definition of the electromagnetic field in terms of a potential field, regarded as the one fundamental electromagnetic quantity, or as a law relating distinct physical fields. Either way, one is committed to a physical field defined at each space-time point—indeed the very same field represented in the earlier formulation by A_μ. The evolution of this field is still radically underdetermined by the dynamics of the theory, so we have made no progress in resolving the problem of indeterminism. Moreover, the fiber bundle connection is just as empirically underdetermined as the vector potential. Since a vertical bundle automorphism transforms one to another connection that shares all the same holonomies, both on the principal fiber bundle and on the associated vector bundle, it represents a gauge transformation between distinct connections which no experiment or observation could discriminate. For, as Wu and Yang say, the empirical content of this theory is exhausted by the Dirac phase factors, which are fixed by the bundle holonomies.

Suppose instead that one takes the holonomies themselves directly to represent electromagnetism and its effects on quantum particles. Since connections that generate the same holonomies are related by a gauge transformation, the holonomies of the principal bundle also fix the electromagnetic field strength (though it is also easy to show this directly—see appendix B). Moreover, the holonomies of curves at one time determine the holonomies of curves at earlier and later times, so taking holonomies as basic restores the determinism of electromagnetism and its influence on the wave-function in a particular gauge.[1] It is still convenient to work with a particular connection as a representative of its class of holonomies, but this may now be treated as mere mathematical surplus structure that does not itself represent any physical field taking values

---

[1] This is most easily shown in the Hamiltonian formulation, in which the holonomies of curves at a time are points in the configuration space of the reduced phase space that results from the original phase space as its quotient under the equivalence relation of lying on the same gauge orbit, as explained by Belot 1998: for details, see appendix C.

at space-time points (over and above the electromagnetic field itself, which is directly represented by its curvature).

Taking the holonomies as directly representing electromagnetism and its effects on quantum particles has interesting implications for locality. In the magnetic Aharonov–Bohm effect, the holonomies of closed curves enclosing a current-carrying solenoid are not zero even if there is no electromagnetic field outside the solenoid. Since this is the spatial region to which the electrons are confined as they traverse the apparatus, the influence of electromagnetism on the electrons is not produced directly by a field in a spatially distant region, as it would be if the electromagnetic field exhausted the content of electromagnetism. But if the holonomies directly represent electromagnetism and its effects, then there is still a sense in which the action of electromagnetism on the electrons is not completely local, since holonomies attach to extended curves rather than points. Moreover, on most interpretations (though not on all), quantum mechanics does not ascribe continuous spatial trajectories to electrons as they pass through the region outside the solenoid, so they could not be acted on at each point, even if electromagnetism did exert its influence "pointwise" there. Whether classical electromagnetism as represented by holonomies acts locally on quantum particles in phenomena like the Aharonov–Bohm effect depends both on how one interprets quantum mechanics and on exactly what is meant by locality. But it is important to note that the fiber bundle formulation has merely introduced a new perspective on, and a new language (holonomies rather than Dirac phase factors) for stating, an issue that could be seen to arise already in the older formulation in terms of vector potentials.

2.2 A gauge-invariant, local explanation?

Some have suggested that the magnetic Aharonov–Bohm effect may be explained without positing any action at a distance by describing electromagnetism by a gauge-invariant field other than the net magnetic field, defined at each point in the region outside the solenoid. I shall consider three such suggestions before arguing against this approach.

Holland (1993, p.196) notes that while $\mathbf{A}$ is gauge dependent, the quantity

$$\mathbf{G}(\mathbf{x}) = \mathbf{A} - \nabla \partial_j \Delta^{-1} A_j \text{ (sum over } j = 1, 2, 3) \quad (2.12)$$

where

$$\Delta^{-1}\mathbf{A}(\mathbf{x}) = -\int_{\text{all space}} d^3 y \frac{\mathbf{A}(\mathbf{y})}{4\pi |\mathbf{x} - \mathbf{y}|} \quad (2.13)$$

is gauge invariant under 1.2. He proposes (ibid.)

to adopt the vector field **G** as the "true" physical degree of freedom generated by solutions of Maxwell's equations

and continues:

The approach suggested here tends to point away from the global formulation of the Wu–Yang type towards a more intuitive local description.

Now the vector field **G** satisfies Laplace's equation. In fact it corresponds to a so-called harmonic one-form ω_h. In unpublished papers, R. J. Kennedy has pointed out that it is a consequence of the Hodge decomposition theorem (see e.g. Nakahara 1990, p. 255) that every one-form A defined on the spatial region outside the solenoid may be expressed *uniquely* as the sum of three terms as follows:

$$A = d\Lambda + d^\dagger \Omega + \omega_h \tag{2.14}$$

where Λ is a function (a zero-form), Ω is a two-form, d is the exterior derivative operator (see appendix B, equation B.35), and $d^\dagger$ is its adjoint. Since ω_h is a harmonic form, it satisfies the conditions $d\omega_h = d^\dagger \omega_h = 0$. And since $d^2 = 0$, the decomposition 2.14 gives

$$dA = dd^\dagger \Omega \tag{2.15}$$

If we identify A as the magnetic vector potential and $dd^\dagger \Omega$ as the magnetic field B expressed as a two-form, then this last equation is just a way of rewriting 1.1, and since the magnetic field is zero outside the solenoid, 2.14 reduces to

$$A = d\Lambda + \omega_h \tag{2.16}$$

This last equation is an elegant expression of the fact that while the vector potential in the region outside the solenoid is gauge dependent, every such potential can be decomposed canonically into a gauge-invariant part and the gradient of an arbitrary smooth function. The harmonic one-form ω_h, or the corresponding vector field **G**, may be thought invariantly to represent a physical aspect of electromagnetism in the region outside the solenoid even though the magnetic field is zero there. Moreover, this would constitute a field defined at each point of space outside the solenoid and would therefore be capable of acting locally on anything located at that point.

In order to offer a completely local account of the magnetic Aharonov–Bohm effect, more is required. It is also necessary to account for variations in this supposed new field in response to changes in the current in the solenoid. But even if ω_h, or the corresponding vector field **G**, describes the instantaneous condition of each point outside the solenoid in equilibrium, it cannot describe transitions between one equilibrium state and another. Consider the vector potential with cylindrical components $A_z = A_r = 0$, that Aharonov and Bohm used in their original paper. This already satisfies $\nabla \times \mathbf{A} = 0$, $\nabla \mathbf{A} = 0$ and is therefore the harmonic vector field **G** for the region outside the solenoid. If the current through the solenoid increases during an interval T, so that the flux increases from Φ to $\bar{\Phi}$, the circumferential component of **G** at the end of the interval will have changed from $A_\theta = \Phi/2\pi r$ to $A_\theta = \bar{\Phi}/2\pi r$, no matter how large r may be.

Of course, a changing magnetic field inside the solenoid will give rise to changing electric and magnetic fields in the vacuum outside it, which will then propagate at the speed of light in accordance with Maxwell's equations. But what will happen to the additional local field that supposedly acts locally on the quantum particles? It is a violation of locality to suppose that at each instant it takes the harmonic form **G** corresponding to the instantaneous flux through the solenoid. Moreover, this supposition is inconsistent with the view that the additional local field gives rise to the holonomies around closed curves in space, since these include those around closed curves in space-time that are functions of a four-vector potential A_μ compatible with the changing field $F_{\mu\nu}$.

One might wish, analogously, to uniquely express an arbitrary one-form on space-time corresponding to such a four-vector potential A_μ as the sum of three components, in the manner of the Hodge decomposition of the spatial one-form A, and to regard the harmonic component as the uniquely natural, gauge-invariant generalization of the spatial harmonic one-form ω_h corresponding to **G**. Granted this wish, one could hope to show that increasing the field through the solenoid would produce changes in this harmonic component that propagate continuously at the speed of light from the solenoid, until a new stable state was reached in which the harmonic component's spatial projection could once more be represented by the harmonic vector field **G** with increased circumferential component $G_\theta = \bar{\Phi}/2\pi r$.

Unfortunately, this wish would be frustrated by the failure of the Hodge decomposition theorem to generalize to manifolds (like Minkowski space-time) whose metric is not Riemannian but pseudo-Riemannian. When the magnetic (or electromagnetic) potential in the Aharonov–Bohm effect is represented by a one-form A on a manifold like the spatial region outside the solenoid, it is the topology (not the metric) that allows it to be closed ($dA = 0$) but not exact ($A \neq d\Lambda$) when there is no magnetic field in that region. But only if the metric is Riemannian does the Hodge decomposition yield a unique harmonic one-form. Thus while ω_h is uniquely defined in the spatial region outside the solenoid, there is no analogous unique one-form in the region of space-time corresponding to that spatial region throughout an interval during which the current in the solenoid is changed. Hence there is no mathematically privileged candidate available to represent an underlying gauge-invariant physical field, defined at each space-time point outside the solenoid during that interval.

If one supposes that the harmonic one-form ω_h or its corresponding vector field **G** represents a real physical field defined at each point outside the solenoid when a current is passing through it, then one can give no local account of how turning on the current in the solenoid gives rise to this field. To say that changes in the current instantaneously affect the value of ω_h is to admit that there is action at a distance in the Aharonov–Bohm effect. But if one is prepared to offer no account at all of how the values of ω_h are affected by changes in the current, then one is essentially abandoning the claim that ω_h or its corresponding vector field **G** represents a real physical field.

Mattingly (2006) has proposed another candidate for a novel field, defined at each point in the region outside the solenoid even when the current is changing. He believes that this could ground a local account of the magnetic Aharonov–Bohm effect. The account generalizes to all other phenomena in which classical electromagnetism affects the phase of quantum particles. The novel field is uniquely determined by the currents and charges (or four-current J_μ) present in the given physical situation, and Mattingly calls it the *current field*. The current field A_μ^m at space-time point x is given by

$$A_\mu^m(x) = 4\pi \int_{\text{all space-time}} d^4x' D_r(x-x') J_\mu(x') \tag{2.17}$$

where D_r is the so-called retarded Green's function familiar from classical electrodynamics (see e.g. Jackson 1999, p. 614). The notation A_μ^m for the current field is intended to reflect the fact that it is one solution to Maxwell's equations for what is usually called the four-vector electromagnetic potential A_μ in a situation in which no sourceless incoming or outgoing fields are present. But of course it is not the only such solution. Because of the gauge symmetry of the equations, if 2.17 represents one solution, then so also does

$$A_\mu^\Lambda(x) = A_\mu^m(x) + \partial_\mu \Lambda \tag{2.18}$$

for arbitrary smooth Λ. In the purely classical context in which electromagnetic potentials are regarded just as calculational aids, there is no reason to suppose that any of these solutions directly represents a physical field; rather, each of them is to be regarded simply as offering a different, indirect representation of the same electromagnetic field $F_{\mu\nu}$ via 1.7.

But Mattingly proposes that the action of classical electromagnetism on quantum particles is mediated by the current field A_μ^m, to be thought of not just as the four-vector electromagnetic potential expressed in one particular gauge, but rather as a real physical field defined at each point in space-time. He claims that once we acknowledge the existence of such a field we can give a local, deterministic account of the Aharonov–Bohm effect. He also says (p. 252)

2.2 A GAUGE-INVARIANT, LOCAL EXPLANATION?

The new field (or newly taken seriously field) that I've described just is how the value of the 4-current on the past light-cone of the electron registers at the position of the electron. And the interaction of the electron's wave function with that field is just the 4-d inner product.

There is a possible confusion here between two fields, namely $A_\mu^m(x)$ and $J_\mu(x')$ (where I have symbolized the arguments x, x' differently to stress the fact that it is the four-current at points x' on the past light-cone of x that contribute to the value of $A_\mu^m(x)$). There is nothing new about the four-current on the past light-cone of the electron $J_\mu(x')$. What *is* new is $A_\mu^m(x)$, regarded as directly representing the value of a physical field at x rather than just as an expression of the four-vector potential at x in one particular gauge. Before we consider whether such a new field could ground a local, deterministic account of the Aharonov–Bohm effect, it is interesting to compare it to the harmonic one-form ω proposed earlier as offering a gauge-invariant representation of a new field at each spatial point outside the solenoid in the equilibrium situation of the magnetic Aharonov–Bohm effect.

Note that $\partial^\mu A_\mu^m = 0$; and, in the equilibrium situation of the magnetic Aharonov–Bohm effect, the electromagnetic field $F_{\mu\nu} = \partial_\mu A_\nu^m - \partial_\nu A_\mu^m = 0$ outside the solenoid. These relations may be rewritten in the notation of differential forms as $d^\dagger A^m = dA^m = 0$, from which it follows that A^m is a harmonic one-form on a four-dimensional pseudo-Riemannian space. If one performs the gauge transformation 2.18, then A^Λ will also be a harmonic one-form on this space provided that Λ is itself harmonic, i.e. $\partial^\mu \partial_\mu \Lambda = 0$. Since 2.18 implies that $A^\Lambda = A^m + d\Lambda$ for arbitrary harmonic Λ, it is clear why A^m is *not* a unique gauge-invariant harmonic one-form on the region of space-time corresponding to the spatial region outside the solenoid during equilibrium. But the choice $\Lambda = 0$ still has a certain salience among the class of harmonic one-forms in this situation, since in that class it is only A^m that is determined solely by the current one-form J.

Granted the existence of a physical field represented by $A_\mu^m(x)$, one could always regard 2.18 as simply a gauge transformation to another way of representing this field. So the fact that the original expression $A_\mu^m(x)$ has some (but not other) nice mathematical properties does not by itself provide a reason to believe that it represents a physical field. What is required is rather empirical evidence for the existence of such a field. The issue then becomes whether experimental demonstrations of the Aharonov–Bohm effect, and more generally of the empirical adequacy of the union of classical electromagnetism and the quantum mechanics of particles, do, or could, supply such evidence. This issue will be taken up later, in chapter 4. For now, let us see whether such a field could provide the basis for a local, deterministic account of the Aharonov–Bohm effect.

Suppose there were a physical field $A_\mu^m(x)$ at x. Then it could certainly act locally on anything present at x—including charged quantum particles like electrons, if an electron could be present at a particular space-time point. The (position representation) wave-function $\Psi(x)$ of a quantum particle is defined at each space-time point x, but it certainly does not follow that the particle is at every point. A classical point particle would follow a continuous trajectory, occupying a single point of space at each instant of time. This is also true of quantum particles according to Bohmian mechanics, but not according to quantum mechanics on most interpretations. Whether a field that takes values at each space-time point in a region could act locally on quantum particles in that region depends on how one interprets quantum mechanics. Without entering that contentious debate at this point, it does seem reasonable to assume that for a physical field present in a region to take on values at each space-time point improves the prospects of a local account of how that field acts on quantum particles in that region.

But can one give a local account of how the hypothetical field $A_\mu^m(x)$ is related to its sources $J_\mu(x')$? Many of the ingredients for such an account are available. The definition of A_μ^m ensures that its value at x is a function only of J_μ, an incontrovertibly physical field; and since this value depends only on the values of J_μ at points on the back light-cone of x, the action of J_μ in generating A_μ^m conforms to the demands of relativistic locality. One can think of A_μ^m as propagating continuously with the speed of light from its sources J_μ to each point in space. Moreover, A_μ^m satisfies a deterministic evolution equation—in a current-free region of space-time, the values of A_μ^m on a space-like hypersurface determine their subsequent values, and indeed in such a way that $A_\mu^m(x)$ is determined by the values of $A_\mu^m(x')$ only on points x' on a space-like hypersurface inside and on the back light-cone of x.

But there is a problem: A_μ^m carries no energy or momentum if $\partial_\mu A_\nu^m - \partial_\nu A_\mu^m = 0$. One of the strongest arguments for the reality of the electromagnetic field is that it can carry energy and momentum (see e.g. Lange 2002). Yet the hypothetical field A_μ^m in the equilibrium situation of the magnetic Aharonov–Bohm effect satisfies $\partial_\mu A_\nu^m - \partial_\nu A_\mu^m = 0$ in the region outside the solenoid and therefore has zero energy and momentum density, and zero energy–momentum flow, in that region. The problem may not be insuperable: one could try to argue that propagating influences on quantum phases do not have to carry energy or momentum to be real. But it does give grounds for skepticism of much the same kind that attached to Newtonian gravitational potentials but were subsequently removed by general relativity's story of propagating gravitational waves that carry (non-localized) energy.

Partly in response to this problem, Mattingly (in press) has proposed an alternative account of the Aharonov–Bohm effect. The phase of the quantum particles' wave-function is modified, on this account, not by the current field, nor by the *net* (electro)magnetic field they experience (which is zero in the ideal case), but rather under the influence due to each individual charge as it makes its own independent contribution to the net (electro)magnetic field. I follow Mattingly (in press) in referring to these independent contributions as *components* of the net field. According to Mattingly (in press), in the magnetic Aharonov–Bohm effect, the electrons' phase is sensitive to component (magnetic) fields not net field quantities. Each such component field arises from the action of an infinitesimal current element in the solenoid and produces an infinitesimal effect on the electron phase. Its total effect on the phase of each separate component of the electrons' wave-function near a point is just the sum of these effects near that point. The total change in the phase of that component of the wave-function is the "sum" of the total phase changes due to all the infinitesimal current elements along its path. This is different for different components of the wave-function in a way that depends on the current through the solenoid. Hence the interference pattern produced when these different component waves overlap at the screen depends on the current through the solenoid, even though at no time is any phase change caused by the *net* magnetic field, either inside or outside the solenoid.

Mattingly (in press) implements this account by writing down the (approximate relativistic) Darwin Lagrangian for the motion of a classical electron (of charge) e acted on by a set of moving charges q, where R_{eq} is the length of a spatial vector $\mathbf{R}_{eq}$ from q to e (in a frame in which their velocities are $\mathbf{v}_e, \mathbf{v}_q$ respectively), and $\mathbf{n}_{eq}$ is a unit vector in the direction of $\mathbf{R}_{eq}$:

$$L_D = \frac{1}{2}m_e v_e^2 + \frac{1}{8}\frac{m_e v_e^4}{c^2} - e\sum_q \frac{q}{R_{eq}} + \frac{e}{2c^2}\sum_q \frac{q}{R_{eq}}\left[\mathbf{v}_e\cdot\mathbf{v}_q + (\mathbf{v}_e\cdot\mathbf{n}_{eq})(\mathbf{v}_q\cdot\mathbf{n}_{eq})\right] \tag{2.19}$$

In the magnetic Aharonov–Bohm effect, the third term on the right of 2.19 vanishes, while the second term may be neglected for non-relativistic electrons. On these assumptions, the Darwin Lagrangian becomes

$$L'_D = \frac{1}{2}m_e v_e^2 + \frac{e}{2c^2}\sum_q \frac{q}{R_{eq}}\left[\mathbf{v}_e\cdot\mathbf{v}_q + (\mathbf{v}_e\cdot\mathbf{n}_{eq})(\mathbf{v}_q\cdot\mathbf{n}_{eq})\right] \tag{2.20}$$

$$\equiv \frac{1}{2}m_e v_e^2 + e\frac{\mathbf{v}_e}{c}\cdot\mathbf{A}_I \tag{2.21}$$

Mattingly (in press) then rewrites the final term as an interaction between e and the component magnetic fields of the other charges as

$$\frac{e}{2c^2}\sum_q B_q|_e\kappa\cdot\left[\mathbf{v}_e + \mathbf{n}_{eq}(\mathbf{v}_e\cdot\mathbf{n}_{eq})\right] \tag{2.22}$$

where

$$\frac{q}{R_{eq}} \mathbf{v}_q = B_q|_c \frac{R_{eq} c \hat{\mathbf{v}}_q}{\sin(\theta_{\mathbf{R}_{eq},\mathbf{v}_q})} \equiv B_q|_c \boldsymbol{\kappa} \qquad (2.23)$$

Note that the resulting Lagrangian

$$L'_D = \frac{1}{2} m_e v_e^2 + \frac{e}{2c^2} \sum_q B_q|_c \boldsymbol{\kappa} \cdot [\mathbf{v}_e + \mathbf{n}_{eq}(\mathbf{v}_e \cdot \mathbf{n}_{eq})] \qquad (2.24)$$

is gauge independent, unlike the usual classical (non-relativistic) Lagrangian L_A for electrons e moving in a magnetic field $\mathbf{B} = \nabla \times \mathbf{A}$

$$L_A = \frac{1}{2} m_e v_e^2 + e \frac{\mathbf{v}_e}{c} \cdot \mathbf{A} \qquad (2.25)$$

Perhaps the easiest way to understand how the electron wave-function's phase is affected by the "magnetic" term in the Darwin Lagrangian is to move to a Hamiltonian formulation and then compare solutions to the Schrödinger equation with the Darwin Hamiltonian corresponding to different currents through the solenoid. The Darwin Hamiltonian arises from the Darwin Lagrangian in the usual way:

$$H_D \equiv \sum_i p_i \dot{q}_i - L_D \qquad (2.26)$$

where $\dot{q}_i = (v_e)_i$ and $p_i \equiv \frac{\partial L_D}{\partial \dot{q}_i}$, ($i = 1, 2, 3$). Under the same assumptions as those leading to L'_D, the Darwin Hamiltonian becomes

$$H'_D = \frac{1}{2m_e} \left(\mathbf{p}_e - \frac{e}{c} \mathbf{A}_I \right)^2 \qquad (2.27)$$

If Ψ is a solution to the Schrödinger equation for a component of the wave-function when $\mathbf{A}_I = 0$ (corresponding to no current through the solenoid), then Ψ' is the corresponding solution for $\mathbf{A}_I \neq 0$, where

$$\Psi' = \exp\left[\frac{ie}{\hbar c} \int^{\mathbf{r}} \mathbf{A}_I \cdot d\mathbf{r}' \right] \Psi \qquad (2.28)$$

The Aharonov–Bohm effect may then be seen to arise from the *difference* in phase shifts for different components associated with the line integral

$$\oint \mathbf{A}_I \cdot d\mathbf{r} \qquad (2.29)$$

2.2 A GAUGE-INVARIANT, LOCAL EXPLANATION? 39

which does not vanish for curves enclosing the solenoid when current flows through it. Moreover

$$\mathbf{A}_I = \frac{1}{2c} \sum_q B_q|_e \left[\boldsymbol{\kappa} + (\boldsymbol{\kappa} \cdot \mathbf{n}_{eq}) \mathbf{n}_{eq} \right] \qquad (2.30)$$

and so the Aharonov–Bohm shift may be thought to arise solely from the cumulative effects on the phase of components of the magnetic field outside the solenoid, associated with all the individual moving charges that constitute the current flowing through the solenoid.

Does Mattingly's new account succeed in giving a gauge invariant, local explanation of the Aharonov–Bohm effect? The component fields to which it appeals seem required by the theory—without them, there could be no net field in the Aharonov–Bohm effect, outside or inside the solenoid; and neither they nor $\mathbf{A}_I$ are gauge dependent. Moreover, the influence of each component field could (it seems) be subjected to independent experimental investigation, by changing the configuration of the solenoid to remove the current elements that generate all the other component fields and observing the effects of the resulting magnetic field. But the account does not square well with a field-theoretic point of view, either classically or (looking ahead) quantum theoretically; and it is not as easy as it seems to experimentally disentangle the component fields from the net field to which they contribute.

Classically, it is the *net* field that acts on charged particles and propagates energy and momentum even through otherwise empty space. Once they have contributed their share, components of the net field have no further role to play and consequently cannot be independently investigated. Mattingly (in press) rejects this view of the relation between the net field and its components when he says "it is sensible to say that the *net* field is fictional while the *component* field is factual." But by denying the reality of the net field one is effectively abandoning a field-theoretic perspective in favor of a particle action-at-a-distance theory. For nothing is gained by interpolating a component field between a charged current element and its effect on another charge (be it production of a corresponding element of a Lorentz force or modification of the phase of a wave-function). Indeed, as has often been remarked (see Essén 1996), the Darwin Lagrangian is just that approximation to the full relativistic Lagrangian for electrodynamics which neglects the independent degrees of freedom of the radiation field and so permits one to treat interactions purely in terms of velocity-dependent particle–particle interactions![2]

---

[2] Note also that a choice of Coulomb gauge ($\boldsymbol{\nabla} \cdot \mathbf{A}_I = 0$) has effectively already been made before one writes down the Darwin Lagrangian (see Essén 1996), though this is no more telling an objection against Mattingly's new account than was the choice of Lorenz gauge $\left(\partial^\mu A_\mu^m = 0 \right)$ for the current field against his previous account.

When matter as well as the electromagnetic field is represented by a quantum field (as it is in the Standard Model), it is unclear how to make sense of interactions among individual moving charges mediated by field components. Mattingly (in press) draws an analogy with a representation of interactions in quantum electrodynamics in terms of exchange of virtual photons between charged particles. But however useful Feynman diagrams may be as a heuristic device in that theory, this representation fails to mesh with any coherent localized particle ontology for quantum field theory. So the analogy does not make it any easier to see how the component field idea may be extended to yield an account of interactions in quantum field theories.

The status of Mattingly's component fields vis-à-vis the net field is in many ways analogous to that of the gravitational and electric forces acting on a charged body (such as one of Millikan's oil drops) that was the topic of an old debate between Cartwright and Creary (see Cartwright 1983). There Cartwright argues for the facticity of the resultant force while denying that of its gravitational and electric components. She interprets Creary as maintaining that only the component forces are real. While acknowledging that Creary may be right about this case, Cartwright argues against his attempt to generalize it to cover all cases in which causes compose to produce a joint effect. Her argument there (pp. 66–7) is that his general strategy "proliferates influences in every case," while she sees no reason to think that these influences can always be found. I think Mattingly's new account faces just this problem. It postulates an influence (a component field) of every individual moving charge. But, to quote Cartwright,

I think every new theoretical entity which is admitted should be grounded in experimentation, which shows up its causal structure in detail. Creary's influences seem to me just to be shadow occurrences which stand in for the effects we would like to see but cannot in fact find.

Despite appearances, Mattingly's component fields cannot be grounded in experimentation. The problem is not the (formidable!) practical one of isolating every single charge that contributes to the current through the solenoid and detecting the field produced by its motion alone. The problem is rather that any experiment on moving charges is sensitive only to the *net* field (and Dirac phase factor!) of those charges, whether they are one or many. We can mathematically analyze the net field into its components, but what we cannot do is experimentally distinguish their individual from their collective effects.

2.3 Geometry and topology in the Aharonov–Bohm effect

It is often suggested that geometry and topology hold the key to the Aharonov–Bohm effect. In a recent paper, Batterman (2003) entitles his concluding section "The Importance of Geometry" and maintains (p. 554) that

2.3 GEOMETRY AND TOPOLOGY IN THE AHARONOV—BOHM EFFECT

> ... for a full understanding of [the holonomies of curves in phenomena like the AB effect], one needs to appeal to the topology and geometry of the base space. The fiber bundle formulation makes that topology explicit.

He argues (pp 554–5) that

> ... appeal to topological features ... can provide different and better explanations of the phenomena than one might otherwise have if one fails to mention them explicitly. ... In the AB effect, it appears that we will need to refer to different nonseparable holonomy properties for each case in which there is different flux running through the solenoid. The different cases are unified by the topological idealization of the solenoid as a string absent from spacetime which renders spacetime nonsimply connected. In this way we can understand why, for a given fixed magnetic flux, a loop that goes n times around the solenoid will have [a holonomy] that is n times that of a loop that goes around once. This topological feature enables us to understand the common behavior in different AB experiments in a way that the individual appeals to nonseparable holonomy properties of closed loops in spacetime do not.

Batterman's main interest here is methodological rather than metaphysical. He considers a fiber bundle formulation of the relevant physics not just in the Aharonov—Bohm effect but also in a variety of apparently unrelated physical phenomena (the paper's title is "Falling Cats, Parallel Parking and Polarized Light"!) in order to show that this facilitates a common explanatory strategy whose application is required for a full or genuine understanding of them all. But he peppers his discussion with remarks that downplay the importance of interpretative projects such as that on which I am engaged in this book. For example, he calls the debate about what structures in the fiber bundle formulation of the Aharonov—Bohm effect should be taken to represent physically real magnitudes or properties "largely a red herring" (p. 552), and concludes his paper as follows:

> Questions about the reality of gauge potentials just do not seem to arise in many/most of the examples we have discussed. The suggestion is that such questions may not matter much either when it comes to understanding such quantum effects as the AB effect.

I agree with Batterman and others that geometrical and topological considerations can enhance our understanding of the Aharonov—Bohm effect and other phenomena. But they cannot by themselves provide a full explanation of these phenomena. For a full explanation must also involve appeal to the (often different) basic physical principles involved in each phenomenon. Where these are themselves in need of interpretation (as gauge potentials are in the Aharonov—Bohm effect), a full explanation cannot be given without supplying, and defending, the necessary interpretation. Interpretations of electromagnetism in the Aharonov—Bohm effect differ as to the causal or quasicausal story they tell of how that effect comes about. Moreover, those differences bear importantly on the acceptability of various possible interpretations. Since a full explanation of the effect will help us understand not only how it is possible but also how it comes about, we cannot give a full explanation without entering into interpretative debates. As we shall see, there is even a certain tension between a topological and a (quasi)causal explanation of the Aharonov–Bohm effect, in so far as the idealizations required to give the former are impermissible in the latter. A full explanation will not resolve this tension by ignoring or rejecting the demand for a (quasi)causal account of the effect but by showing how this demand may be met while still acknowledging the explanatory relevance of topology.

How can geometry and topology help us to understand the Aharonov–Bohm effect? Recall that in their original 1959 paper, Aharonov and Bohm idealized the solenoid as an infinitely long and infinitely thin "string" occupying a one-dimensional region of space. As Batterman says in the first quote above, there is a further topological idealization of the solenoid as a string absent from space-time which renders space-time non-simply connected. (In fact, my own interest in the Aharonov–Bohm effect was initially sparked when a physicist whose lecture I was auditing claimed that the Aharonov–Bohm effect was due to "the topology of the vacuum"—a claim I found puzzling since, unlike a black hole, a current-carrying solenoid seemed unlikely so radically to affect the background geometry!)

As Batterman (2003, p. 542) stresses, in a fiber bundle formulation

If there is nontrivial holonomy ... and if the connection is *flat*, then *the base space must be nonsimply connected*.

Even though there is assumed to be no, or negligible, (electro)magnetic field in the region outside the solenoid in the magnetic Aharonov–Bohm effect, the holonomies of curves encircling the solenoid are not trivial when it is carrying a suitable current. As explained earlier, in the fiber bundle formulation of classical electromagnetism, the (electro)magnetic field is here represented by the curvature of a principal fiber bundle with typical fiber U(1). Since the curvature is zero—the connection is flat—if and only if the field is zero, the connection on this bundle is flat everywhere in the region. It therefore follows that the region where the field is zero cannot be simply connected. The topological idealization makes this explicit by modelling the Aharonov–Bohm effect by a fiber bundle whose base space is a spatial/spatiotemporal manifold with topology $S \times \mathbb{R}^n$ ($n = 1, 2,$ or 3).

The topology of the field-free region explains the Aharonov–Bohm effect by helping to answer the question "How is it possible for electromagnetism in a region to have an effect there even though the electromagnetic field is zero throughout that region?" The question is answerable only if there is more to electromagnetism in the region than the electromagnetic field there. But even if there *is* something more, it is not clear how it can do the trick. If we assume that the "extra" is represented by a fiber bundle connection whose curvature represents the field, then the effect may still seem impossible because of the flatness of

2.3 GEOMETRY AND TOPOLOGY IN THE AHARONOV–BOHM EFFECT

the connection throughout the region, until we appeal to the non-simply connected character of the base space. From this perspective, topology supplies the key to understanding how the Aharonov–Bohm effect is possible, and the topological idealization both highlights that key and shows how it could unlock barriers to our understanding in a whole range of abstractly similar situations in which quite different physical laws may be operative.

But notice that topology can do its work here without any appeal to the fiber bundle formulation. For suppose instead that we assume that what there is to electromagnetism in a region in which there is no electromagnetic field is represented directly by a (co)vector or one-form field A on a spatial/spatiotemporal manifold, whose curl or exterior derivative represents the (electro)magnetic field strength. Then Stokes' theorem seems to imply that the line integral of A around a closed curve in the region must be zero since its curl or exterior derivative is everywhere zero. But now we can appeal to topology to see how this is possible after all. For in the topological idealization Stokes' theorem does not apply, since the manifold $S \times \mathbb{R}^n$ in which A is non-zero is not compact!

Stokes's theorem applies to extended regions, when it applies at all. But there is another local way of putting essentially the same point. The vanishing of the field in the region outside the solenoid may be expressed in the language of differential forms by the equation $dA = 0$. Now Poincaré proved that, in a manifold homeomorphic to $\mathbb{R}^n$, a form is exact if and only if it is closed. This would imply that $A = d\Lambda$ for some zero-form (i.e. some function) Λ, whereas we have seen that in fact $A = d\Lambda + \omega_h$ in the region outside the solenoid, where the harmonic one-form ω_h is not zero when there is a current through the solenoid. This is possible only because the topology of that region is *not* homeomorphic to $\mathbb{R}^n$, so a non-zero ω_h is not excluded by Poincaré's theorem. This is yet another way in which topology helps answer the question "How is it possible for electromagnetism in a region to have an effect there even though the electromagnetic field is zero throughout that region?" But, again, it is a way that does not involve any reference to the fiber bundle formulation.

In sum, while the appeal to the geometry and topology of the fiber bundle formulation helps to explain the Aharonov–Bohm effect, it does not itself fully explain that effect, because it does not answer every important question one might have about how the effect comes about. Rather, it makes a twofold contribution to our understanding of that effect. First, it provides one answer (though not the only answer) to the question of how the effect is possible at all in the light of various deep mathematical results that may appear to rule it out. Second, it provides a general framework within which one can answer similar questions about otherwise unrelated physical phenomena. The framework can explain, for example, how (consistent with conservation of angular momentum) it is possible for a cat to land on its feet when dropped upside down, and how it is possible to parallel park a car which cannot move sideways.

It is a widely shared intuition that a constitutive trait of a scientific explanation is its ability to *unify* otherwise diverse phenomena, and the fiber bundle framework can indeed unify a wide range of phenomena by revealing abstract similarities that may not be apparent at the level of the quite different basic physical laws governing them. But just because the basic laws are different in each case, a complete explanation in each case must at some stage appeal to the theory that issues in those laws. If it is somehow unclear how that theory should be interpreted, then a fully satisfactory explanation cannot be given without entering into debates about how the theory should be interpreted. This is especially true when different interpretations of a theory disagree in what kind of causal or quasicausal account the theory offers of particularly puzzling phenomena.

What initially strikes one as puzzling about the Aharonov–Bohm effect is not that it seems ruled out by some sophisticated mathematical theorem, but that classical electromagnetism seems incapable of giving any account of how it comes about. When electromagnetic fields are interpolated between electrical and magnetic sources and their effects, this theory permits causal explanations compatible with both relativity and local action of a host of phenomena that otherwise smack of action at a distance. But then its marriage with quantum mechanics entails the occurrence of other phenomena like the Aharonov–Bohm effect for which classical electromagnetic fields seem incapable of providing any local causal explanation. I take this to be the central explanatory problem posed by the Aharonov–Bohm effect. It is a problem that appeals to the geometry and topology of fiber bundles leave untouched. I pursue this problem in the next section.

But note in conclusion that a topological idealization that excises the current-carrying solenoid and the magnetic flux it contains from space-time thereby cuts the ground from under any attempt to offer a complete causal or quasicausal account of how the Aharonov–Bohm effect comes about. For any such account will begin with the current and flux as the clear causal antecedents of a shift in the interference fringes, as Mattingly appreciates. The challenge is to come up with a defensible interpretation of the theory of classical electromagnetism interacting with charged quantum particles that can give a coherent account of the missing links that connect these distal causes to the observed effect.

2.4 Locality in the Aharonov–Bohm effect

The Aharonov–Bohm effect challenges a cherished view of how classical electromagnetism acts. In this view, if electric and magnetic sources are present in a restricted region of space, they give rise to an electromagnetic field which extends outside of that region, taking on values at each point of space which vary continuously from point to point. Any change in the electromagnetic field in a region is responsible for a corresponding change in its values outside that region: a net change in values in a given region may be the resultant of many earlier changes there or elsewhere. In a vacuum, such changes propagate continuously at the speed of light. The effect of electromagnetism on an object located in a region at a time is a function of the electromagnetic field in that region at that time: the effect of electromagnetism on a particle located at space-time point x is a function of the electromagnetic field $F_{\mu\nu}(x)$ at x.

In this view, there are several distinct senses in which classical electromagnetism may be said to act locally. The electromagnetic field strength $F_{\mu\nu}$ is local in so far as its value is defined at each point x of space-time. Moreover, the Lorentz force law 1.5 describes a local interaction between electromagnetism and classical particles, since the force exerted on a charged particle located at x is a function of the value of the field $F_{\mu\nu}$ just at x. Each of these features is related to a different general conception of locality. The first conceives of locality in terms of constitution, while the second understands it as a causal notion. The latter conception may be more familiar, so I begin by stating and explaining two related conditions that attempt to capture its content before returning to the first conception.

Here is one locality condition, due to Einstein (1948, p. 322):

Local Action

If A and B are spatially distant things, then an external influence on A has no immediate effect on B.

The idea behind *Local Action* is that if an external influence on A is to have any influence on B, that effect cannot be immediate; it must *propagate* from A to B via some continuous physical process. Any such mediation between A and B must occur via some (invariantly) temporally ordered and continuous sequence of stages of this process. In the purely classical context, electromagnetism conforms to *Local Action* because an external electromagnetic influence on A (e.g. moving a magnet or completing a circuit) in region R immediately affects charged particles and the field only in region R. If B (e.g. a charged particle or current loop) is located outside of R, then this has no immediate influence on B: rather, the influence must be carried from A to B's location via the continuous propagation of an electromagnetic wave, which is not instantaneous but travels at the speed of light. B is directly affected only by the changes in the electromagnetic field at its location that occur when the wave gets there.

Local Action is closely connected to another locality condition, namely

Relativistic Locality

All the causes of any event lie inside or on its back light-cone.[3]

---

[3] Recall that the back light-cone of an event consists of those space-time points that are connected to that event by a continuous, future-directed, null curve. The name is appropriate in so far as the following condition holds: in a vacuum, a pulse of light emitted in the right direction at any of these points traces out such a curve and so comes into coincidence with the event.

The connection may be seen as follows. In accordance with *Local Action*, the only immediate causes of an event are spatiotemporally coincident with it, and hence (trivially) occur on its back light-cone. Moreover, any spatially distant causes must propagate to it via some (invariantly) temporally ordered sequence of stages of some physical process. Assume that all features of this process are determined by what happens arbitrarily close to the space-time points composing the regions where these stages occur. (This assumption accords with a constitutive conception of locality, and follows from a condition—weak separability—which I shall shortly introduce as an explication of that conception.) According to relativity, all such points lie in or on the event's back light-cone. Hence any such process that conforms to *Local Action* is also in accord with *Relativistic Locality*. But note that *Relativistic Locality* might hold even if *Local Action* failed, as long as all the unmediated causes of an event lay in or on its back light-cone. In the purely classical context, electromagnetism conforms to *Relativistic Locality* because any electromagnetic causes of an event are either coincident with it or connected to it by a process involving the continuous propagation of an electromagnetic wave, which is not instantaneous but travels at the speed of light. All such causes therefore lie in or on the back light-cone of the event.

On a constitutive conception of locality, what happens anywhere is wholly determined by what happens locally. Philosophers have adopted the active term "supervene upon" in preference to the passive "is determined by" to express this kind of determination. In a space-time context, "anywhere" and "locally" are terms referring to a spatiotemporal rather than a spatial region, and "happens" is understood atemporally. Elsewhere (Healey 1997, 2004) I have proposed the following condition as an attempt to capture a constitutive conception of locality.

(Weak) Separability

Any physical process occupying space-time region R supervenes upon an assignment of qualitative intrinsic physical properties at space-time points in R (and/or in arbitrarily small neighborhoods of those points).

The condition of (weak) separability requires some further explanation. According to this principle, whether a process is non-separable depends on what qualitative, intrinsic properties there are. Deciding this involves both conceptual and scientific difficulties. The conceptual difficulty is to say just what it means for a property to be qualitative and intrinsic.

Intuitively, a property of an object is *intrinsic* just in case the object has that property in and of itself, and without regard to any other thing. This contrasts with *extrinsic* properties, which an object has in virtue of its relations, or lack of relations, to other things. Jupiter is intrinsically a body of mass 1.899×10^{27}

kilograms, but only extrinsically the heaviest planet in the solar system.[4] But however intuitive it may be, philosophers continue to disagree about the further analysis of the distinction between intrinsic and extrinsic properties. This is true also of the distinction between qualitative and individual properties, where a property is *qualitative* (as opposed to *individual*) if it does not depend on the existence of any particular individual. Having a mass of 1.899×10^{27} kilograms is a qualitative property of Jupiter, while both the property of being Jupiter and the property of orbiting our sun are individual properties of Jupiter (and the last is also one of its extrinsic properties).

After such an inconclusive resolution of the conceptual difficulty, it may seem premature to consider the scientific difficulty of discovering what qualitative, intrinsic properties there in fact are. But this is not so. Whatever a qualitative, intrinsic property is *in general*, it seems clear that science, and in particular physics, is very much in the business of finding just such properties.

Physics proceeds by first analyzing the phenomena with which it deals into various kinds of systems, and then ascribing states to such systems. To classify an object as a certain kind of physical system is to ascribe to it certain, relatively stable, qualitative intrinsic properties; and to further specify the state of a physical system is to ascribe to it additional, more transitory, qualitative intrinsic properties. Fundamental physics is concerned with the basic kinds of physical systems, and it seeks to characterize the states of these systems so completely as to determine all the physical properties of all the systems these constitute. A physical property of an object will then be both qualitative and intrinsic just in case its possession by that object is wholly determined by the underlying physical states and physical relations of all the basic systems that compose that object. Of course, physics has yet to achieve, and indeed may never achieve, true descriptive completeness in this sense. But to the extent that it is successful, it simultaneously defines and discovers an important class of qualitative intrinsic properties.

What is meant by a process being supervenient upon an assignment of qualitative intrinsic properties at space-time points (or their neighborhoods) in a space-time region R? The idea is familiar. It is that there is no difference in that process without some difference in the assignment of qualitative intrinsic physical properties at space-time points (or their neighborhoods) in a space-time region R. I take the geometric structure of R itself to be independently specified by means of its spatiotemporal properties, where if R is closed it may be necessary to add information on how points in R are related to points just outside it. The supervenience claim is that if one adds to this geometric

---

[4] Note that I follow philosophers' usage rather than physicists' here. I take Jupiter's mass to be intrinsic to it even though Jupiter's mass may vary, or indeed might always have been different, from 1.899×10^{27} kilograms. Physicists tend to use the term 'intrinsic' differently, to refer only to unchanging, or even essential, properties (where an essential property is one that an object could not have lacked while remaining that very object).

structure an assignment of qualitative intrinsic physical properties at space-time points in R (or arbitrarily small neighborhoods of them), then there is physically only one way in which that process can occur.

As it is usually understood, in a wholly classical context, electromagnetism describes processes that are everywhere at least weakly separable. For a complete specification of the electromagnetic state of any region is given by the field $F_{\mu\nu}$ and the current density J_μ at each space-time point x in any region R. And this specification reflects an underlying assignment of qualitative intrinsic physical properties—either at each space-time point x in R, or (more plausibly) in arbitrarily small neighborhoods of each such point. There are a couple of reasons why it is more plausible to suppose that it is not the points but their arbitrarily small neighborhoods at which these properties attach. Any vector or tensor field such as J_μ or $F_{\mu\nu}$ implicitly codes information about *directions* in R as well as scalar magnitudes at points of R; but the specification of a direction at a point in R cannot be given without saying how that point is related to other points in its arbitrarily small neighborhoods (via the tangents to curves connecting it to them). And it is not clear how to understand any density, such as the charge density $J_0(x)$, except as a limit of the ratio of the charge contained in a neighborhood of x to the volume of that neighborhood for smaller and smaller neighborhoods of x. For these reasons, and in order to capture the intended concept of locality in its application to a classical field theory like electromagnetism, it appears necessary to add the parentheses that convert the simple condition of *Separability* into the slightly weaker condition of *Weak Separability*. With that addition we may say that, in a wholly classical context, all electromagnetic processes are weakly separable, as well as conforming to both *Local Action* and *Relativistic Locality*.

But the classical electromagnetic field $F_{\mu\nu}$ does not act locally on charged particles in the Aharonov–Bohm effect, since the interference pattern they produce depends not only on its value (zero) in the region to which they are confined, but also on its values elsewhere (in the magnetic effect it contributes to the flux inside the solenoid). If $F_{\mu\nu}$ exhausts the content of classical electromagnetism, then *Local Action* is violated: changing the current in the solenoid directly affects the spatially distant electrons. This effect is not mediated by the propagation of any intervening continuous physical process. It is true that while the current is changing, a non-zero field $F_{\mu\nu}$ will be produced that propagates outside the solenoid in conformity to *Relativistic Locality*. But this field cannot itself mediate the effect, since the effect is present even when the field outside the solenoid returns to its previous (zero) value. It is as if the region outside the solenoid "retains the memory" of changes in $F_{\mu\nu}$, even though there is no residual electromagnetic field there to store this "memory".

What does remain outside the solenoid in the magnetic effect is whatever "extra" content of classical electromagnetism is represented by the magnetic vector potential **A**—the spatial part of the four-vector potential A_μ. Aharonov and Bohm argued that it was this potential in the region outside the solenoid that affects the electrons as they pass through that region. Feynman (Feynman, Leighton, and Sands 1965a, vol. II, 15-12) was even more explicit:

> In our sense then, the **A**-field is "real." You may say: "But there was a magnetic field." There was, but remember our original idea—that a field is "real" if it is what must be specified *at the position* of the particle in order to get the motion. The **B**-field in the [solenoid] acts at a distance. If we want to describe its influence not as action-at-a-distance, we must use the vector potential.

But even though A_μ takes on values at each space-time point x outside the solenoid during the experiment, there are reasons to doubt that this permits a local account of the effect.

The problem is that it is difficult to interpret the value of A_μ at a point x as representing any qualitative intrinsic physical properties of x, of arbitrarily small regions enclosing x, or of anything located at, or arbitrarily near to, x. But failing such an interpretation, there is no reason to suppose one can give an account of the Aharonov–Bohm effect in terms of the action of $A_\mu(x)$ on something else (like a charged particle or its wave packet) located at or near x. Certainly no properties represented by $A_\mu(x)$ are *observable*, if the combination of quantum mechanics and classical electromagnetism used to account for the Aharonov–Bohm effect exhausts the empirical content of A_μ. For the gauge symmetry of the theory implies that if A_μ and its gauge transform in accordance with 1.6 are physically distinct potentials, representing different distributions of qualitative intrinsic physical properties, no observation or experiment could ever tell them apart. Only gauge-invariant magnitudes are observable, including the electromagnetic field strength $F_{\mu\nu}$ and the Dirac phase factor $\exp\left(\frac{ie}{\hbar}\oint_C A_\mu dx^\mu\right)$. But we have already seen that $F_{\mu\nu}$ alone cannot provide a local account of the action of electromagnetism in the Aharonov–Bohm effect. And we shall soon see that the Dirac phase factor does not permit a local account either.

Now even if the value of $A_\mu(x)$ is not observable, it does not automatically follow that the vector potential has no value at x, nor that this value cannot represent any qualitative intrinsic physical properties at or near x. Only a positivist or instrumentalist would be prepared to make such an inference without further ado. But a scientific realist should also be concerned by the epistemic inaccessibility of any such hypothetical properties, even given the rest of the theory. This concern will be explored and shown to be well founded in chapter 4, which compares and evaluates alternative interpretations of classical gauge theories, including electromagnetism.

Even if electromagnetism in the Aharonov–Bohm effect is separable since the value of the vector potential A_μ at each space-time point x in a region does represent some qualitative intrinsic physical properties in the vicinity of x, this may not permit a local explanation of the effect. Such an explanation requires also some account as to how these properties affect something else (like a charged particle or its wave packet) located there; and it further requires an account of how variations in these properties propagate continuously from the solenoid in accordance with *Local Action* and *Relativistic Locality*. Once more the gauge dependence of A_μ makes it hard to give these accounts.

Any account of the propagation of electromagnetic influences represented by values of A_μ must reckon with the fact that the equations of the theory do not prescribe a unique time evolution for these values: prescribing the values of A_μ everywhere at some initial time (or on an initial space-like hypersurface) determines its values at a later time (on a later space-like hypersurface) only up to an arbitrary gauge transformation 1.6 on the intermediate space-time region that diverges smoothly away from the identity. In an appropriate (or inappropriate!) gauge, these values fail to conform to *Relativistic Locality*. In this way the radical indeterminism of the theory bars the way to a local account of the propagation of electromagnetic influences represented by the vector potential to the region outside the solenoid.

But note that one can always choose a gauge corresponding to Mattingly's "current field" A_μ^m, and in this gauge the propagation of the vector potential is both deterministic and in conformity to *Relativistic Locality*! It is open to one who seeks to give a local account of the Aharonov–Bohm effect, in terms of qualitative intrinsic physical properties in the vicinity of each space-time point represented by the vector potential, to argue that these properties propagate deterministically and in conformity to *Relativistic Locality* in a way that is made apparent by the choice of gauge A_μ^m in which to represent them. Alternative choices of gauge are then equally legitimate, as long as one realizes that each merely offers an alternative representation of the very same local, deterministic evolution.

To complete a local account of the Aharonov–Bohm effect in terms of qualitative intrinsic electromagnetic properties in the vicinity of each space-time point x in an Aharonov–Bohm experiment represented (in some gauge) by $A_\mu(x)$, it is necessary to say how these properties affect charged particles in a region of space-time in which the field strength $F_{\mu\nu}$ is everywhere zero. Quantum mechanically, the particles are assigned a wave-function $\Psi(x)$ whose value at each point $x \equiv (\mathbf{x}, t)$ is a complex number, where the probability of observing a particle in a volume V at time t is proportional to

$$p = \int_V |\Psi(\mathbf{x}, t)|^2 \, d^3x \qquad (2.31)$$

To decide whether the particles are acted on locally, it is necessary to say more about how the particles' wave-function is related to their spatiotemporal location. Note that one assumption about that relation is implicit in the discussion to this point. I have assumed, along with all other commentators of

2.4 LOCALITY IN THE AHARONOV–BOHM EFFECT

whom I am aware, that charged particles are not located in a volume V of space at time t if the probability of observing a particle in that region at that time is zero. This seems to be quite a weak assumption whose denial would render observation remarkably impotent. In accordance with this assumption, it is customary to represent the fact that charged particles do not experience any magnetic field in the (idealized) magnetic Aharonov–Bohm effect by the vanishing of their wave-function inside the solenoid—the only region where the magnetic field is present. This is then taken to justify the claim that particles which form the interference pattern arrive at the screen after passing through the region outside the solenoid. One cannot say more about how the properties of the particles change as they pass through this region without discussing old and fierce debates about the interpretation of quantum mechanics, and especially about the relation between the particles' wave-function and their properties. It is best to postpone this discussion while considering an alternative account of the action of electromagnetism in the Aharonov–Bohm effect.

If the value of the vector potential A_μ at each space-time point x in a region does not represent some qualitative intrinsic physical properties in the vicinity of x, it may be that some function of its integral around each closed curve C in that region does represent such properties of or at (the image of) C. Recall Wu and Yang's discussion of electromagnetism (Wu and Yang 1975), in which they focus on two such functions, the *phase* $\frac{e}{\hbar} \oint_C A_\mu dx^\mu$ and the *Dirac phase factor* $\exp\left(\frac{ie}{\hbar} \oint_C A_\mu dx^\mu\right)$. To these one may add the line integral $\oint_C A_\mu dx^\mu$ itself, which is independent of the charge e of the particles. This last quantity is gauge invariant and so, therefore, are the phase, the phase factor, and all other functions of the line integral of the vector potential around a closed curve C. As a special case, in the equilibrium condition of the magnetic Aharonov–Bohm effect, the line integral $I(C)$ of the magnetic vector potential $\mathbf{A}$ around a closed curve in space is similarly invariant under the gauge transformation 1.2. Since the gauge dependence of the vector potential made it hard to accept Feynman's view that it is a real field that acts locally in the Aharonov–Bohm effect, there is reason to hope that a gauge-invariant function of its line integral around closed curves might facilitate a local account of the action of electromagnetism on quantum particles in the Aharonov–Bohm effect and elsewhere.

But suppose $\oint_C A_\mu dx^\mu$ (or some function of this line integral) does represent qualitative intrinsic physical properties of, or at, the space-time loop R_C that is the image of C, while the value of the vector potential A_μ at each space-time point x in an open space-time region S containing R_C does not represent a qualitative intrinsic physical property of, or at, x. Then classical electromagnetism involves processes that violate even the condition of *Weak Separability*. For an electromagnetic process occupying S will involve qualitative intrinsic physical properties of or at R_C that do not supervene upon an assignment of qualitative intrinsic physical properties at space-time points in R_C or arbitrarily small neighborhoods of them. This move to non-localized properties cannot fulfill the hope of a completely local account of how classical electromagnetism acts on quantum particles in the Aharonov–Bohm effect and elsewhere.

But even if such an account is not separable, it may still satisfy *Local Action* and *Relativistic Locality*! Consider, for example, what happens when the current through the solenoid is increased in the magnetic Aharonov–Bohm effect. This will initially affect the line integral $I(C)$ of the magnetic vector potential **A** only for curves C that tightly circle the solenoid. Changes in $I(C)$ then propagate out (and in) from the surface of the solenoid with the speed of light (see Peshkin and Tonomura 1989, p. 14). Hence the process by which changes in the current in the solenoid affects $I(C)$ may conform both to *Local Action* and to *Relativistic Locality*, even though it violates *Weak Separability*.

To establish conformity to these principles more generally, one can appeal once more to Mattingly's "current field gauge" A_μ^m for the four-vector potential A_μ. In this gauge, the (mathematical) field A_μ propagates continuously at the speed of light. Moreover, the gauge-invariant value of $\oint_C A_\mu dx^\mu$ for closed space-time curve C is determined by the values of A_μ^m on points on R_C. It follows that the value of $\oint_C A_\mu dx^\mu$ for any closed space-time curve C is a function only of the value of $\oint_{C'} A_\mu dx^\mu$ for closed space-time curves C', each point in whose image lies on the back light-cone of a point on R_C. This establishes conformity to *Relativistic Locality*. Conformity of classical electromagnetism to *Local Action* follows as a special case for the propagation of electromagnetic influences between closed curves on space-like hypersurfaces, since only these can be said to have a spatial location. To answer the remaining question as to whether quantum particles are themselves acted on locally by (non-separable) electromagnetism, it is necessary to take up the discussion about the relation between the particles' wave-function and their properties that has been postponed until now.

While there is no consensus on how quantum mechanics should be understood, most interpretations agree that quantum particles do not have continuous trajectories or even occupy a precise position at each moment.[5] Orthodox interpretations take the assignment of a wave-function to an ensemble of similarly prepared particles to offer a complete representation of their properties. But what does this mean? It may be understood as the radical claim that the wave-function has no descriptive significance—that it has the purely instrumental role of permitting statistical predictions of the results of measurements

---

[5] Bohm's (1952) hidden-variable interpretation and its modern developments (see Holland 1993, for example) constitute a significant exception to this generalization. My (1997) paper discusses problems faced by an attempt to give a local Bohmian interpretation of the Aharonov–Bohm effect. Appendix F contains a brief sketch of a number of interpretations of quantum mechanics, including Bohmian mechanics.

on the particles, and wholly fulfills that role (in the sense that there is no supplementary characterization of the particles that would permit more definite predictions of such results). This understanding goes along with a strong version of the Copenhagen interpretation, according to which quantum mechanics simply has nothing to say about a system when it is not being observed. Those who adhere to this version of the Copenhagen interpretation will not ascribe even a non-localized position to quantum particles under the influence of classical electromagnetism under most circumstances: in the magnetic Aharonov–Bohm experiment depicted in figure 2.1 particles would not be ascribed even an indefinite position until observed at the detection screen. One can say nothing at all about the position of the particles between their emission and their detection at the screen. Consequently, no local account of the action of electromagnetism on these particles can be given, even if electromagnetism is localized and propagates in accordance with *Relativistic Locality*.

But there is another way of understanding the completeness claim which goes along with a weaker version of the Copenhagen interpretation. On this version, an individual system may be described by a wave-function somewhat as follows: if the wave-function at some moment is non-negligible only for some set Δ of possible values of some dynamical variable Q, then the system has the dynamical property Q *is restricted to* Δ. For example, even though a hydrogen atom whose wave-function is a superposition of its ground state and first excited state wave-functions has no precise energy, it does have the property *energy is not greater than* -3.4 *electronvolts*. Applied to position, this interpretation implies that a particle may have an imprecise location, being localized only within a region in which its wave-function is non-negligible. This does not, of course, imply that the particle has any component parts or internal spatial structure: it may still be called a "point particle."

On this understanding, quantum mechanics describes the passage of a single charged particle through the apparatus sketched in figure 2.1 as a non-separable process. Specifically, each particle is confined at each moment to one of a continuous sequence of regions, each of which overlaps *both* paths 1 and 2 indicated in the figure. In that sense, each particle's trajectory actually covers loops encircling the solenoid, and so may interact locally with qualitative intrinsic electromagnetic properties on such a loop, whether or not those properties supervene on an assignment of properties at its constituent points. The same is true on other interpretations of quantum mechanics that ascribe such a non-localized position to a particle, including many modal interpretations (including that in my 1989 book), collapse interpretations like that of (Ghirardi, Rimini, and Weber 1986), and some versions of the Everett and consistent histories interpretations. Any such interpretation permits an account of the interaction of classical electromagnetism with quantum particles that conforms to *Local Action* and *Relativistic Locality* even though it violates *Separability*. Arguably, this is an account that satisfies the wish (expressed by

Feynman) to describe the influence of electromagnetism on quantum particles not as action at a distance, in the Aharonov–Bohm effect or elsewhere.

2.5 Lessons for classical electromagnetism

What can be learnt about classical electromagnetism from the Aharonov–Bohm effect? The first, and most important, lesson is that there are three general ways of understanding classical electromagnetism in its application to the quantum mechanics of charged particles (cf. Belot 1998, Lyre 2004). Each would require significant changes from how it is usually understood when applied to classical mechanical systems, but the changes are different in each case.

One can continue to maintain that the only qualitative intrinsic electromagnetic properties are those that may be represented in one or more of the following ways: by the values of the electric and magnetic fields in a given frame, covariantly by the tensor $F_{\mu\nu}$ on the space-time manifold, or by the curvature two-form Ω on a principal U(1) bundle over this manifold. These are localized properties, in the sense that they are predicated of, or at, arbitrarily small neighborhoods of space-time points. On this view, magnetic and electric potentials in a given frame, the four-vector field A_μ on the space-time manifold, and a connection one-form ω on a principal U(1) bundle over this manifold are all elements of surplus mathematical structure that play basically the same calculational role in the theory in slightly different ways; but none of these represents any additional qualitative intrinsic electromagnetic properties. I shall call this the *no new EM properties* view.

The Aharonov–Bohm effect is widely taken to provide powerful reasons to reject this interpretation of classical electromagnetism. Even though the alteration in the interference pattern is a function only of the (electro)magnetic field strength, to suppose that this field is directly responsible for the effect is to accept electromagnetic action at a distance. Such action at a distance has been widely regarded as physically or metaphysically problematic. The replacement of Newton's theory of gravity by Einstein's general relativity completed a program of eliminating action-at-a-distance theories from classical physics that had been greatly advanced by Maxwell's formulation of classical electromagnetism. But the acceptance of the independent reality of electromagnetic fields is intimately connected to their role in eliminating electromagnetic action at a distance, as noted by Feynman in the quote in the previous section. So acceptance of electromagnetic action at a distance in the Aharonov–Bohm effect looks like a step backward in natural philosophy.

Moreover, as Aharonov and Bohm themselves noted, whereas in classical mechanics the fundamental equations of motion (including the Lorentz force law) can always be expressed directly in terms of the electromagnetic fields alone, the potentials cannot be eliminated from basic quantum mechanical equations like the Schrödinger equation. (Electro)magnetic potentials play a vital, and apparently ineliminable, role in the theoretical account of the Aharonov–Bohm effect.[6]

It is on the basis of such considerations that Aharonov and Bohm, Feynman, and many others have rejected this first way of understanding classical electromagnetism in its application to the quantum mechanics of charged particles in favor of an approach that acknowledges that electromagnetic potentials have independent significance and can have physical effects. The most straightforward way of implementing such an approach is to adopt the view that magnetic and electric potentials in a given frame, the four-vector field A_μ on the space-time manifold, and/or a connection one-form ω on a principal U(1) bundle over this manifold themselves represent qualitative intrinsic electromagnetic properties. These properties are still taken to be localized, in the sense that they are predicated of, or at, arbitrarily small neighborhoods of space-time points. But only some of them are represented by localized values of the electromagnetic field. There are taken to be additional "new" localized electromagnetic properties, and it is these that are responsible for the Aharonov–Bohm effect as they affect charged particles even as they pass through a region throughout which the electromagnetic field is zero. I shall call this the *new localized EM properties* view.

This view has been developed and ably defended in different ways recently by several philosophers, including Leeds (1999) and Maudlin (2007). I take Mattingly (2006) also to have advocated a version of this view, though he seems to prefer a description of his current field A_μ^m as newly taken seriously rather than simply new. The view generalizes to an approach to other gauge field theories that I will call the *new localized gauge properties* view. It is a view that deserves the serious consideration it will receive later (in chapter 4) after it becomes clear how to extend it to other Yang–Mills gauge theories and general relativity. Until then it will suffice to point out the kind of difficulties inherent in such a view.

If there are new localized gauge properties, then neither theory nor experiment gives us a good grasp on them. Theoretically, the best we can do is to represent them either by a mathematical object chosen more or less arbitrarily from a diverse and infinite class of formally similar objects related to one another by gauge transformations, or else by this entire gauge-equivalence

---

[6] In fact, several formulations have since been given wholly in terms of electromagnetic fields (see in particular Mandelstam 1962). But the resulting accounts of the Aharonov–Bohm effect are not as simple or natural as the standard account and seem to hide rather than clarify the nature of electromagnetic action at a distance. Mattingly (in press) seems to give a local account in terms of (component) electromagnetic fields. The account requires action at a distance unless these exert physical influences over and above those exerted by the net field. But if they do exert such influences, then they, in effect, represent new local electromagnetic properties. The epistemic status of such new local electromagnetic properties would be highly suspect, as we shall see.

class. This is very different from the way the electromagnetic field is represented by a tensor field on the space-time manifold, where (excluding manifold diffeomorphisms) the only latitude involved in representing its strength by one rather than another tensor field comes from the arbitrary choice of a scale of units. Two vector potentials A_μ, $\bar{A}_\mu$ in a region may have such different forms that it may not be at all obvious at first sight whether they are related by a gauge transformation 1.6: if that region is not simply connected, leading to the same field $F_{\mu\nu}$ is not a sufficient condition for this, as the Aharonov–Bohm effect makes clear. Even if two vector potentials A_μ, $\bar{A}_\mu$ in a region are gauge equivalent, it is not clear whether we can or should take them both to represent the same new localized gauge properties. If they represent different new localized gauge properties, then it is not clear how each can succeed in representing the particular properties it does represent. While if they do represent different new localized gauge properties, then no experiment that can be modeled within the present framework could ever provide evidence relevant to discriminating between their different property assignments. Moreover, if two vector potentials A_μ, $\bar{A}_\mu$ are gauge equivalent everywhere and coincide in their values of **A** on some space-like hypersurface but represent different new localized gauge properties off that hypersurface, then the radically indeterministic evolution of **A** means that the theory can say nothing about whether A_μ rather than $\bar{A}_\mu$ correctly represents the actual new localized gauge properties off that hypersurface.

While agreeing that the Aharonov–Bohm effect shows that there are new electromagnetic properties, the third general way of understanding classical electromagnetism in its application to the quantum mechanics of charged particles maintains that these are non-localized rather than localized. I call it the *new non-localized EM properties* view, and its generalization the *new non-localized gauge properties* view. On the new non-localized EM properties view, while magnetic and electric potentials in a given frame, the four-vector field A_μ on the space-time manifold, and/or a connection 1-form ω on a principal U(1) bundle over this manifold can be used to represent new qualitative intrinsic electromagnetic properties, they do not do so directly. Moreover, the new properties they represent are non-localized not localized, since they are not predicated of, or at, arbitrarily small neighborhoods of space-time points. Rather, only gauge-invariant functions of these mathematically localized fields directly represent new electromagnetic properties; and these are predicated of, or at, arbitrarily small neighborhoods of *loops* in space-time—i.e. oriented images of closed curves on the space-time manifold. There are various candidates as to just which gauge-invariant functions of electromagnetic potentials should be taken directly to represent new non-localized electromagnetic properties. Among the candidates are line integrals of vector potentials including $I(C)$ and $\oint_C A_\mu dx^\mu$, phases, Dirac phase factors, and holonomies. It will be important to consider which candidate offers the best prospects for developing this view, and I will pursue this point in chapter 4. But what is common to all such developments is the striking fact that on this view, while it may turn out to act locally and develop deterministically, classical electromagnetism is not even weakly separable. In my opinion, this is one of the deepest lessons of the Aharonov–Bohm effect for classical electromagnetism. In chapter 4 I shall argue that the empirical evidence provided by experiments verifying the Aharonov–Bohm effect like those of Chambers (1960) and Tonomura *et al.* (1986) provides good reasons to adopt the new non-localized EM properties view, and to accept that classical electromagnetism is not even weakly separable.

3

Classical gauge theories

3.1 Non-Abelian Yang–Mills theories

Although classical electromagnetism was the first gauge theory to be developed, its structure is rather simpler than that of the gauge theories that were subsequently found to describe the weak and strong interactions. It was with the work of Yang and Mills (1954) that the full power and mathematical richness of theories generalizing electromagnetism became apparent. They developed a gauge theory with structure group SU(2) to describe the (approximate) isospin symmetry of the strong force between nucleons (neutrons and protons). This theory became the paradigm for the later empirically successful theories of weak and strong interactions enshrined in what came to be known as the Standard Model of particle physics. Because of this history, gauge theories that share essential elements of common structure with the theory developed by Yang and Mills are now known as Yang–Mills theories. Because classical electromagnetism also shares so much of this structure, I shall include it in the class of Yang–Mills theories. But in one respect it is not typical of this class. Unlike other Yang–Mills theories, the structure group U(1) of electromagnetism is Abelian, i.e. elements g_1, g_2 of the group obey the commutative law

$$g_1 \circ g_2 = g_2 \circ g_1 \tag{3.1}$$

It is only non-Abelian Yang–Mills theories—those whose structure group contains elements that do not obey this law—which exhibit the full richness of the class. It will be important to appreciate this richness, since it must be encompassed in any adequate interpretation of Yang–Mills gauge theories.

A classical Yang–Mills gauge theory describes a field whose effect on particles may be represented by altering the "base line" for comparing the generalized phases of their wave-function at different points. The value of the wave-function at a space-time point is now a vector in a complex inner product space, and its generalized phase corresponds to a direction in that abstract space. A gauge transformation applied to the wave-function constitutes a rotation in this phase direction that varies smoothly from point to point. The phase transformation 1.11 generalizes to

$$\psi' = \mathbf{U}\psi \tag{3.2}$$

where $\mathbf{U}(x)$ is an appropriate $n \times n$ matrix at each space-time point x. If this is to correspond to a generalized rotation, it cannot change the length of the vector $\psi(x)$, and so $\mathbf{U}(x)$ must be a unitary matrix—$\mathbf{U}\mathbf{U}^\dagger = \mathbf{U}^\dagger\mathbf{U} = \mathbf{1}$. A general $n \times n$ unitary matrix $\mathbf{U}$ may be written as

$$\mathbf{U} = \exp[-(ig/\hbar)\mathbf{\Lambda}] \tag{3.3}$$

where $\mathbf{\Lambda}$ is an $n \times n$ Hermitian matrix, and an arbitrary constant g has been inserted that will turn out to play the role of a coupling constant for the theory, generalizing the electric charge e.[1] So, for an infinitesimal gauge transformation, we have

$$\psi' = (\mathbf{1} - (ig/\hbar)\mathbf{\Lambda})\psi \tag{3.4}$$

This is a linear transformation, and so the matrix $\mathbf{\Lambda}$ may be expanded in terms of a basis of linearly independent $n \times n$ Hermitian matrices T_a as

$$\mathbf{\Lambda} = \sum_a \Lambda^a T_a \tag{3.5}$$

This represents a generalized infinitesimal rotation if and only if $\mathbf{U}$ is an element of a representation of a Lie group and the T_a represent the generators of the Lie algebra of this group, with commutation relations[2]

$$[T_a, T_b] = f_{abc} T_c \tag{3.6}$$

where the f_{abc} are the *structure constants* of the algebra. For example, the three independent generators of the Lie algebra of the group SU(2) satisfy the relations

$$[T_i, T_j] = i\epsilon_{ijk} T_k \quad \text{(for } i, j, k = 1, 2, 3) \tag{3.7}$$

---

[1] The *adjoint* $M^\dagger$ of an $n \times n$ matrix M with components M_{ij} is an $n \times n$ matrix with components M_{ji}^*, where the * operation represents complex conjugation. M is *Hermitian* just in case $M^\dagger = M$. The *trace* of M is the sum of its diagonal elements: $\text{Tr}(M) \equiv \sum_{i=1}^n M_{ii}$.

[2] $[T_a, T_b]$ means $T_a T_b - T_b T_a$. A Lie group is a continuous group of transformations. Appendix B details the relation between a Lie group and its Lie algebra. Any element of the group may be reached from the identity element by a sequence of infinitesimal transformations, each specified by a linear combination of the generators of its Lie algebra.

where $\epsilon_{ijk} = +1(-1)$ if $\{ijk\}$ is an even(odd) permutation of $\{123\}$ and $\epsilon_{ijk} = 0$ otherwise. These generators are represented by traceless Hermitian matrices. Acting on two-component spinor[3] wave-functions ψ, **T** equals $\frac{1}{2}\sigma$, where the components of σ (in one representation) are the Pauli matrices

$$\sigma_1 = \begin{pmatrix} 0 & 1 \\ 1 & 0 \end{pmatrix}, \quad \sigma_2 = \begin{pmatrix} 0 & -i \\ i & 0 \end{pmatrix}, \quad \sigma_3 = \begin{pmatrix} 1 & 0 \\ 0 & -1 \end{pmatrix}$$

By analogy with electromagnetism, a gauge potential is associated with a Yang–Mills gauge theory. Its role is to define what is to count as the same generalized phase at different space-time points. This it does by specifying parallelism of the wave-function in an infinitesimal neighborhood of each point. Since the wave-function is now an n-component object, the generalization of the electromagnetic four-vector potential A_μ is now an $n \times n$ matrix $\mathbf{A}_\mu$. The generalized phases at x and $x + dx$ are to be regarded as parallel just in case they differ by an amount $\frac{g}{\hbar}\mathbf{A}_\mu(x)dx^\mu$. So a wave-function ψ will have a value at $x + dx$ that is parallel to the value of ψ at x just in case

$$\bar{\psi}(x + dx) = \exp((ig/\hbar)\,\mathbf{A}_\mu(x)dx^\mu)\psi(x) \tag{3.8}$$

Here $\mathbf{A}_\mu$ is a Hermitian matrix of the form

$$\mathbf{A}_\mu = \sum_a A_\mu^a T_a \tag{3.9}$$

For a theory with structure group $SU(n)(n = 2, 3, \ldots)$ the T_a are traceless.

For this definition of parallelism of generalized phases to be consistent with the symmetry of the theory under variable generalized phase transformations 3.2, it must be applicable whatever gauge one uses to represent the wave-function. Under a gauge transformation 3.2, the value of $\bar{\psi}(x)$ at $x + dx$ becomes

$$\bar{\psi}'(x + dx) = \mathbf{U}(x + dx)\bar{\psi}(x + dx) \tag{3.10}$$

This is still to be regarded as parallel to $\psi'(x)$, so if $\mathbf{A}_\mu$ transforms to $\mathbf{A}'_\mu$, then

$$\bar{\psi}'(x) = \exp((ig/\hbar)\,\mathbf{A}'_\mu(x)dx^\mu)\psi'(x) \tag{3.11}$$

---

[3] A spinor is just a vector in a two-dimensional complex vector space; and so each of its components in a basis for that space is a complex number.

Hence, for all $\boldsymbol{\psi}(x)$,

$$\mathbf{U}(x+dx)\exp\left((ig/\hbar)\,\mathbf{A}_\mu(x)dx^\mu\right)\boldsymbol{\psi}(x) = \exp\left((ig/\hbar)\,\mathbf{A}'_\mu(x)dx^\mu\right)\mathbf{U}(x)\boldsymbol{\psi}(x) \tag{3.12}$$

Therefore

$$\left(\mathbf{U}(x)+\partial_\mu\mathbf{U}(x)dx^\mu\right)\left(\mathbf{1}+(ig/\hbar)\,\mathbf{A}_\mu(x)dx^\mu\right) = \left(\mathbf{1}+(ig/\hbar)\,\mathbf{A}'_\mu(x)dx^\mu\right)\mathbf{U}(x) \tag{3.13}$$

and so

$$\mathbf{A}'_\mu = \mathbf{U}\mathbf{A}_\mu\mathbf{U}^\dagger - \left(\frac{i\hbar}{g}\right)(\partial_\mu\mathbf{U})\mathbf{U}^\dagger \tag{3.14}$$

Equation 3.14 consequently specifies how the generalized four-vector gauge potential $\mathbf{A}_\mu$ must transform when the wave-function is subjected to a variable generalized phase transformation of the form 3.2 in order for these joint transformations to constitute a symmetry of the theory. It is the generalization of 1.6, to which it reduces in the case of an Abelian theory.

What is the corresponding generalization of the electromagnetic field tensor $F_{\mu\nu}$? The easiest way to answer this question is to explain how the field is related to phase changes around infinitesimal closed curves. So consider the infinitesimal parallelogram $ABCD$ shown in figure 3.1.

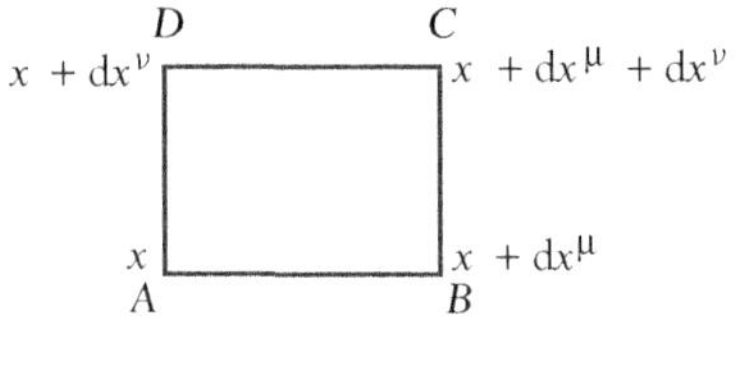

```
        D                    C
   x + dx^ν ┌──────────┐ x + dx^μ + dx^ν
            │          │
            │          │
            │          │
            └──────────┘
         x              x + dx^μ
         A              B
```

Figure 3.1.

If the wave-function $\boldsymbol{\psi}(x)$ at A is parallel-transported to B the result is

$$\boldsymbol{\psi}_{AB}(x) = \exp\left((ig/\hbar)\,\mathbf{A}_\mu(x)dx^\mu\right)\boldsymbol{\psi}(x) \tag{3.15}$$

or, since dx^μ is infinitesimal,

$$\boldsymbol{\psi}_{AB}(x) = \left(\mathbf{1}+(ig/\hbar)\,\mathbf{A}_\mu(x)dx^\mu\right)\boldsymbol{\psi}(x) \tag{3.16}$$

continuing to C gives

$$\boldsymbol{\psi}_{ABC}(x) = \left(\mathbf{1}+(ig/\hbar)\,\mathbf{A}_\nu(x+dx^\mu)dx^\nu\right)\boldsymbol{\psi}_{AB}(x) \tag{3.17}$$

continuing around the parallelogram to D,

$$\psi_{ABCD}(x) = (1 - (ig/\hbar) \mathbf{A}_\mu(x + dx^\nu)dx^\mu)\psi_{ABC}(x) \qquad (3.18)$$

and returning finally to A,

$$\psi_{ABCDA}(x) = (1 - (ig/\hbar) \mathbf{A}_\nu(x)dx^\nu)\psi_{ABCD}(x) \qquad (3.19)$$

Hence we have

$$\psi_{ABCDA}(x) = (1 - (ig/\hbar) \mathbf{A}_\nu(x)dx^\nu)(1 - (ig/\hbar) \mathbf{A}_\mu(x + dx^\nu)dx^\mu) \qquad (3.20)$$
$$\times (1 + (ig/\hbar) \mathbf{A}_\nu(x + dx^\mu)dx^\nu)(1 + (ig/\hbar) \mathbf{A}_\mu(x)dx^\mu)\psi(x)$$

which implies, to second order in infinitesimals,

$$\psi_{ABCDA}(x) = \{1 - (ig/\hbar)(\partial_\nu \mathbf{A}_\mu - \partial_\mu \mathbf{A}_\nu) + g^2/\hbar^2 [\mathbf{A}_\mu, \mathbf{A}_\nu]\} dx^\mu dx^\nu \psi(x) \qquad (3.21)$$

Now if the phase change around $ABCDA$ is written as $\frac{g}{\hbar}\mathbf{F}_{\mu\nu}dx^\mu dx^\nu$ then

$$\psi_{ABCDA}(x) = (1 + (ig/\hbar) \mathbf{F}_{\mu\nu}dx^\mu dx^\nu) \psi(x) \qquad (3.22)$$

from which it follows that

$$\mathbf{F}_{\mu\nu} = (\partial_\mu \mathbf{A}_\nu - \partial_\nu \mathbf{A}_\mu) - (ig/\hbar) [\mathbf{A}_\mu, \mathbf{A}_\nu] \qquad (3.23)$$

or, if we write

$$\mathbf{F}_{\mu\nu} = \sum_a F^a_{\mu\nu} T_a \qquad (3.24)$$

$$F^a_{\mu\nu} = \partial_\mu A^a_\nu - \partial_\nu A^a_\mu - (ig/\hbar) f_{abc} A^b_\mu A^c_\nu \qquad (3.25)$$

In the Abelian case this reduces to 1.7,

$$F_{\mu\nu} = \partial_\mu A_\nu - \partial_\nu A_\mu \qquad (3.26)$$

thus confirming in this case the relation between the gauge field strength $\mathbf{F}_{\mu\nu}(x)$ and the phase change around a closed infinitesimal curve at x.

In an Abelian theory like classical electromagnetism, the field $F_{\mu\nu}$ is invariant under gauge transformations. But this is not so for a non-Abelian theory. Under the transformation 3.14 the field $\mathbf{F}_{\mu\nu}$ transforms as follows:

$$\mathbf{F}'_{\mu\nu}(x) = \mathbf{U}(x)\mathbf{F}_{\mu\nu}(x)\mathbf{U}^\dagger(x) \tag{3.27}$$

as is most easily seen from 3.22 and 3.2, but can be confirmed also using 3.14 and 3.23.

So far only infinitesimal parallel transport of generalized phases has been considered. What is the generalization of the phase factor $\exp\left(\frac{ie}{\hbar}\int_C A_\mu dx^\mu\right)$ for electromagnetism that defines sameness of phase along a (possibly open) curve C? Equation 3.8 cannot be simply extended to a finite curve by exponentiating the line integral along the curve, since the values $\mathbf{A}_\mu(x)$ at different points x along the curve do not commute for a non-Abelian theory. Instead, one defines a so-called path-ordered exponential

$$\wp \exp\left\{(ig/\hbar)\int_C \mathbf{A}_\mu(x)dx^\mu\right\} \tag{3.28}$$

as follows. Suppose that the matrices $\mathbf{A}_\mu(x_i)$ ($i = 1,..,n$) all lie on the parametrized curve $C(s)$ at parameter values $\{s_1, s_2, \ldots, s_n\}$. Let σ be a permutation of these parameter values such that each element in the sequence $\{s_{\sigma(1)}, s_{\sigma(2)}, \ldots, s_{\sigma(n)}\}$ is no bigger than the preceding element. Then the path-ordered product of n matrices $\wp\{\mathbf{A}_{\mu_1}(x_1)\mathbf{A}_{\mu_2}(x_2) \ldots \mathbf{A}_{\mu_n}(x_n)\}$ is their product permuted so that $\mathbf{A}_{\mu_i}(x_i)$ precedes $\mathbf{A}_{\mu_j}(x_j)$ only if $s_{\sigma(i)} \leq s_{\sigma(j)}$. Now the integral

$$\int_{s \geq s_1 \geq s_2 \geq \cdots \geq s_n \geq 0} \mathbf{A}_{\mu_1}(x_1)\mathbf{A}_{\mu_2}(x_2) \ldots \mathbf{A}_{\mu_n}(x_n) dx^{\mu_1} dx^{\mu_2} \ldots dx^{\mu_n} \tag{3.29}$$

is equal to

$$\frac{1}{n!}\int_{s_i \in [0,s]} \wp\{\mathbf{A}_{\mu_1}(x_1)\mathbf{A}_{\mu_2}(x_2) \ldots \mathbf{A}_{\mu_n}(x_n)\} dx^{\mu_1} dx^{\mu_2} \ldots dx^{\mu_n} \tag{3.30}$$

which may be written as

$$\frac{1}{n!}\wp\left\{\int_0^s \mathbf{A}_\mu(x)dx^\mu\right\}^n \tag{3.31}$$

Finally, the path-ordered exponential is defined by

$$\wp \exp\left\{(ig/\hbar)\int_C \mathbf{A}_\mu(x)dx^\mu\right\} \equiv \sum_0^\infty \frac{1}{n!}\wp\left\{(ig/\hbar)\int_C \mathbf{A}_\mu(x)dx^\mu\right\}^n \quad (3.32)$$

This transforms under the gauge transformation 3.14 as follows:

$$\wp \exp\left\{(ig/\hbar)\int_{x_1}^{x_2} \mathbf{A}_\mu(x)dx^\mu\right\} \Rightarrow \mathbf{U}(x_2)\wp \exp\left\{(ig/\hbar)\int_{x_1}^{x_2} \mathbf{A}_\mu(x)dx^\mu\right\}\mathbf{U}^\dagger(x_1)$$
$$(3.33)$$

Hence the non-Abelian generalization of the Dirac phase factor is not invariant but transforms more like the gauge field $\mathbf{F}_{\mu\nu}$ under gauge transformations.

3.1.1 The fiber bundle formulation

The formulation of classical Yang–Mills gauge theories offered in the previous section did not exploit the elegant geometric framework of fiber bundles that has become increasingly common since the work of Wu and Yang (1975), Trautman (1980), and many others. The fiber bundle formulation presents an illuminating perspective on these theories even though it is not indispensable, and it has influenced philosophical reactions to gauge theories as well as giving rise to a great deal of interesting mathematics. A fiber bundle formulation of classical electromagnetism was sketched in section 1.2. The generalization to other classical Yang–Mills gauge theories is straightforward; further details may be found in appendix B.

One illustration of the power of the formalism is provided by the following more transparent derivation of the transformation law 3.14. This follows from the requirement that a theory involving derivatives of the wave-function ψ be symmetric under the variable generalized phase transformation 3.2. For this to hold, the derivatives must transform in the same way as ψ, i.e. they must appear in the equations as *covariant derivatives* D rather than the corresponding coordinate derivatives ∂. In the fiber bundle formulation, the wave-function ψ is represented by a section of a vector bundle, whose covariant derivative may be expressed as

$$D_\mu \psi = (\partial_\mu + \mathbf{A}_\mu)\psi \quad (3.34)$$

(see appendix B). This equation will be covariant under 3.2 just in case

$$D_\mu(\psi') = \mathbf{U}(D_\mu\psi) \quad (3.35)$$

3.1 NON-ABELIAN YANG–MILLS THEORIES

But the covariant derivative of the section representing the transformed wavefunction may be expressed directly in terms of the transformed components $\mathbf{A}'_\mu$ as

$$D_\mu(\psi') = \left(\partial_\mu + \mathbf{A}'_\mu\right)\psi' \qquad (3.36)$$

Equating these last two expressions, we get

$$\mathbf{U}\left(\partial_\mu \psi + \mathbf{A}_\mu \psi\right) = \partial_\mu (\mathbf{U}\psi) + \mathbf{A}'_\mu \mathbf{U}\psi \qquad (3.37)$$

From which it follows that

$$\mathbf{A}'_\mu = \mathbf{U}\mathbf{A}_\mu \mathbf{U}^\dagger - \left(\partial_\mu \mathbf{U}\right)\mathbf{U}^\dagger \qquad (3.38)$$

which is 3.14, except for a numerical factor in the second term on the right, stemming from differing definitions of the "components" of $\mathbf{A}_\mu$.

The curvature $F(X, Y)$ is defined by its action on an arbitrary section s, namely

$$F(X, Y) = D_X D_Y - D_Y D_X - D_{[X,Y]} \qquad (3.39)$$

It components are therefore

$$\mathbf{F}_{\mu\nu} = D_\mu D_\nu - D_\nu D_\mu - D_{[\partial_\mu, \partial_\nu]} \qquad (3.40)$$
$$= D_\mu D_\nu - D_\nu D_\mu \text{ (since } [\partial_\mu, \partial_\nu] = 0) \qquad (3.41)$$
$$= \partial_\mu \mathbf{A}_\nu - \partial_\nu \mathbf{A}_\mu + [\mathbf{A}_\mu, \mathbf{A}_\nu] \qquad (3.42)$$

which is 3.23, except for the now familiar numerical factor. Under the transformation 3.2 this implies that

$$\mathbf{F}_{\mu\nu}\psi' = (D_\mu D_\nu - D_\nu D_\mu)\mathbf{U}\psi \qquad (3.43)$$
$$= [D_\mu(\mathbf{U}D_\nu) - D_\nu(\mathbf{U}D_\mu)]\psi \qquad (3.44)$$
$$= \mathbf{U}(D_\mu D_\nu - D_\nu D_\mu)\psi \qquad (3.45)$$
$$= \left(\mathbf{U}\mathbf{F}_{\mu\nu}\mathbf{U}^\dagger\right)\psi' \qquad (3.46)$$

where 3.35 has been used repeatedly. Hence $\mathbf{F}_{\mu\nu}$ transforms as follows under the joint transformations 3.2, 3.35:

$$\mathbf{F}_{\mu\nu} = \left(\mathbf{U}\mathbf{F}_{\mu\nu}\mathbf{U}^\dagger\right) \qquad (3.47)$$

Thus regarding the gauge potential as determining the covariant derivative corresponding to a connection on a vector bundle whose curvature represents the gauge field strength illuminates both the relation between the gauge potential $\mathbf{A}_\mu$ and the field strength $\mathbf{F}_{\mu\nu}$, and also what transformation properties these must have under a variable generalized phase transformation in ψ for this to be a symmetry of a Yang–Mills gauge theory.

But the fiber bundle formulation provides a more basic representation of the gauge potential by means of a connection ω—a Lie-algebra-valued one-form—on a principal fiber bundle $P(M, G)$ to which such a vector bundle may be associated. Here M is the space-time manifold (or some submanifold of it) and G is an abstract Lie group (for example, SU(2) or SU(3)). The gauge field strength is then represented by a Lie-algebra-valued two-form Ω on $P(M, G)$—the covariant derivative of ω. The wave-function of quantum particles is a section of an associated vector bundle (E, M, G, π, V, P), where the typical fiber V is a vector space on which acts a representation of G. If G is SU(2), this may be the two-dimensional space of spinors—two-component complex vectors. But SU(2) (for example) also has representations on vector spaces of different dimensionality. By representing the gauge field on a principal bundle, one leaves open what vector representation of the abstract structure group G is appropriate for the wave-functions of any particles that may or may not be interacting with the gauge field. This is important not just because it makes the principal bundle representation more flexible, but because of its significance for understanding the nature of the gauge field in itself, whether or not it is thought of as interacting with other matter.

As we shall see in more detail in subsequent chapters, the Yang–Mills gauge fields of the Standard Model are taken to interact not with particles represented by wave-functions, but with quantized matter fields. Such a matter field may be represented by a section of a vector bundle which is associated to a principal bundle whose connection and curvature represent the gauge field. But what kinds of matter fields there are, and so what these associated vector bundles are like, in no way affects the representation of a gauge field on its principal fiber bundle. There is a perspective that views matter fields as somehow primary, or more basic than gauge fields. This perspective is one motivation for the so-called gauge argument to be examined in chapter 6. It is implicit even in the terminological distinction between matter and force that is typically used to contrast matter fields and their associated particles (electrons, quarks, etc.) with gauge fields and their associated particles (photons, gluons, etc.). There are important physical reasons for drawing a distinction, many stemming from the fact that particles of the first sort are fermions, whose quantized field operators obey *anti*commutation relations, whereas particles of the second sort are bosons, whose quantized field operators obey commutation relations. But both kinds of entities are fundamental to the empirical success of theories of the Standard Model, and there is no more sense contemplating a world of

3.1 NON-ABELIAN YANG–MILLS THEORIES

matter acted on by no forces than a world of forces acting on nothing but themselves. Indeed, because the "carriers" of forces described by quantized non-Abelian gauge theories themselves interact directly through those forces, the latter fantasy world seems a bit closer to home.[4,5]

A connection ω—a Lie-algebra-valued one-form—on a principal fiber bundle $P(M, G)$ may be represented on (an open subset of) the space-time manifold M by its pull-back with a (local) section σ—a Lie-algebra-valued one-form $\mathcal{A}$ on M,

$$\mathcal{A} = \sigma^* \omega \tag{3.48}$$

If σ_1, σ_2 are each local sections defined on (an open subset of) M, and related by $\sigma_2(x) = \sigma_1(x) g(x)$, then the corresponding representatives of ω will be related by

$$\mathcal{A}_2 = g^{-1} \mathcal{A}_1 g + g^{-1} dg \tag{3.49}$$

which may be written as

$$\mathcal{A}_{2\mu} = g^{-1} \mathcal{A}_{1\mu} g + g^{-1} \partial_\mu g \tag{3.50}$$

and again compared to 3.14. Since $\mathcal{A}$ is a Lie-algebra-valued one-form, it may be expanded in terms of a basis $\{T_a\}$ of the Lie algebra as follows:

$$\mathcal{A} = \sum_a A_\mu^a(x) T_a dx^\mu \tag{3.51}$$

In the case of an infinitesimal gauge transformation, the group element $g(x)$ near the identity may be written in terms of this basis as

$$g(x) = (\mathbf{1} - \theta(x)) = (\mathbf{1} - \theta^a(x) T_a) \tag{3.52}$$

in which case 3.50 becomes

$$\begin{aligned} A_{2\mu}^a T_a &= (\mathbf{1} + \theta) A_{1\mu}^a T_a (\mathbf{1} - \theta) + (\mathbf{1} + \theta) - \partial_\mu \theta \\ &= A_{1\mu}^a T_a - \theta^b A_{1\mu}^a f_{abc} T_c - \partial_\mu \theta^a T_a \end{aligned} \tag{3.53}$$

---

[4] A colleague suggested to me that for a God to create such a world would be analogous to creating a world with Eve but not Adam. Perhaps—but it is not only feminists who will take that analogy to make my point!

[5] Impatient readers may wish to skip the rest of this section, at least on a first reading.

giving the following transformation law for the components of A^a:

$$A^a_{2\mu} = A^a_{1\mu} + f_{abc}\theta^b A^c_{1\mu} - \partial_\mu \theta^a \quad (3.54)$$

which may be compared to the corresponding transformation law for the Abelian case 1.6.

Similarly, Ω may be represented on (an open subset of) M by its pull-back with a (local) section σ—a Lie-algebra-valued two-form $\mathcal{F}$ on M,

$$\mathcal{F} = \sigma^* \Omega \quad (3.55)$$

This is related to the corresponding $\mathcal{A}$ by

$$\mathcal{F} = d\mathcal{A} + \mathcal{A} \wedge \mathcal{A} \quad (3.56)$$

where d is the exterior derivative on M (see appendix B, equation B.35). $\mathcal{F}$ may be written as

$$\mathcal{F} \equiv \frac{1}{2}\mathcal{F}_{\mu\nu} dx^\mu \wedge dx^\nu \quad (3.57)$$

in which case 3.56 becomes

$$\mathcal{F}_{\mu\nu} = \partial_\mu \mathcal{A}_\nu - \partial_\nu \mathcal{A}_\mu + [\mathcal{A}_\mu, \mathcal{A}_\nu] \quad (3.58)$$

Under the change of section $\sigma_1(x) \Rightarrow \sigma_2(x) = \sigma_1(x)g(x)$, $\mathcal{F}_{\mu\nu}$ transforms as follows:

$$\mathcal{F}_{\mu\nu 2} = g^{-1} \mathcal{F}_{\mu\nu 1} g \quad (3.59)$$

If $\mathcal{F}_{\mu\nu}$ is also expanded in terms of the basis $\{T_a\}$ of the Lie algebra as

$$\mathcal{F}_{\mu\nu} = \sum_a F^a_{\mu\nu} T_a dx^\mu dx^\nu \quad (3.60)$$

then 3.58 gives

$$F^a_{\mu\nu} = \partial_\mu A^a_\nu - \partial_\nu A^a_\mu + f_{abc} A^b_\mu A^c_\nu \quad (3.61)$$

an equation that differs from 3.25 only through the same alternative numerical choice in how to define A^a_μ (and so also $F^a_{\mu\nu}$) that we encountered in connection with the transformation law for $\mathbf{A}'_\mu$. $F^a_{\mu\nu}$ transforms under change of section as follows (cf. 3.47):

$$F^a_{\mu\nu 2}(x) = g^{-1}(x) F^a_{\mu\nu 1}(x) g(x) \tag{3.62}$$

If C is a closed curve in M with base point m, then its horizontal lifts $\tilde{C}$ map $\pi^{-1}(m)$ onto itself, defining the holonomy map

$$\tau_m : \pi^{-1}(m) \to \pi^{-1}(m) \tag{3.63}$$

corresponding to the connection ω on the principal fiber bundle $P(M, G)$. This is given explicitly in a given section with $\sigma(m) = u$ by the action of a group element called the holonomy of C (see appendix B)

$$\mathcal{H}_m(C) = \wp \exp\left\{-\oint_C \mathcal{A}_\mu(x) dx^\mu\right\} \tag{3.64}$$

where the action is compatible with the action of the group on $\pi^{-1}(m)$

$$\mathcal{H}_m(C)(ug) = \mathcal{H}_m(C)(u)g \tag{3.65}$$

Under a gauge transformation corresponding to a change of section $\sigma_1(x) \to \sigma_2(x) = \sigma_1(x) g(x)$ the holonomy transforms as

$$\mathcal{H}_{\sigma_2(m)} = g^{-1}(m) \mathcal{H}_{\sigma_1(m)} g(m) \tag{3.66}$$

Note that the holonomies of all closed curves with base point m transform in the same way under change of sections.

If ω is a connection on $P(M, G)$, and h is a vertical automorphism of P, then $h^*\omega$ is also a connection on P. Their holonomy maps $\tau_\omega, \tau_{h^*\omega}$ are related as follows. Let σ be a section on M. The holonomy maps are defined by their holonomies in this section, acting on $u = \sigma(m)$. As appendix B shows, they transform as

$$\mathcal{H}_{m,h^*\omega} = g^{-1}(m) \mathcal{H}_{m,\omega} g(m) \tag{3.67}$$

for an element $g(m)$ of the structure group G which is independent of the curves C on which they act. This is the transformation rule for a holonomy under a vertical bundle automorphism, corresponding to what Trautman (1980) called a gauge transformation of the second kind. Holonomies related in this way form an equivalence class. It follows that, for an Abelian structure group, holonomies are invariant under such transformations.

70 3 CLASSICAL GAUGE THEORIES

Two curves $C_{m'}, C_m$ with the same image on M but different base points m, m' have related holonomies:

$$\mathcal{H}_{m',\omega} = g^{-1}(\gamma_{mm'})\mathcal{H}_{m,\omega} g(\gamma_{mm'}) \qquad (3.68)$$

where $\gamma_{mm'}$ is a curve from m to m', and $g(\gamma)$ is defined by

$$g(\gamma) = \wp \exp\left\{-\int_\gamma \mathcal{A}_\mu(x)\mathrm{d}x^\mu\right\} \qquad (3.69)$$

It follows that, for an Abelian structure group, holonomy maps as well as holonomies are gauge independent and independent of choice of base point m.

3.1.2 Loops, groups, and hoops

One can view the holonomies of a principal fiber bundle as representations of an underlying group which is sometimes referred to as a group of loops. This representation permits a more intrinsic formulation of a Yang–Mills gauge theory whose conceptual significance will be a subject of investigation in later chapters. Since the term 'loop' has multiple uses in this field, a prerequisite for this investigation will be to make my own terminology clear.

Begin with the idea of a continuous path in space or space-time without any "kinks." A theory models such a path by means of an appropriately smooth curve in a manifold M representing space-time. A *curve* C is a map from the interval $I = [0, 1]$ of real numbers into M: elements of I are called the *parameters* of C. Its *beginning point* is the image of 0, and its *end point* is the image of 1. A curve is *closed* if its beginning point and its end point are a single point m of M—its *base point*: otherwise it is *open*. The image of a *trivial* curve coincides with its base point. A curve intersects itself at a point $m \in M$ when m is the image of more than one element of I. An open curve is non-self-intersecting if there is no point of its image at which it intersects itself; but a closed curve is *non-self-intersecting* if it intersects itself only at its base point, and this is the image of no point of I except $0, 1$. A curve is *continuous* (C) if and only if the coordinates of the image of $C(s)$ are continuous functions of the parameter s, in the domain of any coordinate chart on M; it is n times *differentiable* (C^n) if and only if these functions are n times differentiable—a (C^∞) curve is said to be *smooth*.

Suppose C_1, C_2 are curves, and that the end point $m \in M$ of C_1 is also the beginning point of C_2. Then one can *compose* C_1, C_2 by defining a curve $C_2 \circ C_1$ whose image is the union of the images of C_1, C_2; $C_2 \circ C_1$ traces out that image by first tracing out the image of C_1, then tracing out the image of C_2. Closed curves with the same base point may always be composed in this way. A curve is continuous and *piecewise smooth* if and only if it is continuous and composed of a finite number of segments, each of which is a smooth curve. There is an equivalence class of curves associated with each curve C that includes all other closed curves that share its image and result from it by orientation-preserving reparametrizations.[6] Such an equivalence class constitutes an unparametrized curve. Whenever it is defined, the composition of two parametrized curves extends naturally to the corresponding unparametrized curves.

The image of a (non-trivial) continuous and piecewise smooth, non-self-intersecting closed curve C is a one-dimensional region of the manifold M on which the curve is defined. It inherits an orientation from the way that C is traced out by the parameter s. A space-time theory uses this region to represent a corresponding oriented one-dimensional region L of space or space-time. It is this closed, oriented, one-dimensional region of space or space-time that I shall call a *loop*.

What has sometimes been called the group of loops (e.g. by Gambini and Pullin 1996) does not concern loops in this sense, but something more abstract. Return to the idea of a closed curve on M. Note that the set of closed curves as defined does not form a group under the composition operation, since a curve has no natural inverse under this operation (retracing a curve in the reverse direction simply results in a longer (self-intersecting) curve with the same image). To impose a group structure, it is therefore necessary to form appropriate equivalence classes of closed curves.

Restrict attention now to unparametrized closed curves with a common base point o of M. These form a semi-group under the operation $\circ$ of composition of curves, where the identity element is the trivial curve whose image is simply o. But there is still no inverse, since the image of a curve $C \circ \overline{C}$ that results from composing an arbitrary curve C with its retracing in the opposite orientation $\overline{C}$ is not just the trivial curve at o.

To arrive at a group structure, it is necessary to widen the equivalence classes so that $C \circ \overline{C}$ and the trivial curve at o are both elements of the identity. This may be achieved by identifying unparametrized curves that differ only on a finite number of "trees"—(self-intersecting) curves that enclose no area.

Each resulting equivalence class $[C] = \gamma$ now has an inverse $\gamma^{-1} = [\overline{C}]$. These equivalence classes are often dubbed loops, though to avoid confusion I shall not refer to them by that term. Fortunately a more descriptive term is available. For, as we will see, in the context of a Yang–Mills gauge theory there is a natural way to associate curves on a manifold M with holonomies of a connection on a fiber bundle over M such that the holonomies of every pair of curves C_1, C_2 in γ are equal: $H(C_1) = H(C_2) \equiv H(\gamma)$. This is what motivated Ashtekar and others to abbreviate the phrase "holonomy-equivalent loop" by the simple word *hoop*, a usage I shall follow here. With these definitions, the

---

[6] If $C: I \to M$ is a curve, then $C_r: I \to M$ is an orientation-preserving reparametrization of C if and only if there is a continuous, monotonically increasing, function $f: I \to I$ such that $C_r(f(i)) = C(i)$ for all $i \in I$.

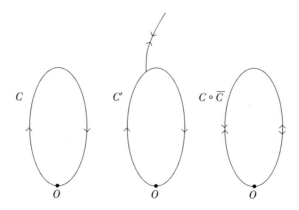

Figure 3.2. Curves C, C' differ by a tree. C∘C̄ is a tree, not the trivial curve O.

hoops with base point o constitute a group L_o under composition (uniquely defined by reference to composition of their constituent curves). I shall call L_o the *group of hoops* at o. Since the holonomy of every $C \in \gamma$ is equal, one can write $H(\gamma)$ instead of $H(C)$.

One can view the holonomies of a Yang–Mills gauge theory as representations of such an underlying group of hoops. The representation is not in a vector space, but in a Lie group—the gauge group of the theory, i.e. the structure group of its fiber bundle representation. This follows from representation theorems due to Anandan (1983), Barrett (1991), and others. Any sufficiently smooth homomorphism from the group of hoops into a suitable Lie group may be represented by the holonomies of a connection on a principal fiber bundle with that Lie group as structure group.

As shown in the previous section, a connection on a principal fiber bundle determines an associated set of holonomies. Consider a horizontal lift $\tilde{C}$ of a closed curve C with base point o. For each point u in the fiber above o, the horizontal lift through u will trace out a corresponding curve in the bundle that returns to a point v in the fiber above o. In this way the connection maps the fiber above o isomorphically onto itself. The map is called the holonomy map, and its action on a point u is an element of the bundle's structure group called the holonomy of C at u. Conversely, the holonomies of a gauge theory (relative to an arbitrary base point o) determine the connection, but only up to a principal bundle automorphism—a vertical automorphism that smoothly maps the fiber above each point onto itself.

Following earlier work of Anandan (1983), Barrett in 1985 proved two representation theorems:

(1) If M is a connected manifold with base point o and $H: L_o(M) \to G$ is a smooth homomorphism from the hoop group of M into a Lie group

G, then there exists a differentiable principal fiber bundle $P = P(M, G)$, a point $u \in \pi^{-1}(o)$, and a connection Γ on P such that H is the holonomy mapping of (P, Γ, u).

(2) If M is a connected, Hausdorff manifold, then this bundle is unique up to bundle isomorphisms that act as the identity on M. The notion of smoothness invoked here depends on a topology on L_o that renders the maps H continuous.

These representation theorems show why fiber bundles provide a convenient, but not fundamental, framework for formulating Yang–Mills gauge theories. But the existence of a smooth homomorphism H is a poor candidate for a "brute fact". The representation theorems apply to a manifold with a distinguished point, and it is unclear what features of these base-pointed hoops are represented by the abstract Lie group G.

There is a perspective which makes these problems seem more tractable. It is provided by properties of what are known as *Wilson loops*—i.e. traces of holonomies.[7] These are now holonomies of a connection defined on a vector bundle associated to the principal fiber bundle, so the operation of taking the trace makes sense—it is applied to matrices that act on representations of the structure group G rather than to the abstract group itself. The Wilson loop of a closed curve C with base point m and holonomy $\mathbf{H}_m(C)$ corresponding to gauge potential $\mathbf{A}_\mu$ is (cf. equation B.46 of appendix B)

$$W(C) = \text{Tr}\left[\wp \exp\left\{ -\oint_C \mathbf{A}_\mu(x)\mathrm{d}x^\mu \right\} \right] \quad (3.70)$$

As we saw in the previous chapter, in the case of electromagnetism, the holonomy of a closed curve C in the base manifold M of a vector bundle associated to the principal $P(M, \text{U}(1))$ bundle is just (the complex conjugate of) the gauge-invariant Dirac phase factor $\exp[(ie/\hbar) \oint_C A_\mu(x)\mathrm{d}x^\mu]$, where e is the charge of particles subject to an interaction represented (in an arbitrary gauge) by the four-vector potential A_μ. The structure group U(1) of this theory is Abelian and the typical fiber of the vector bundle is the (one-dimensional) complex vector space $\mathbb{C}$, and so the Wilson loops simply equal the corresponding holonomies (which are gauge invariant and independent of base point). In contrast, the holonomies of a non-Abelian Yang–Mills theory depend on a choice of base point in M, and transform by a common similarity transformation under changes of gauge at that base point (cf. equations B.49,

---

[7] Note that Wilson loops are therefore *not* loops in the sense in which I am using the term *nor* in the sense in which it is used by those who refer to the group of hoops as the loop group! Perhaps it is unnecessary to add that *none* of these usages has anything to do with the "loops" that figure in Feynman diagrams as graphical representations of certain terms which crop up when perturbation theory is used to extract predictions from the quantized field theories of the Standard Model.

74 3 CLASSICAL GAUGE THEORIES

B.48). But their Wilson loops are still both independent of base point and gauge invariant. While the holonomies take values in (representations of) the group G, their Wilson loops are complex valued. Theorems due to Giles (1981) and others show how to reconstruct all the gauge-invariant information contained in the gauge potential from the properties of its Wilson loops. Provided these satisfy a set of equations known as the *Mandelstam identities*, one can use them to find a set of holonomy matrices defined *modulo* similarity transformations of which the Wilson loops are the traces.

The gauge-invariant physical content of a classical Yang–Mills theory is therefore wholly contained in an infinite set of complex-valued functions $W(\gamma)$ of hoops γ that satisfy Mandelstam identities such as the following, all of which hold for SU(2) Wilson loops in the two-dimensional representation:

$$W(\gamma_1 \circ \gamma_2) = W(\gamma_2 \circ \gamma_1)$$
$$W(\iota) = 2, \text{ where } \iota \text{ is the identity hoop}$$
$$W(\gamma_1)W(\gamma_2) = W(\gamma_1 \circ \gamma_2^{-1}) + W(\gamma_1 \circ \gamma_2)$$
$$W(\gamma) = W^*(\gamma^{-1}) = W(\gamma^{-1})$$

This makes it plausible to maintain that what an SU(2) Yang–Mills theory ultimately describes is not a localized field represented by a gauge potential, but a set of intrinsic properties of what I have simply called loops—closed, oriented, one-dimensional regions of space-time—where each such loop is represented in the manifold M by the oriented image of a non-self-intersecting closed curve from a hoop. (*Warning*: such a loop must be sharply distinguished both from any Wilson loops whose values represent its properties and from a hoop an element of which is an unparametrized closed curve whose oriented image represents the loop.) In a classical isospin gauge theory, the coupling to matter would depend not only on the values of W, but also on properties of that matter (different for a proton and a neutron, say). But the theory could be taken to describe all this without even implicit mention of gauges, which would enter only in our more abstract mathematical representations of gauge-free reality.

3.1.3 Topological issues

Some of the most interesting features of Yang–Mills gauge theories may be seen to have their origins in topological features of fiber bundles. These include magnetic monopoles and the Gribov ambiguity. In each case, the topological features arise because the relevant principal fiber bundle lacks a global section and is therefore non-trivial.

Monopoles

In their seminal paper introducing the fiber bundle formulation of gauge theories, Wu and Yang (1975) demonstrated the power of this formalism by applying it to magnetic monopoles in electromagnetism and other gauge theories. When classical Maxwellian electromagnetism is formulated in terms of the connection and curvature of a principal fiber bundle as described in section 1.2, the base space of the bundle is the space-time manifold M, which is simply connected—every closed curve in M may be continuously contracted to a point. But if the theory is modified to include models of situations featuring isolated magnetic charges as well as electric charges, then the principal fiber bundle is no longer trivial.

Dirac (1931) showed that the presence of a single isolated magnetic charge m implied the quantization of electric charge in units $e = \frac{\hbar c}{2m}$, but his analysis involved an unphysical infinite "string" emanating from the monopole where its magnetic potential became infinite. Wu and Yang (1975) proved that even though the modified theory of electromagnetism may still be written in terms of four-vector potentials, there is no non-singular, single-valued function $A_\mu(x)$ that represents this everywhere (except at the location of the monopole). But one can represent electromagnetism by more than one such function, whose values are related by appropriate transition functions on the overlap of their domains, provided that Dirac's quantization condition is met. Any two such functions, A_1, A_2, may be regarded as deriving from a single connection ω on a non-trivial fiber bundle as pull-backs $\sigma_1^*\omega, \sigma_2^*\omega$ with respect to local sections σ_1, σ_2 as $\mathcal{A} = \sigma^*\omega = iA_\mu(x)dx^\mu$. The domain of any one such section cannot extend to all of M minus the monopole's world-line, though the union of their domains does cover this region. As in the case of coordinate patches on the sphere, two local sections suffice.

Wu and Yang (1975) extended their analysis to the case of magnetic monopoles in classical non-Abelian Yang–Mills gauge theories. In each case, the fiber bundle formulation facilitated an elegant explanation of the kinds of monopole fields that may be present in such a theory. For example, in the case of an SU(2) Yang–Mills theory, the principal fiber bundle does have a global section and all monopole fields are of the same type; while in the case of an SO(3) theory with spontaneous symmetry breaking, t'Hooft (1974) and Polyakov (1974) exhibited magnetic monopole solutions arising from a non-trivial U(1) sub-bundle of the trivial SO(3) bundle. Further details are available in Chan and Tsou's book (1993).

Despite their theoretical interest, the experimental search for monopoles has so far proved fruitless.

The Gribov ambiguity

Fiber bundles that do not have a space or space-time manifold as base space have also proved useful in understanding the features of gauge theories. An interesting example is provided by Singer's (1978) analysis of the so-called Gribov ambiguity. As shown in an earlier section (3.1.1), a Yang–Mills gauge field theory may be formulated as a theory of the connection and curvature of a principal fiber bundle over a space(-time) manifold. Assuming this bundle is trivial, the connection ω may be represented on the manifold by a Lie-algebra-valued one-form $\mathcal{A} = \sigma^*\omega$, for some global section σ, or the associated four-vector fields $A_\mu^a(x)$, where

$$\mathcal{A} = \sum_a A_\mu^a(x) T_a dx^\mu \tag{3.71}$$

A vertical bundle automorphism of this bundle represents a gauge transformation. Whether the transformed connection and vector fields are to be considered as merely offering an alternative theoretical representation of the same physical situation, or rather a representation of a transformed physical situation, is an issue to be taken up later (in chapter 6). When considering quantized gauge theories in the path-integral formulation (see chapter 5, section 5.8), physicists take the former alternative, and to avoid massive over-counting of possible physical situations, they attempt to select a single representative of each distinct possible situation by picking just one connection ω to represent that situation, from an infinite set of connections related to it by vertical bundle automorphisms. This is equivalent to picking a unique set of functions $A_\mu^a(x)$ representing ω in the (arbitrarily) chosen section σ. This is called fixing the gauge, and it is typically done by imposing some condition of the form $f(A_\mu^a(x)) = 0$, where the function f is required to be a solution to some differential equations, together with boundary conditions. A simple example of these for electromagnetism would be the conditions defining the Coulomb gauge (see 5.12), together with the condition that $A_\mu(x) \to 0$ as $x \to \infty$.

Gribov (1977) realized that an analog to the Coulomb condition failed to lead to an unambiguous specification of gauge in certain non-Abelian gauge theories. Singer (1978) then generalized this point by proving that no gauge-fixing meeting certain conditions is possible for a non-Abelian gauge theory. Singer's result may be explained as follows. Two connections ω, ω' that are gauge equivalent because they are related by a vertical bundle automorphism are said to lie on the same gauge orbit. A successful gauge-fixing condition would pick out a unique element of each gauge orbit. One can form a fiber bundle, the points of whose base space are gauge orbits and whose typical fiber is isomorphic to the gauge group—the group of gauge transformations. A gauge-fixing condition would constitute a global section of this fiber bundle—a continuous selection of connection from each gauge-equivalence class. Suppose that one specifies a gauge-fixing condition by imposing restrictions on the functions $A_\mu^a(x)$ representing a connection ω on the space-time manifold; and suppose further that one such restriction is the non-Abelian generalization of the Coulomb boundary condition $A_\mu^a(x) \to 0$ as $x \to \infty$. This is equivalent to requiring that the $A_\mu^a(x)$, or equivalently ω, be defined on an extension of the space-time manifold (with topology $\mathbb{R}^4$) by addition of a point at infinity, to form a manifold with topology S^4, the four-sphere. So a gauge-fixing condition meeting this further restriction would correspond to a global section of a fiber bundle over a base space of gauge orbits of connections on a principal bundle over S^4, whose typical fiber is a connection on this bundle. Singer (1978) proved that there are no such gauge-fixing conditions. Neither the generalized Coulomb condition nor any similar condition corresponding to $A^a_\mu(x) \to$ constant as $x \to \infty$ succeeds in fixing the gauge in a non-Abelian Yang–Mills theory.

3.2 A fiber bundle formulation of general relativity

Classical general relativity may also be formulated as a theory of a connection and curvature of a principal fiber bundle. There is a variety of choices for the structure group in such a formulation, but in no case is the structure group compact. Intuitively, this means that applying group operations successively can take you further and further away from your starting point. By contrast, successive operations in a compact group like U(1) or SU(2) correspond to generalized rotations, which can only take you so far away from the orientation in which you started.[8]

To bring out a conceptually significant disanalogy between general relativity and Yang–Mills gauge theories, it will be helpful to consider a formulation of general relativity in terms of a principal bundle $A(M)$ of affine frames. A frame is a linearly independent set of basis vectors of the tangent space at a point, and an affine transformation is an invertible inhomogeneous linear transformation applied to this space, considered as an affine space. (Transformations in the Poincaré group are a subgroup of the group of affine transformations, namely those that preserve the Minkowski metric.) The gravitational gauge potential is now represented by a connection on this bundle. It is important not to confuse this with the usual linear connection in general relativity, which is defined instead on the bundle of linear frames, $F(M)$—especially since the latter is usually *called* the affine connection! But though distinct, the two connections are intimately related.

For there is a natural homomorphism from $F(M)$ onto $A(M)$ which induces a mapping of the connection on $A(M)$ onto a Lie-algebra-valued one-form on $F(M)$ that decomposes into the sum of two parts. One part is simply the usual general relativistic connection. The other part is called the solder(ing) form on $F(M)$, since it "solders" $F(M)$ to M. According to Trautman (1980, p.306),

---

[8] This is closely related to another significant technical difference between general relativity and a Yang–Mills gauge theory. While the first-class constraint functions in a constrained Hamiltonian formulation of a Yang–Mills gauge theory form a Lie algebra under the Poisson bracket operation, for general relativity they do not (see appendix C).

The most important difference between gravitation and other gauge theories is due to the soldering of the bundle of frames [$F(M)$] to the base manifold M.

This occurs as follows. The solder form Θ yields a unique way of "projecting" a vector $\tilde{v}$ in the tangent space to $F(M)$ at point u of $F(M)$ down onto a vector v in the tangent space to M at the point m "below" u. Since Θ is itself a Lie-algebra-valued one-form on $F(M)$, it acts like a connection on $F(M)$. So it defines the "Θ-horizontal lift" of all the vectors in the frame u into a basis $\tilde{u}$ for the horizontal subspace of the tangent space to $F(M)$ at u. The components of $\tilde{v}$ with respect to $\tilde{u}$ now define a vector v, in the tangent space to M at m, of which they are the components with respect to the basis u.

The solder form is important because principal bundle automorphisms that preserve it simply correspond to diffeomorphisms of the base space M. They do not "rotate" the elements around within the fiber above each point but just carry the horizontal lift of a curve in the base space into the horizontal lift of its image under such a diffeomorphism. This means that the only vertical automorphism is the identity! The next chapter will argue that this implies that general relativity, unlike Yang–Mills gauge theories, is separable.

3.2.1 A gravitational analog to the Aharonov–Bohm effect

Suppose that classical electromagnetism is non-separable in its action on quantum particles. Since general relativity, our best classical theory of gravitation, may also be formulated in terms of the connection and curvature of a principal fiber bundle, one might suspect that gravity also acts non-separably on quantum particles. Moreover, the striking quantum phenomena of electromagnetically induced phase shifts in a field-free region used to illustrate the non-separability of electromagnetism in the previous chapter have an apparent analog in general relativity. For there is a model of general relativity that represents a situation bearing a strong resemblance to the circumstances of the magnetic Aharonov–Bohm effect. But a closer examination of this model will show that the formal differences between general relativity and Yang–Mills gauge theories pointed out earlier in this section give rise to significant physical disanalogies between the Aharonov–Bohm effect and its purported general relativistic analog. Unlike Yang–Mills theories, general relativity is separable in the quantum domain.

The Aharonov–Bohm effect comes about because the effects of electromagnetism on the phase (and subsequent behavior) of charged quantum particles that pass through a region of space are not always wholly determined by the electromagnetic field there while they pass. General relativity predicts an analogous effect in which the effects of gravity on the phase (and subsequent behavior) of quantum particles that pass through a region of space are not always wholly determined by the space-time curvature in that region while they pass. The fiber bundle formulations described earlier make the analogy even clearer. In each case, the bundle curvature is zero everywhere in the region through which the particles pass. Nevertheless, the bundle connection defines parallel transport around a closed curve in that region, resulting in a phase change around any curve that encloses a separate central region in which the curvature is non-zero.

As Anandan (1993) notes, Marder (1959) described a static general relativistic space-time with cylindrical symmetry which is empty of matter except for an infinite central cylinder. Everywhere outside the cylinder the Riemann curvature tensor is zero and the metric is Minkowski. The two-dimensional spatial geometry of each normal section through the cylinder is the same and may be represented in a hypothetical embedding space by the surface of a truncated cone with a smooth cap. If a tangent vector is parallel-transported along a closed curve once around the cylinder, it is rotated through the angle of the cone, even though the curvature is zero everywhere along the curve.

In the Aharonov–Bohm effect, the interference pattern produced by charged particles passing through a region is not determined solely by the electromagnetic field there while they pass: it depends also on the field in regions from which they are excluded. In particular, the position of the intensity maxima produced by interfering two coherent beams that have passed on either side of an infinite solenoid while traversing a region throughout which the field is zero depends on the current passing through the solenoid. The phase shift between the two beams along a closed curve C is given by the expression

$$F(C) = \exp[-(ie/\hbar) \oint_C A_\mu(x) dx^\mu] \qquad (3.72)$$

In the gravitational analog, the interference pattern produced by spinning particles passing through a region is not determined solely by the space-time curvature in that region: it depends also on the space-time curvature in regions from which they are excluded. In particular, the position of the intensity maxima produced by interfering two coherent beams that have passed on either side of an infinite "string" while traversing a completely flat region depends on the energy and momentum flow inside the "string." The phase shift between the two beams along a closed curve C is given by the expression

$$F(C) = \wp\exp[-(i/\hbar) \oint_C (e_\mu^a P_a + \frac{1}{2}\Gamma_\mu^{ab} M_{ab}) dx^\mu] \qquad (3.73)$$

where M_{ab} are (representations of) the generators of the Lorentz group and P_a of the translation group, e_μ^a is dual to the frame field e_a^μ, Γ_μ^{ab} are the components of the linear (Levi-Civita) connection ∇ with respect to e_a^μ, and the integral is path ordered.[9] The second summand in 3.73 represents the phase shift due to the particles' spin. These expressions for the phase change can both be seen as instances of the general equation

$$F(C) = \wp\exp[-(ig/\hbar) \oint_C A^a_\mu T_a dx^\mu] \qquad (3.74)$$

where the T_a generate the Lie algebra of the structure group of an arbitrary gauge field, and A^a_μ represents the corresponding gauge potential.

But despite these analogies between the effects of gravity on quantum particles in Marder space-time and the Aharonov–Bohm effect, Anandan (1993) has also pointed out a significant disanalogy. While electromagnetic phase differences given by 3.72 are physically significant and measurable only around closed curves, gravitational phase differences given by 3.73 are physically significant, and potentially measurable, also along open curves. For example, for spinless particles, the gravitational phase acquired by a locally plane wave along a curve C corresponding to a classical trajectory is given approximately by

$$\varphi = (1/\hbar) \int_C e^a_\mu p_a dx^\mu \qquad (3.75)$$

where p_a are the eigenvalues of the energy–momentum operator $\hat{P}_a$. Unlike the electromagnetically induced phase change in the Aharonov–Bohm effect, this gravitationally induced phase change is physically significant, and potentially measurable, along open curves C, as well as closed curves. It would be observable by the Josephson effect for a path across the Josephson junction, or by the strangeness oscillations in the neutral K meson system for an open time-like path C along the K meson beam.

The solder form described earlier in this section is the formal reason for this physical difference between the gravitational and electromagnetic interactions. As one transports the phase of quantum particles along a curve C, the vector representing their phase lies in a vector bundle associated with the principal affine frame bundle $A(M)$. The solder form effectively "locks" the phase vector to a corresponding space-time vector in the tangent bundle of M. It is the linear connection on $F(M)$ that defines parallel transport of this latter vector along C; and, because of the solder form, it therefore also uniquely specifies how the phase changes from point to point along C. In the case of gravitation, it is the solder form that privileges a unique connection in

---

[9] Note that the e^a_μ are simply the components of the solder form Θ in a basis for the Lie algebra of the translation subgroup of the affine group.

the affine frame bundle $A(M)$. The class of symmetries of the system is restricted to those that preserve the solder form. This excludes non-trivial "vertical" bundle automorphisms, and leaves only automorphisms of $A(M)$ that simply correspond to diffeomorphisms of M. The absence of the solder form in electromagnetism corresponds to the absence of any such privileged connection in the U(1) bundle.

The existence of a uniquely privileged connection on the principal bundle $A(M)$ implies that classical gravitation is separable in its action on quantum particles. The intrinsic geometric properties in (an infinitesimal neighborhood of) each point in any region of space-time determine the effects of gravitation on quantum particles in that region, including effects on their spins. By contrast, the intrinsic electromagnetic properties pertaining to (even an infinitesimal neighborhood of) each point in an extended region of space-time do not determine the effects of electromagnetism on charged particles in that region: those effects further depend on holonomy properties pertaining to extended loops in the region.

4

Interpreting classical gauge theories

After chapter 3's outline of the structure of some classical gauge theories, it is now time to ask how these theories should be interpreted. What beliefs about the world are (or would be) warranted by the empirical success of a classical gauge theory? Note the conditional character of this question. Non-Abelian gauge theories have contributed to the empirical success of the Standard Model in quantum rather than classical guise. And while classical electromagnetism and general relativity have indeed enjoyed considerable empirical success, neither is believed to be empirically adequate, and indeed the former is now known not to be. But even if none of our classical gauge theories is empirically adequate, we may still hope to learn something about the world by viewing it through the lens of these theories. The history of science is replete with empirically inadequate theories that have taught us a great deal about what kind of world we find ourselves in. Indeed, it is arguable that we owe most, if not all, of our scientific knowledge to empirically inadequate theories. Be that as it may, I bracket the question of empirical adequacy in this chapter, proceeding on the working assumption that there is substantial empirical evidence for, and none against, the theories under discussion.

But there is still a question as to the scope of that positive evidence. We are dealing with theories of interactions, whose empirical consequences are manifested by how they affect the behavior of objects subject to them. A theory of mechanics is required to describe the behavior of these objects, and that theory may be either classical or quantum. As chapter 2's discussion of the Aharonov–Bohm effect made plain, a classical gauge theory may have empirical consequences in conjunction with quantum mechanics that are qualitatively different from those it has in conjunction with classical mechanics. This is an important instance of the familiar fact that the empirical consequences of a theory are a function of what other theories are used in deriving them. If the evidence supporting a classical gauge theory is forthcoming only in conjunction with classical mechanics, then that may favor an interpretation which would be rendered much less plausible by evidence for the theory obtained in conjunction with quantum mechanics. While this makes any judgment as to how a gauge theory should be interpreted provisional, it does not mean that all such judgments are merely relative to an arbitrary choice of auxiliary theory. A proposed interpretation of a classical gauge theory should be evaluated on the basis of the evidence supporting it when taken in conjunction with the best available auxiliary theory; i.e. the known theory in conjunction with which it is not merely (assumed to be) empirically adequate but also of widest scope.

Some may not be comfortable with this talk of comparing and evaluating alternative interpretations of a theory with a view to deciding which to adopt. For our working assumption is that the theory under scrutiny is empirically adequate independent of any such decision. Positivists will then likely claim that there is no substantive difference between the supposed alternatives, so that any decision would be a decision among what Reichenbach called equivalent descriptions—merely different ways of saying the same thing. A constructive empiricist like van Fraassen (1980) will not agree that alternative interpretations offer equivalent descriptions, since he adheres to a literal interpretation of the language the theory uses to describe unobservable structures. But van Fraassen maintains also that it is not the business of science to choose among alternative interpretations of an empirically adequate theory; and he advocates a voluntarist epistemology in accordance with which it is up to anyone reflecting on such a theory in science to adopt whatever interpretation he or she likes best.

Against such antirealists, I maintain that evidential considerations of the sort that figure in science itself operate also at the level of choice among rival interpretations of an empirically successful theory. Such considerations can give us (defeasible) reasons to adopt one interpretation rather than another. Indeed, they provide reasons to adopt a particular interpretation of each of the classical gauge theories presently under consideration. Because of the differences explained in chapter 3 between the structure of general relativity and that of Yang–Mills gauge theories, we should interpret these theories differently—or so I shall argue.

4.1 The no gauge potential properties view

Despite phenomena like the Aharonov–Bohm effect, one may deny that the successful application of classical electromagnetism in the quantum domain requires one to accept the reality of anything represented by electromagnetic potentials, over and above the electromagnetic field. In chapter 2 I called this the no new EM properties view. Its adherents maintain that the empirical success achieved by conjoining classical electromagnetism with quantum rather than classical mechanics should not lead one to accept that electromagnetic potentials somehow represent additional qualitative intrinsic EM properties.

Each of the gauge theories described in chapter 3 may be viewed in a similar way. For each theory was there formulated in terms of what may be called a gauge field and a gauge potential.[1] On the *no gauge potential properties* view it is only the former that directly represents gauge properties, even in the quantum domain, while the latter is mere mathematical surplus structure, of use in the theory only as a way of calculating the former.

The main objection lodged against the no new EM properties view in chapter 2 was based on the fact that, in the context of the Aharonov–Bohm effect, it implied electromagnetic action at a distance. This generalizes into a similar objection to the no gauge potential properties view. But additional objections to this view arise, one in the case of non-Abelian Yang–Mills gauge theories and another in the case of general relativity.

Wu and Yang (1975a) have shown that in the case of an SU(2) gauge theory there may be physically distinct situations in a simply connected region of space-time, represented by inequivalent gauge potentials, even though the gauge field is the same throughout the region in each situation. There are two respects in which this goes beyond the Aharonov–Bohm effect. In the case of electromagnetism, analogous situations can occur only in multiply connected regions. More importantly, in the case described by Wu and Yang (1975a), there is a *locally detectable* difference between the distinct physical situations: in one situation sources are present at each point in the region, while in the other it contains no sources! This can happen because of a crucial structural difference between the roles of potentials in Abelian and non-Abelian Yang–Mills gauge theories (see appendix A). Maxwell's inhomogeneous equations relate the electromagnetic field alone directly to its sources. But a vector field representing the non-Abelian gauge potential appears independently and ineliminably in their non-Abelian generalizations. Alterations in the potential that preserve the associated field may consequently be balanced by corresponding alterations in the sources. Not only the gauge field but also the gauge potential in a region is affected by sources, and (unlike the Aharonov–Bohm effect) it is the sources *in the region* that have these independent effects. This provides a powerful reason to accept that a classical non-Abelian gauge potential indeed represents qualitative intrinsic gauge properties over and above those represented by its associated gauge field.

Now consider chapter 3's formulation of general relativity as the theory of a connection and curvature on a principal fiber bundle—the affine bundle $A(M)$. Recall that the natural homomorphism h from the affine bundle $A(M)$ onto the frame bundle $F(M)$ maps the connection on $A(M)$ onto a one-form on $F(M)$ that decomposes uniquely into the solder form and the usual linear connection

---

[1] Confusingly, a different terminology is sometimes employed in the case of Yang–Mills theories, according to which it is a non-Abelian generalization of the electromagnetic *potential* that is referred to as the gauge field. The non-Abelian generalization of the electromagnetic field may then be referred to as the field strength.

on the space-time manifold. This last is the Levi-Civita connection—the unique symmetric connection that is compatible with the metric. The Levi-Civita connection certainly represents qualitative intrinsic properties of the space-time manifold, since it defines parallel transport of vectors along curves in the manifold, and thereby specifies the class of geodesics through each point. It follows that the gauge potential, as represented by the affine connection on $A(M)$, also represents these same properties, via the homomorphism h. These include the property that through each point outside the central region in Marder space-time there are distinct geodesics that meet again at another point after passing around opposite sides of the central region—a property that would be readily observable by standard geometric techniques already in the classical domain.

This property is in no way represented by the gauge field, which in this formulation corresponds to the curvature of $A(M)$. The curvature at a point u of $A(M)$ will represent the curvature of the space-time manifold at the point $\pi(u)$ "below" u, by virtue of the homomorphism h. But the space-time manifold is flat everywhere except over the central region, and the curvature of $A(M)$ cannot represent the fact that there are geodesics that meet after passing around opposite sides of the central region. In contrast to the case of electromagnetism, it is not appropriate to call this a "new" gauge property, since it is apparent already in the classical domain. But it does provide an illustrative example of a qualitative intrinsic gravitational/geometric property of a region that is represented by the gauge potential but not the gauge field in that region, in a formulation of general relativity which represents potential and field respectively by the connection and curvature on a principal affine bundle $A(M)$.

These objections provide compelling reasons for the following conclusion. The empirical success of a classical Yang–Mills theory (including electromagnetism) or general relativity cannot be understood unless one accepts that there are indeed gauge potential properties—i.e. qualitative intrinsic physical properties that are not represented in that theory by the gauge field but only by the gauge potential. But should these properties be understood as localized or non-localized?

4.2 The localized gauge potential properties view

According to the localized gauge potential properties view, the gauge potential of a theory represents qualitative intrinsic properties predicated of, or at, space-time points (or their arbitrarily small neighborhoods)—properties in addition to those represented by the gauge field. There are several ways in which this may come about, depending not only on what mathematical object plays the role of the gauge potential in a particular formulation of the theory, but also on how this object is described. As we have seen in chapter 3, the gauge potential of a non-Abelian Yang–Mills theory may be represented by a Lie-algebra-valued one-form on a principal fiber bundle over space-time, by a Lie-algebra-valued one-form on (an open set of) the space-time manifold itself, or by a set of vector fields on (an open set of) space-time that transform among themselves according to a representation of the Lie group that is the bundle's structure group. Which mathematical object one takes to represent qualitative intrinsic gauge potential properties may affect the details of how the view is applied. But whatever the details, what makes these properties localized is that the view takes the mathematical object in question to represent gauge potential properties that attach locally—at space-time points or their arbitrarily small neighborhoods. It does this either directly by itself being defined at points (in open sets) of the space-time manifold, or indirectly by being defined at each bundle point u, with projection $\pi(u)$ onto a corresponding point of the space-time manifold.

Before we can assess the localized gauge potential properties view, we need to answer a number of questions about it. What are these properties supposed to be like? Assuming there are such properties, what can we say about them? Could we observe them, and if so, how? How are they supposed to act on quantum and/or classical particles? One might expect the gauge theory itself to provide the answers to these questions; but, prior to interpretation, it merely constrains them.

Some constraints are formal. Any localized gauge potential properties in classical electromagnetism must be faithfully representable by a vector at each point in (an open set of) the space-time manifold. The vector at a point would partially determine the electromagnetically induced rate of change of the phase of the wave-function representing quantum particles at that point: the rate would be proportional also to the particles' charge, so neutral particles suffer no electromagnetically induced phase shift. In the general case, the vector in question would be the four-vector potential A_μ, representing rate of change of $\Psi(p)$ in any space-time direction at space-time point p. So any localized gauge potential properties in classical electromagnetism would be properties of, or at, arbitrarily small neighborhoods of p.

To each space-time direction at p, there would correspond a mutually exclusive and jointly exhaustive set of these properties, one and only one of which would be possessed in any given situation. The relations between properties and directions, and between the properties of, or at, nested neighborhoods of p would be just those required in order to secure a faithful representation by A_μ. A change of gauge from A_μ to $\bar{A}_\mu$ could be thought of passively as a mere "relabeling" of the very same distribution of localized gauge potential properties, and in that sense there would be no ONE TRUE GAUGE. But, having chosen A_μ to represent their actual distribution, one could then regard $\bar{A}_\mu$ as representing a *distinct* possible distribution, as follows. Instead of changing the way values of vector potential are mapped onto gauge potential properties, one would keep the *same* correspondence between possible values of a component of a vector potential at a point and possible gauge potential properties thereabouts as for A_μ, so that (for example) if $A_2(p) = 5$ represented property P of a small neighborhood of p, then $\bar{A}_2(p) = 10$ would represent a different property $\bar{P}$ of that neighborhood—intuitively, a property that would double the electromagnetically-induced rate of change in the phase of any quantum particles there as compared to P.

Localized gauge potential properties in a non-Abelian gauge theory would be governed by more complex formal constraints. Here, the value of the four-vector potential A_μ^a at a point would partially determine the gauge-field-induced rate of change of the phase of *the a-component of* the wave-function representing quantum particles at that point. The wave-function is now itself a multi-component vector (or spinor) in an "internal" space—an element of a representation of the structure group of the theory. Hence the value of $A_\mu^a(p)$ would now represent many related localized gauge properties of, or at, arbitrarily small neighborhoods of p. For each space-time direction at p, and each "direction" in the internal gauge space, one out of a range of mutually exclusive and jointly exhaustive localized gauge potential properties would attach to each such neighborhood. This would act on the phase of the component of the wave-function in that "direction" so as to determine its gauge-field-induced phase shift along a curve in that neighborhood, where particles with different "gauge charges" experience different shifts. Again, a gauge transformation may be thought of either passively—as merely relabeling the same distribution of localized gauge potential properties, or actively—as representing a distinct possible distribution.

Localized gauge potential properties in the formulation of general relativity as the theory of the connection and curvature of the affine frame bundle $A(M)$ could be represented by the one-form field e_μ^a dual to the frame field e_a^μ, and the Levi-Civita connection ∇ with components $\Gamma^\lambda_{\mu\nu}$, on the space-time manifold. e_μ^a defines the space-time metric $g_{\mu\nu}$ by

$$g_{\mu\nu} = \eta_{ab} e_\mu^a e_\nu^b \tag{4.1}$$

where η_{ab} is the Minkowski metric. ∇ is the unique torsion-free, linear connection that is compatible with $g_{\mu\nu}$ in the sense that the covariant derivative of $g_{\mu\nu}$ vanishes (so lengths of vectors and angles between them are preserved on parallel transport)

$$D_\lambda g_{\mu\nu} = \partial_\lambda g_{\mu\nu} - \Gamma^\rho_{\lambda\mu} g_{\rho\nu} - \Gamma^\rho_{\lambda\nu} g_{\mu\rho} = 0 \tag{4.2}$$

This one-form field and connection represent familiar geometric properties of the space-time manifold. These include lengths of curves and angles at which they intersect, geodesics and parallel transport, the light cone structure, durations of time-like curves, areas and volumes, etc. These are properties of a space-time region that are not determined by the curvature, either of that region or of the bundle $A(M)$ "above" it, as the example of Marder space-time brings out vividly. They are, or supervene on, localized geometric properties—qualitative intrinsic geometric properties of arbitrarily small neighborhoods of each point of the region. They affect the behavior of classical particles as well as classical rods and clocks: and they also affect quantum particles by affecting the phase of their wave functions. A gauge transformation could correspond either to a new choice of section on $A(M)$ or to a principal bundle automorphism of $A(M)$. Either way, the effect would be to alter e^a_μ, $g_{\mu\nu}$, and $\Gamma^\lambda_{\mu\nu}$ so as to correspond to a diffeomorphic model of general relativity. As is now familiar from discussions of the hole argument (Earman and Norton 1987), this may readily be considered as a passive transformation from one to another representation of the same state of affairs. Some (e.g. Butterfield 1989, Maudlin 1989, Healey 1995) maintain that it could also be interpreted as a transformation to a representation of a distinct state of affairs in which all fields (including e^a_μ, $g_{\mu\nu}$, and $\Gamma^\lambda_{\mu\nu}$) are distributed differently on space-time points. But they then argue in various ways that this is not a *possible* state of affairs, so really no active interpretation of the gauge transformation is available.

Assuming that there are localized gauge potential properties, what more can we say about them beyond the bare assertion that they may be represented by one of the mathematical objects we have considered? Mattingly (2006) has suggested that in the case of electromagnetism we can say that they amount to what he calls the *current field*, given by the expression

$$A^m_\mu(x) = 4\pi \int_{\text{all space-time}} d^4x' D_r(x - x') J_\mu(x') \qquad (4.3)$$

This expression gives a canonical way of representing the supposed localized electromagnetic potential properties in a region; but does it say anything more about them than a representation by means of a gauge transform of this expression? The name 'current field' and the expression for it do indeed refer to currents—a concept whose meaning and reference may be assumed to be unproblematic in this context. And the current field is naturally thought to be caused by these currents alone, whereas any gauge transform of this expression would introduce some mathematical field $\Lambda(x)$ with no similarly clear physical content, "masquerading" as an additional cause of the supposed localized electromagnetic potential properties in the region. But we already knew that in cases like the Aharonov–Bohm effect the behavior of quantum particles in a certain region may be affected just by distant currents: what we want to know more about is the nature of the modifications *in that region* these distant currents bring about in order to have these effects. To admit that there are no such modifications is to revert to the *no gauge potential properties* view with all its difficulties. Just referring to the supposed mediating localized

4.2 THE LOCALIZED GAUGE POTENTIAL PROPERTIES VIEW

electromagnetic potential properties in the region by the expression 'current field' does not tell us any more about them.

If there were any localized gauge potential properties, then we could say that they are interrelated in such a way as to permit their representation by vector potentials or one of the other more abstract mathematical objects recently considered, and that they are caused by (possibly distant) source currents. In order to say anything more interesting about them, we would need to be able to discriminate among them—to say that *these* are the properties in this region, and not *those*. But the gauge symmetry of the theory is a barrier to any such discriminating reference.

To see the problem, suppose one were to truly state that the localized electromagnetic potential properties in a region are representable by a particular four-vector potential A_μ. This assertion is incompatible with the claim that they are representable by any potential A'_μ that fails to be gauge equivalent to A_μ. But it is compatible with the assertion that they are representable also by a distinct, but gauge-equivalent, potential $\bar{A}_\mu$. Nevertheless, one may wish to discriminate the actual electromagnetic state of the region from the state it would be in if its localized electromagnetic potential properties were systematically "shuffled around" in a such a way that $\bar{A}_\mu$ rather than A_μ represents the "shuffled" properties. This idea may be made precise.

Suppose the localized electromagnetic potential properties in a region are representable by a particular four-vector potential A_μ. Then there is some structure-preserving map f from the set of allowed property distributions in that region to the set of four-vector potentials whose value for this particular distribution is A_μ. It would take considerable effort to spell out in detail exactly what conditions a function like f must meet to preserve the relevant structure in this case. A minimal requirement would be that they ensure that if the localized electromagnetic potential properties in the immediate neighborhoods of points p and q are the same, then $A_\mu(p) = A_\mu(q)$. There is a large literature in what has come to be known as measurement theory devoted to the task of spelling out the conditions that must be met by a relational structure in order that it may be represented mathematically. When the conditions have been spelled out successfully for a particular relational structure, it is possible to prove a (representation) theorem that shows that a faithful representation exists, and a further (uniqueness) theorem that circumscribes the range of faithful representations of a given relational structure. In the present case, the representation theorem would show that localized electromagnetic potential properties in a (sufficiently small) open set of the space-time manifold are always representable by some four-vector potential A_μ; while the uniqueness theorem would show that this representation is unique up to a gauge transformation 1.6, i.e.

$$\bar{A}_\mu = g(A_\mu), \text{ where } \bar{A}_\mu(x) = A_\mu(x) + \partial_\mu \Lambda(x) \tag{4.4}$$

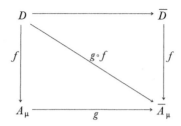

Figure 4.1.

Now suppose that the distribution D of localized electromagnetic potential properties in a region is faithfully represented by A_μ in accordance with a representing function f. Then (it is a consequence of the uniqueness theorem that) the same distribution D would also be faithfully represented by $\bar{A}_\mu$ in accordance with the different representing function $g \circ f$. But a *different* distribution $\bar{D}$ of localized electromagnetic potential properties in the region would be faithfully represented by $\bar{A}_\mu$ in accordance with the *same* representing function f, as depicted in figure 4.1.

The point is that what is represented by a given four-vector potential depends on a choice of representing function. Any of a class of gauge-equivalent four-vector potentials could be used to represent any of an infinite class of distinct distributions of localized electromagnetic potential properties (if there were such properties). And a gauge transformation 1.6 from one to another could be taken to correspond either to adoption of a different way of representing the same distribution in accordance with an alternative choice of representing function, or to the representation of a different distribution in accordance with the same choice of representing function.

So what a particular four-vector potential represents is in large part up to the representor. If there are localized electromagnetic potential properties, then to use a four-vector potential to say what their distribution is in a region one needs to adopt a particular representing function. But how can one do that?

Suppose one were somehow in a position to know that a particular four-vector potential $\mathring{A}_\mu$ in the region yields correct predictions for the results of all possible observations on particles within that region. One might convince oneself, for example, that Aharonov and Bohm's choice $A_t = A_z = A_r = 0$, $A_\theta = \Phi/2\pi r$ correctly predicts the results of all interference experiments that could be performed on charged particles passing through the region surrounding the solenoid in the magnetic Aharonov–Bohm effect. Assuming that there are localized electromagnetic potential properties in that region, one could then simply adopt whatever representing function f was required in order for this four-vector potential to represent these properties in accordance with f, secure in the knowledge that there must be such a function even though one could not explicitly give it.

But does this really say what the distribution of localized electromagnetic potential properties is in the region? What it says is just that this distribution is whatever it had to be in order to be represented by $\overset{\circ}{A}_\mu$. The problem is that a continuous infinity of different distributions satisfies that description, and one has not succeeded in saying which of these in fact obtains.

The proponent of the new localized EM properties view here faces an instance of a quite general problem. The problem deserves an extended discussion since it can afflict any new fundamental scientific theory with certain symmetries.[2] It is also closely related to influential views of the philosopher David Lewis (1970, 2001/2007) on the meaning of scientific terms and our acquaintance with their referents. Since Lewis has created a framework within which it is easy to pose the general problem, it will be best to begin by sketching that framework before showing how to apply it to theories with gauge symmetry.

4.2.1 Problems defining theoretical terms

Lewis (1970) outlines a method for using the Ramsey sentence of a theory to construct explicit definitions for terms newly introduced in a theory T in a previously understood language. It may seem surprising that anything like this is possible, given the perceived failure of attempts by the logical positivists to define theoretical terms in (what they called) an observation language, and work of Suppes (1957) and others establishing (against claims of Mach) the indefinability of a term like 'mass' within classical particle mechanics. The success of Lewis's method depends not only on a liberal understanding of the logical and linguistic resources available to provide the required definitions, but also on a certain substantive assumption that the theory be *uniquely realized*. It is this substantive assumption which is called in question by theories with a certain kind of symmetry. If it fails, then so does Lewis's method for defining theoretical terms. So also does the attempt to say that supposed localized gauge potential properties are distributed one way rather than another in a region of space-time, in the context of a Yang–Mills gauge theory like classical electromagnetism. To understand why, it is necessary to explain Lewis's method and to say just what the assumption of unique realization amounts to.

Lewis begins by assuming that, given a successful new theory T, it is possible to formulate it by means of a single (finite or infinite) postulate of the form '$T[\tau_1 \ldots \tau_n]$', where $\tau_1 \ldots \tau_n$ are all the terms newly introduced by T, while all the other terms appearing in the postulate are assumed to be already understood, with determinate sense and reference. In defense of this assumption, he argues that if new terms appear in the theory as predicates, functors, etc. one can replace them in these occurrences by

---

[2] For further discussion of this general problem, see Healey (2006).

names of corresponding properties, functions, etc. by making use of already understood copulas. For example, a new monadic predicate '$\Phi\_\_$' could be replaced by a term '$\_\_$ has the property ϕ' where 'ϕ' purports to name the property (necessarily) shared by all and only objects of which 'Φ' is true, and '$\_\_$ has $\_\_$' is a previously understood relational predicate. He calls the result '$T[x_1 \ldots x_n]$' of replacing the terms $\tau_1 \ldots \tau_n$ by distinct variables $x_1 \ldots x_n$ that do not already occur in T the *realization formula for T*, so that any n-tuple of entities that satisfies it (keeping the interpretation of all other terms in T fixed) *realizes* or is a *realization* of T. So the postulate T says that T is realized by the n-tuple of entities denoted (respectively) by $\tau_1 \ldots \tau_n$. The *Ramsey sentence R*, i.e. '$\exists x_1 \ldots \exists x_n T[x_1 \ldots x_n]$,' on the other hand, merely says that T is realized by at least one n-tuple of entities. But any consequence of T in language available before the introduction of the new terms $\tau_1 \ldots \tau_n$ by T is still a consequence of R, and so the logically weaker R shares all the same predictions, and therefore predictive success, as T, while containing none of its new terms. But of course, R is still committed to the existence of entities corresponding to these new terms, since it will not be true unless some n-tuple of entities realizes T.

If a theory is uniquely realized, then the following identities will serve to define its newly introduced terms in previously understood language:

$$\tau_1 = \imath\lambda_1 \exists y_2 \ldots \exists y_n \forall x_1 \ldots \forall x_n (T[x_1 \ldots x_n] \equiv (y_1 = x_1) \& \ldots \& (y_n = x_n))$$
$$\ldots \tag{4.5}$$
$$\tau_n = \imath\lambda_n \exists y_1 \ldots \exists y_{n-1} \forall x_1 \ldots \forall x_n (T[x_1 \ldots x_n] \equiv (y_1 = x_1) \& \ldots \& (y_n = x_n))$$

where the symbol '$\imath$' is the description operator: '$\imath x$' stands for 'the object x such that.' The identities fix the references of the newly introduced terms by saying what they actually denote: and they specify their senses by saying what they would denote if T were uniquely realized by some other n-tuple of entities. Lewis (1970, p. 433) says this about the assumption of unique realization:

A uniquely realized theory is, other things being equal, certainly more satisfactory than a multiply realized theory. We should insist on unique realization as a standard of correctness unless it is a standard too high to be met. Is there any reason to think that we must settle for multiply realized theories? I know of nothing in the way scientists propose theories which suggests that they do not hope for unique realization. And I know of no good reason why they should not hope for unique realization. Therefore I contend that we ought to say that the theoretical terms of a multiply realized theories (*sic*) are denotationless.

In a more recent paper (2001/2007), he again deploys the same basic mechanism for a different purpose, namely to argue that

Quite generally, to the extent that we know of the properties of things only as role-occupants, we have not yet identified those properties. No amount of knowledge about what roles are occupied will tell us which properties occupy which roles.

This time the mechanism is applied not just to a particular scientific theory, but rather to a "true and complete 'final theory'" capable of delivering "a true and complete inventory of those fundamental properties that play an active role in the actual workings of nature." My concern is of course limited to gauge theories, which are presumably neither true nor complete. But there is still a certain commonality of argumentative strategy here, since I wish to argue that a proponent of the localized gauge potential properties view is caught in the very similar predicament of being unable to say which such properties occupy which roles in any actual situation.

In contrast to his 1970 paper, Lewis (2001/2007) now thinks that he can secure unique realization for his true and complete "final theory" by a simple move. His new thesis is that there is no way to distinguish this from its multiple *possible* realizations.

Though our theory T has a unique actual realization, I shall argue shortly that it has multiple possible realizations. Suppose it does indeed have multiple possible realizations, but only one of them is the actual realization. Then no possible observations can tell us which one is actual, because whichever one is actual, the Ramsey sentence will be true. There is indeed a true contingent proposition about which of the possible realizations is actual, but we can never gain evidence for this proposition, and so can never know it.

The new thesis and Lewis's argument for it are not my concern, which is the possibility of multiple actual realizations of a gauge theory when viewed as a theory of localized gauge potential properties. But this makes it important to consider Lewis's (2001/2007) new reason for dismissing such a possibility. Here is what he says:

We have assumed that a true and complete final theory implicitly defines its theoretical terms. That means it must have a unique actual realization. Should we worry about symmetries, for instance the symmetry between positive and negative charge? No: even if positive and negative charge were exactly alike in their nomological roles, it would still be true that negative charge is found in the outlying parts of atoms hereabouts, and positive charge is found in the central parts. O-language has the resources to say so, and we may assume that the postulate mentions whatever it takes to break such symmetries. Thus the theoretical roles of positive and negative charge are not purely nomological roles; they are locational roles as well.

The idea seems to be to secure unique realization for the terms 'positively charged' and 'negatively charged' in face of the assumed symmetry of the fundamental theory in which they figure by adding one or more sentences stating what might be thought of as "initial conditions" to the laws of that theory. These sentences S would be formulated almost exclusively in what Lewis calls the O-language—i.e. the language that is available to us without benefit of the term-introducing theory T. But they would also use one or more of the terms 'positively charged' and 'negatively charged' to break the symmetry of how these terms figure in T. They would do this by applying further constraints that must be met by the denotations of these terms in order that $S\&T$ be true. Those constraints would then fix the actual denotations of 'positively charged' and 'negatively charged' in T so that, subject to these further constraints, T is indeed uniquely realized.

Lewis expresses an important insight here. While a fundamental theory in physics is concerned to capture universal laws governing the workings of the world, to apply this theory to a particular situation it must be possible to use the theory to describe or represent that situation. If this were not possible, the theory would be useless. Moreover, we could have no reason to believe it, since observations of particular situations could provide no evidence for the theory. Applications of the theory provide the resources to set further constraints on the denotations of its newly introduced terms—constraints that may suffice to break the symmetries of its laws and so secure its unique realization. Note that such constraints need not involve descriptions in an O-language, though they typically will. But they will involve demonstration or ostension, as does Lewis's own suggestion when it includes the term 'hereabouts.' One could, for example, simply point to a cathode and say "That is negatively charged."

But it may still be possible successfully to apply a theory that introduces new terms while leaving some of their denotations indeterminate. Here I am indebted to Tim Maudlin, who suggested a toy theory like this in correspondence, though I have modified his example for my own purposes.

Suppose that physicists in a possible world not too different from our own try to account for the properties of the strong nuclear force in their world. They arrive at a classical (not quantum) theory modeled on classical electrodynamics, but resembling chromodynamics in that it postulates three different "color" charges (along with their opposites). The physicists postulate that the building blocks of matter are quarks, each of which bears a smallest unit of color charge. They formulate detailed dynamical laws governing the behavior of particles under the strong force. These laws are completely symmetric under permutations of color charge, and also imply that quarks will always be confined within color-neutral combinations. Nucleons are taken to consist of three confined quarks, each of a different "color," while mesons are composed of an oppositely colored quark and antiquark pair. Confinement is very strong in this theory. For example, the three quarks in a nucleon are point particles that always occupy exactly the same point of space. The theory can model the dynamics of free quarks, including how appropriate combinations would "collapse" into color-neutral point combinations. But in fact there have never been any free quarks; and, because of strong confinement, there never will be.

This theory could be applied to explain detailed properties of nuclei in this world, as well as predicting cross-sections for various scattering processes, such as the production of pi-mesons in proton–proton collisions. It could prove very successful in such applications, and could come to be believed on the basis of that success. But because of its color symmetry, the theory would not have a unique realization in that world. Moreover, because of permanent, strong confinement, there would be no way in that world to say or demonstrate which quarks are "green," which "blue," and which "red," even though (for example) every nucleon was known to consist of precisely one quark of each color. So Lewis's move would be to no avail: multiple realization would be unavoidable.

The lessons of the toy theory extend to some actual classical gauge theories, if the localized gauge potential properties view is right. Consider classical electromagnetism. On the localized EM potential properties view, a gauge transformation 1.6 has an active interpretation, so that the transformed potential $\bar{A}_\mu$ may be taken to represent a distinct distribution $\bar{D}$ of localized gauge potential properties to that (D) represented by A_μ. But, because of their gauge symmetry, if D is consistent with the laws of the theory, then so also is $\bar{D}$. It would be difficult to state those laws in a language capable of explicitly describing, rather than simply representing, the distributions D, $\bar{D}$. But suppose the theory of classical electromagnetism were somehow formulated as a postulate EM in a Lewisian language with the full complement of names for localized EM potential properties. Then EM would be multiply realized on a massive scale.

For let D, $\bar{D}$ be any pair of distributions of localized gauge potential properties in some space-time region, represented by gauge-equivalent potentials A_μ, $\bar{A}_\mu$. Now suppose that, for each r, τ_r is a term in EM that purports to name a localized gauge potential property. If EM is true, then there must be an assignment of extensions to every predicate *has τ_r* in D under which EM is true of D. But corresponding to the gauge transformation $A_\mu \Rightarrow \bar{A}_\mu$ there is a systematic permutation of this assignment that yields another assignment of extensions to every predicate *has τ_r* in $\bar{D}$ under which EM is true of $\bar{D}$. This is true for every such pair of distributions D, $\bar{D}$ in every space-time region. If EM is true, then it is realized, so there is an assignment of extensions to every predicate *has τ_r* under which it comes out true in every region. But the same systematic permutation of this assignment will also make it come out true in every region. Every such systematic permutation of the extensions of the predicates *has τ_r* in every region corresponds to a transformation in the properties named by the τ_r. Hence EM is multiply realized. Moreover, the class of available permutations is non-denumerably infinite, so there would be massive multiple realization in the *actual* world, if classical electromagnetism were a true theory purporting to describe how localized EM potential properties are distributed in it.

The argument of the last two paragraphs applies equally well to any Yang–Mills gauge theory. It establishes a clear and worrying sense in which, even if there were localized gauge potential properties, one could not use such a gauge theory to say, or represent, what they are. Of course, one cannot exclude the possibility of additions to the theory that would make this possible. But since it is the gauge symmetry of the theory that keeps it silent, this could only happen if such additions violated gauge symmetry. That would be a radical development indeed.

This conclusion does not apply to general relativity, when it is formulated as a theory of localized gauge potential properties represented by the connection and curvature of a principal fiber bundle $A(M)$, as in chapter 3. It is important to see why not, for this points to a significant difference between general relativity and classical Yang–Mills gauge theories directly relevant to their interpretation. There is a close connection to the "hole" argument (Earman and Norton 1987, etc.) here. It turns out that the very assumption that enables one to deny the radical indeterminism of general relativity despite the hole argument also permits one to assert the unique realization of a Lewisian postulate GR corresponding to this fiber bundle formulation of general relativity. In contrast to the case of a Yang–Mills gauge theory like EM, no more needs to be said.

Assume that in such a formulation of general relativity the connection ω on $A(M)$ represents localized gauge potential properties. A non-trivial principle bundle automorphism transforms this into a different connection $\overline{\omega}$. Just as for the case of classical electromagnetism, one can argue that this transformed potential may be taken to represent a distinct distribution $\overline{D}$ of localized gauge potential properties to that (D) represented by the original objects. And again, because of gauge symmetry, if D is consistent with the laws of general relativity, then so also is $\overline{D}$. It then appears that one can establish the multiple realization of GR by exactly the same argument that established the multiple realization of EM. But appearances prove deceiving!

To see why, note first that since ω is naturally homomorphic to the sum of the solder-form Θ and the Levi-Civita connection ∇, defined in the usual way on the bundle of frames $F(M)$, it follows that $\overline{\omega}$ will be naturally homomorphic to the sum of transformed solder-form $\overline{\Theta}$ and Levi-Civita connection $\overline{\nabla}$. So transforming the gauge potential ω implies transforming the geometric properties of all points in the manifold representing space-time. The resulting distribution of geometric objects (frame-field, metric, linear connection) over manifold points is simply the drag-along of the original objects associated with some manifold diffeomorphism h. Because of the diffeomorphism invariance of general relativity, the new distribution will be a model of general relativity (in its usual formulation) if and only if the original distribution was. But do these models actually represent distinct possible states of affairs, according to the theory?

4.2 THE LOCALIZED GAUGE POTENTIAL PROPERTIES VIEW

The key move in responding to the "hole" argument is to answer "No" to this question. By accepting Leibniz equivalence, one accepts that diffeomorphic models of general relativity represent the *same* state of affairs, in which the same physical objects and events have the same geometric (and indeed all other) properties. Some maintain that this refutes space-time substantivalism—or at least the most natural form of that view. Others (including me—see Healey 1995) have argued that a plausible version of substantivalism is immune to the hole argument. But all agree that acceptance of Leibniz equivalence is the key to "defanging" the hole argument. So we should accept it, and join the company of all those careful writers on general relativity who point out that the theory permits one to represent a state of affairs by any of an equivalence class of diffeomorphically related models.

But now Leibniz equivalence entails that $\bar{D}$ is actually the *same* distribution of localized gauge potential properties as D! $\bar{D}$ and D distribute these properties differently over points of the *manifold* representing a region of space-time, but they still agree on how those properties are distributed over that region *itself*. They can do this because they disagree on *how* the manifold M represents space-time. More precisely, if point $m \in M$ represents space-time location p according to D, but $f(m)$ represents p according to $\bar{D}$ (where f is a diffeomorphism of the manifold M onto itself), then whatever localized gauge potential properties D attaches at (or near) m, $\bar{D}$ attaches at (or near) $f(m)$: hence they both attach the same properties at (or near) p. So when general relativity is formulated as a theory of the connection on a principal bundle $A(M)$ that is taken to represent localized gauge potential properties, there is no active interpretation of the transformation $\omega \Rightarrow \bar{\omega}$ according to which $\bar{\omega} \neq \omega$ represent distinct distributions of localized gauge potential properties. It is true that the theory is gauge symmetric in the sense that a transformation $\omega \Rightarrow \bar{\omega}$ maps models into models. But distinct models related by such a transformation simply represent the same state of affairs, so there is no symmetry that remains to be broken by a move like Lewis's (2001/2007).

Notice that in the case of a Yang–Mills gauge theory like electromagnetism one *cannot* maintain that a gauge transformation $\omega \Rightarrow \bar{\omega}$ induced by a vertical bundle automorphism leaves unchanged the supposed localized gauge potential properties represented by the bundle connection. For this transformation leaves fixed not only the manifold point m representing each space-time location p, but also the metric, frame-field, and linear connection on M. It follows that m still represents the very same location p before and after the gauge transformation. So the gauge transformation changes (only) the supposed EM properties (or other "internal" gauge potential properties) at the same space-time location (p). Whereas a principal automorphism applied to the general relativistic connection on $A(M)$ typically *changes* what space-time point m represents. A (non-trivial) principal automorphism applied to the general relativistic connection on $A(M)$ also "repaints" the metric, frame-field, and linear connection on M in accordance with a diffeomorphism f, in such a way that if $m = f(n)$ then m comes to represent not space-time location p, but rather the space-time location formerly represented by n.

Notice that this asymmetry between gravity and other interactions arises from the fact that there is no analog to Leibniz equivalence in the case of other interactions. Even though all our gauge theories may be formulated as theories of a connection on a principal fiber bundle over a manifold representing space-time, it is the (narrowly) geometric, not the electromagnetic or other gauge, properties that constrain the way manifold points represent space-time locations. But geometric properties *are* gravitational properties within general relativity. This identification was Einstein's key innovation, whose significance he himself only finally realized when he saw how to use Leibniz equivalence to respond to the hole argument that had for so long held back his progress toward the final formulation of his theory (see the papers by Norton and Stachel (Howard and Stachel 1989)).

What are we to conclude from this extended discussion of the problem faced by a theory with gauge symmetry on the localized gauge potential properties view? We can restate the problem by echoing an earlier quote from Lewis (2001/2007). To the extent that we know of these supposed localized gauge properties only as role-occupants, we have not identified those properties. No amount of knowledge of what roles are occupied will tell us which properties occupy which roles. Even if a classical Yang–Mills gauge theory were known to accurately represent or truly describe the distribution of localized gauge potential properties in a given physical situation, this representation or description would be massively ambiguous. The gauge symmetry of the theory would prevent us from being able to say or otherwise specify which among an infinity of distinct distributions so represented or described is realized in that situation. This is not, however, the case for general relativity, when formulated as a theory of the connection on a principal bundle of affine frames.

This conclusion is semantic, not epistemological: it concerns the limits of what we can say, not of what we can know. But it is closely connected to another question it is important to ask about the localized gauge potential properties view: Assuming there are such properties, could we observe them, and if so, how? This question has an important bearing on how we should interpret a gauge theory. If nothing could count as observing localized properties allegedly represented by the potential of a classical gauge theory, then no empirical success achieved by that theory would provide any reason to believe that there are such properties—or so I shall argue in section 4.5. Semantics and epistemology come together here for a very simple reason. If there is no way for us to say how the supposed localized gauge potential properties are distributed, then we cannot even describe the results of a hypothetical observation or experiment that would reveal their distribution to us. The most that we could hope to learn from any observation or experiment is that there *are* such properties distributed in such a way as to be representable by a gauge potential. But section 4.5 argues that even this hope proves vain.

Recall the final question posed at the start of this section: How are localized gauge potential properties supposed to act on quantum and/or classical particles? That question was broached already in section 2.4 for the case of electromagnetism. Leeds (1999, p. 606) defends "the view that the Aharonov–Bohm effect is caused by an interaction between the electron and the vector potential". His defense is based on a sophisticated articulation of the new localized EM properties view—the localized gauge potential properties view applied to classical electromagnetism. It is interesting to examine Leeds's defense to see whether it provides better reasons to believe in new localized EM properties.

4.2.2 Leeds's view

As Leeds (1999) says, his defense depends on taking the fiber bundle formulation of electrodynamics literally, or almost literally. His paper concludes with a discussion of how the quantized electromagnetic field acts in quantum electrodynamics—a topic to be discussed in later chapters (5–8). But most of his paper concerns the theory of classical electromagnetism acting on quantum particles whose phase is represented on a fiber bundle on which the electromagnetic potential acts as a covariant derivative. This is a little different from the usual fiber bundle formulation of classical electromagnetism, since the bundle Leeds describes is neither a principal fiber bundle nor its associated vector bundle, but a sort of hybrid object which has been called the bundle of phases (by Bernstein and Phillips 1981, amongst others).[3] The base space is physical space, or space-time. The typical fiber consists of the set of "directions" in an (abstract) plane, each of which may be indicated by a corresponding angle—its "compass bearing." The fiber "above" a point represents phases for the particles' wave-function at that point. A bundle section yields the actual phase of the wave-function in a particular position representation. A choice of section corresponds to a choice of position representation and at the same time to a choice of gauge. A change of section corresponds to a variable phase transformation 1.11.

The electromagnetic potential is represented by a connection on the phase bundle: it determines parallel transport of "directions" in the bundle. This permits one to compare the phase of the wave-function at "neighboring" points, and so to say how fast the wave-function's phase is changing in a particular direction in the base space, and how much it changes along a particular path in

---

[3] Mathematical details of principal fiber bundles and their associated vector bundles are provided in appendix B.

the base space, all in a particular gauge/position representation. The form of the covariant derivative is therefore gauge dependent. It is related to the ordinary derivative by the addition of a constant multiple of the vector potential A. The vector potential that appears in the expression for the covariant derivative in a particular gauge consequently changes under a change of section, in accordance with a gauge transformation 1.2 (or 1.6 in the general case). But the connection itself is a geometric object on the bundle that remains invariant under changes of section.

According to Leeds (1999, pp. 612–13), the vector potential

... plays a crucial role in the dynamics: in the Aharonov–Bohm experiment, the phases of the wave at two different points on either of the paths enclosing the solenoid are related by parallel transport along that path; it is because of the path-dependence of parallel transport that the two components of the wave acquire a phase difference (in addition to that accounted for by difference in path length) between the point of emission and the point they are brought back together. The relation w' *is the parallel transport of* w *along path P* is gauge-independent: like the relation $v = wg$ within fibers, it holds between w' and w independently of, and prior to, any choice of gauge. By contrast, the particular labels, say a and a', which we give to w and w', depend of course on our choice of gauge. And so, too, with our representation of the vector potential: we typically represent this by choosing a gauge, and then giving at each point x in space, and for each fiber element w above x, the rates of change of w in this gauge when parallel transported in each of three spatial directions; these are (a constant multiple of) the components of the classical vector potential **A**. The passage from vector potential qua geometric object in the fiber bundle to classical vector potential is as unique as we could wish it to be: given a fixed notion of lift or parallel transport in the fiber bundle, the different choices of gauge for the fiber bundle (or, if the fiber bundle is non-trivial, for neighborhoods within the fiber bundle) lead to a family of classical vector potentials, all related to each other by classical gauge transformations (and closed under all classical gauge transformations).

... In the fiber bundle picture (at least, in our very literal-minded reading of that picture), the vector potential does exactly the work that Aharonov and Bohm claim it does: it is a field in space, acting on the electron wave by affecting its phase.

Does this defense shed light on how new localized EM properties act on quantum particles? There are several reasons for denying that it does. On Leeds's "very literal-minded reading" of his fiber bundle formulation, all that a gauge transformation comes to is a change of bundle section, leading to a different representation of the same invariant structures—the bundle connection representing the EM potential, and (what we might call) the intrinsic phase of the wave-function, i.e. for each space point x, an assignment of an element w from the fiber above x. In fact Leeds quickly backs away from this "very literal-minded reading" when he suggests on the next page (614) that

4.2 THE LOCALIZED GAUGE POTENTIAL PROPERTIES VIEW

... we might try to see how much of the fiber bundle picture we can hold on to, if we jettison the idea of the fiber elements as representing properties of electrons.

After abandoning this idea he argues (p. 618)

... for taking the gauge-dependent phase factor (or its derivatives) and the gauge-dependent rate of phase change, as representing certain "elements of reality" for which we have no coordinate-free representation.

and concludes that

... we do have an argument which applies to all the cases that we have reason to think arise in nature, and which gives us both a reality represented by the vector potential and local features of the electron for it to act upon. And we continue to have reasons for thinking of the vector potential as a field in space: it is well-defined independent of the entrance of any particular electron on the scene, it is given locally as a vector defined at each point in space, and, of course, it moves in space as a superposition of waves moving at speed c [the speed of light *in vacuo*].

Leeds's main argument for these conclusions will be scrutinized later (section 4.5). Meanwhile I shall criticize his understanding of the fiber bundle formalism, and especially of the significance of gauge transformations within that formalism. And I shall claim that his account of how the reality allegedly represented by the vector potential acts on charged quantum particles answers none of the concerns about the new localized EM properties view raised in 2.4.

Leeds backs away from his "very literal-minded reading" of his fiber bundle formulation because it implies that a vertical bundle automorphism that rotates the fiber above every point by the *same* angle will induce a constant phase change in the wave-function in the same section: $\Psi \Rightarrow \exp(i\lambda)\Psi$ with constant λ. He accepts that such an automorphism merely results in an alternative representation of the same state of the electrons, in accordance with the usual understanding that the quantum state is represented not by a single *vector*, but rather by a *ray* in Hilbert space—i.e. an equivalence class of vectors $[\exp(i\lambda)|\Psi\rangle]$ rather than a single vector $|\Psi\rangle$. But by an extension of this thought, an *arbitrary* smooth vertical bundle automorphism that induces a position-dependent phase change in the wave-function in the same section: $\Psi \Rightarrow \exp(i\lambda(\mathbf{x}))\Psi$ with variable $\lambda(\mathbf{x})$ will *also* result in an alternative representation of the same state of the electrons, provided that the automorphism is applied simultaneously to the bundle connection. Thus we arrive at Trautman's understanding of a "local" gauge transformation as corresponding to just such an automorphism. Admittedly, for Trautman, this was to be applied to a principal fiber bundle, but the bundle of phases may be regarded as associated to a U(1) principal fiber bundle, since the U(1) structure group is isomorphic to the group of rotations in a plane.

This observation undercuts Leeds's defense of his view in the following passage (Leeds 1999, p. 613):

And so what was in classical electromagnetism the major objection to thinking of the vector potential as real—our inability to see the different, but physically equivalent vector potentials as alternative coordinatizations of a single quantity—simply does not arise in the present context. The existence of many physically equivalent vector potentials *does* turn out to be the result of our choice of coordinatization.

For Leeds, choice of coordinatization corresponds to choice of section. If that were the only way to understand gauge transformations in the fiber bundle formalism, then it would be natural to suppose that it is the single quantity—the bundle connection—that may appear differently when differently coordinatized—i.e. represented by distinct vector fields in different sections. But now we see that the bundle connection representing the (electro)magnetic potential is not unique either—it changes also under gauge transformations, understood as vertical bundle automorphisms. A natural extension of Leeds's reasoning would therefore lead him away from the new localized EM properties view to a new *non-localized* EM properties view like that to be described in the next section and subsequently defended.

Suppose one were to accept with Leeds that we have a reality represented (in a particular section) by the vector potential, and localized features of the electron for it to act upon, represented by the gauge-dependent phase factor (or its derivatives) and the gauge-dependent rate of phase change. Would that constitute, or permit, an account of how localized EM gauge properties act on charged quantum particles, in the Aharonov–Bohm effect and elsewhere?

Leeds does not address the controversies in the interpretation of quantum mechanics that arise when one asks what are the intrinsic properties of quantum particles, and how are these represented by, or related to, their wave-function. He claims without argument that the vector potential propagates on the light cone. Such arguments may be found elsewhere (for example, see Peshkin and Tonomura 1989 p. 14). *If* we had reason to adopt a localized gauge potential properties view of classical electromagnetism, then these could contribute to an account of their causal propagation. We have yet to find such reasons in Leeds (1999): the search will continue in section 4.5. But nothing Leeds (1999) says about the localized gauge potential properties he takes to be represented by the (electro)magnetic vector potential has made it easier to understand how any such properties could act on charged quantum particles.

4.2.3 Maudlin's interpretation

Maudlin (2007) believes that gauge potentials act locally on quantum particles. But it may not be strictly accurate to call his view a localized gauge potential *properties* view. For he takes gauge theories to teach philosophers a new metaphysical lesson by postulating a category in a sense intermediate between general properties and particular individuals. According to Maudlin,

4.2 THE LOCALIZED GAUGE POTENTIAL PROPERTIES VIEW 103

the localized items a classical gauge theory attaches at a space-time point are neither properties nor individuals but something more like tropes—i.e. individual instances that are parts, rather than exemplifications, of a property. Here's the idea.

On one venerable philosophical view, properties are universals—abstract entities that may (or may not) be exemplified by particular individuals. When distinct individuals are similar in some respect, this is explained by their both participating in, or exemplifying, the same abstract property. The property itself exists independently of any and all such exemplifications; it is, in a timeless sense, prior to them. But there is a rival view which takes properties to exist only by virtue of being instantiated, so that it is property-instances, i.e. tropes, that come first, while properties are just collections of similar tropes. The challenge for the rival view is to say what this similarity comes to if it cannot be understood as jointly instantiating the same abstract universal. One response is simply to take the relevant similarity relation as primitive—as an external relation among tropes that binds them together into collections so that they compose (concrete and so supposedly less mysterious) properties.

The metaphysical innovation Maudlin sees in gauge theory is to go one stage further, and to consider the "binding" relation to be not merely external (i.e. not determined by the intrinsic properties of the relata, like the relation of *being on top of*) but extrinsic (so that whether or not it obtains is determined in part by intrinsic properties of things *other than* the relata, like the relation of *being married to*). So what a gauge theory attaches at a point is not an instance of the very same localized gauge property that may or may not be instantiated elsewhere, but something more like a trope that *could not* be instantiated anywhere else. Starting with these trope-like entities, the connection defines an extrinsic, since *path-dependent*, similarity relation that defines what it is for two trope-like entities at different points to count as bits of the same property.

Each trope-like entity at a point is represented on a phase bundle by an element of the fiber above that point: this in turn is represented in a particular gauge by the phase of the wave-function at that point. And the gauge potential represented by the bundle connection defines a path-dependent similarity relation among these entities by saying what constitutes parallel transport of fiber elements in the bundle. Two entities count as similar—as parts of the same localized gauge property—relative to a path, just in case parallel transport along the horizontal lift of that path takes one into the other. This is the only meaning that can be attached to the notion of a localized gauge property. It follows that the question as to whether the localized gauge properties of two points are the same or different is simply not well formed. It is rather like the question as to whether two separated Antarctic explorers, each pointing towards the South Pole, are pointing in the same direction.

Maudlin's view should interest the metaphysician. But does it help us to understand what gauge theories are telling us about our world? Does it help to address worries about whether we could say that things are one way rather than another at various space-time points in our world if a classical gauge theory were true of it? No: if anything, it intensifies those worries. On Maudlin's view, if a classical gauge theory were true, then at each point there attaches one out of a non-denumerable infinity of "localized phase tropes." Since there is a non-denumerable infinity of points, and no trope attaches at more than one point, specifying which trope attaches at which point will require a capacious vocabulary. One might think to press mathematics into service at least to represent these facts. A "phase field" $\lambda(p)$ may take any of a non-denumerable infinity of values, one value at each space-time point p. Rather than representing "localized phase tropes" by a single field, one could instead employ a parametrized family of magnitudes λ_p, one for each point p. And one could try to use this family to represent, for every point p, which of the possible tropes-at-point-p is attached at p by assigning a value $\lambda_p = x$ to the relevant member of the family at p. Then, for $p \neq q$, $\lambda_p = \lambda_q$ would state that distinct magnitudes happened to have the same value, not that a single magnitude had the same value at different points.

But the attempt would fail for a familiar reason. The gauge symmetry of the theory implies that, for each point p, there will be a massive systematic ambiguity as to which trope-at-p is represented by the value of the magnitude λ_p. Even if we could form the words to say how localized phase tropes are distributed, uttering those words would not help to describe that distribution in such a way as to discriminate it from a non-denumerable infinity of others.

Nor does Maudlin's view help us to understand the action of gauge potentials on quantum particles. That would require an account of the physical role of the extrinsic relation represented by the bundle connection in affecting localized phase tropes. But we have been told only that it plays a kind of semantic role, by grouping localized phase tropes at different points into properties, relative to a connecting path. This may make for interesting new metaphysics, but it does not give us a better grasp of the physical processes involved in the action of a classical Yang–Mills gauge field on quantum particles.

4.3 The non-localized gauge potential properties view

According to the non-localized gauge potential properties view, the gauge potential of a theory does represent qualitative intrinsic properties in addition to those represented by the gauge field; but these are predicated only of, or at, extended *regions* of space-time—not their constituent points. It is not part of the view that such regions are represented by open sets of the space-time manifold, nor that they are four-dimensional in their inherited topology. On the version of the view that I shall develop and defend in this chapter, the

4.3 THE NON-LOCALIZED GAUGE POTENTIAL PROPERTIES VIEW

relevant regions are oriented images of closed curves in space-time—and so are represented by closed subsets of the space-time manifold M. Topologically, the subsets are images of immersions of the circular manifold S^1 in M: when the closed curve is non-self-intersecting, its image is a topological embedding in M. In accordance with the usage introduced in the previous chapter (section 3.1.2) I shall call the oriented region of space-time represented by such a submanifold a *loop*. So a loop corresponds to the oriented image of a continuous, and piecewise smooth, non-self-intersecting closed curve in the space-time manifold.

The non-localized gauge potential properties view is motivated by the idea that the structure of gauge potential properties is given by the gauge-invariant content of a gauge theory. The most direct way to implement this idea would be to require that the gauge potential properties are just those that are represented by gauge-invariant magnitudes. But we shall see that a more subtle implementation is required in the case of a non-Abelian gauge theory.

Consider first the case of classical electromagnetism. While the vector potential A_μ is gauge dependent, its line integral $S(C) = \oint_C A_\mu dx^\mu$ around a closed curve C is gauge invariant. So, therefore, are functions of $S(C)$ including the Dirac phase factor $\exp\left(\frac{ie}{\hbar}\oint_C A_\mu dx^\mu\right)$. When Wu and Yang (1975) say (on p. 3846) "What provides a complete description that is neither too much nor too little is the phase factor," they appear to take the view that classical electromagnetism represents non-localized gauge potential properties by the Dirac phase factor. The other two options they have rejected before making this Goldilocks choice are the (localized) field strength $F_{\mu\nu}$ (because it underdescribes electromagnetism, i.e. different physical situations in a region may have the same $F_{\mu\nu}$) and the phase $\frac{e}{\hbar}\oint_C A_\mu dx^\mu$ (because it overdescribes electromagnetism, i.e. different phases in a region may describe the same physical situation). Chapter 2's discussion of the Aharonov–Bohm effect made it clear why the field strength in a region does not suffice to account for the behavior of charged quantum particles in that region; while the observed behavior of quantum particles of charge e in a region would be the same if the phase around each closed curve were increased by $\frac{nh}{e}$, for integral n, thus keeping constant the phase factor for every such curve. But does this justify the claim that the Dirac phase factor correctly represents qualitative intrinsic properties of loops?

It appears that it does not. The phase does not describe electromagnetism itself in a region, but rather the effects of electromagnetism on quantum particles with the particular charge e. While the observed behavior of quantum particles of charge e in a region would be the same if the phase around each closed curve were increased by $\frac{nh}{e}$, the observed behavior of quantum particles of charge $e' \neq e$ would change, unless e' were an integral multiple of e. So different phases may reflect the same electromagnetic state of affairs in a region only if there is some minimal quantity of electric charge e_0 of which all charges are integral multiples: $e = me_0$ ($m = 0, \pm 1, \pm 2, \ldots$).

Similarly, the Dirac phase factor for arbitrary charge e cannot describe electromagnetism alone, but at most the effects of electromagnetism on particles of charge e. It is only if all charges are integral multiples of some elementary quantity e_0 that the Dirac phase factor $\exp\left(\frac{ie_0}{\hbar}\oint_C A_\mu dx^\mu\right)$ may plausibly be claimed to describe electromagnetism itself, rather than the effects of electromagnetism on particles of a particular charge. Now, as it happens, it is currently believed that all charges *are* integral multiples of a certain minimum charge, namely the charge on the down quark, $-1/3$ of the electronic charge. But charge quantization is not usually taken to be a consequence of classical electromagnetism. And while it does follow from non-Abelian gauge theories with a simple structure group, including SU(5) grand unified theories, these appear to be empirically inadequate.

So we face a choice in developing the non-localized gauge potential properties view of classical electromagnetism. We may take these properties to be represented by $S(C)$—the line integral of a vector potential around closed curve C—or, if there is a minimal quantum of electric charge e_0, by the Dirac phase factor $\exp\left(\frac{ie_0}{\hbar}\oint_C A_\mu dx^\mu\right)$ that corresponds to it. In this and other interpretative choices, we should be guided by the evidence, while remaining ready to revise our interpretation in the light of new evidence. The evidence is in favor of a minimal quantum of electric charge, so we should adopt the latter interpretation. By a suitable choice of units we can then absorb the quantities e_0 and $\hbar$ into A_μ. Having done so, we arrive at the view that non-localized EM potential properties in a region are represented by the holonomies $\exp\left(-i\oint_C A_\mu dx^\mu\right)$ of all closed curves in the region (the minus sign in the exponential is a consequence of a conventional choice in the definition of the Dirac phase factor). This is the interpretation of classical electromagnetism I shall defend.

The view extends naturally to other Yang–Mills gauge theories, though there are some complications for non-Abelian theories. In each case, it is reasonable to assume that there is some minimal unit of generalized charge associated with the interaction. The first complications appear in the expression for the non-Abelian holonomy of a closed curve, namely

$$\mathbf{H}(C) = \wp \exp\left(-\oint_C A_\mu^a T_a dx^\mu\right) \tag{4.6}$$

In this expression, A_μ^a is a component of the gauge potential corresponding to the a-axis in a basis for the Lie algebra of the non-Abelian structure group of the theory, while T_a is a matrix representation of the corresponding generator of the algebra. This holonomy is an element of a representation of the structure group of the theory. It acts on the value of the particles' wave-function at the base point m of the curve C to produce a generalized phase shift in the function at that space-time point. The value of the wave-function at a point is

4.3 THE NON-LOCALIZED GAUGE POTENTIAL PROPERTIES VIEW

no longer simply a complex number, but rather a vector or spinor in an abstract "internal" space with the dimensionality of the representation of the structure group, so a generalized phase shift means a rotation in this "internal" space. If the structure group is non-Abelian, distinct matrices T_a, T_b generally fail to commute. So the exponential in the expression for the holonomy cannot be understood as a simple power-series expansion. The symbol $\wp$ indicates that it is a path-ordered exponential, a notion that was explained in chapter 3 (section 3.1).

The next complication is that the holonomy of a closed curve C depends on its base point m, so it should really be written $\mathbf{H}_m(C)$ (cf. equation B.46). This makes it difficult to see how it could represent qualitative intrinsic properties of a loop traced out by the image of a closed curve, since that image is clearly independent of the base point of the curve. But this difficulty is readily overcome, as follows.

A choice of representation of the Lie algebra is effected by a choice of basis e^d for the fiber above the base point m in the associated vector bundle. Suppose the base point is changed to m', and the same choice of basis is made, but this time at m'. By parallel transport of e^d back along a curve $\gamma_{mm'}$ linking m' to m in accordance with some connection compatible with the holonomy $\mathbf{H}_m$, this will correspond to a transformed basis $\bar{e}^d$ for the fiber above m for some unitary transformation $\bar{e}^d = \mathbf{U}^{-1}_{mm'}(C)e^d$, where

$$\mathbf{U}_{mm'} = \wp \exp\left\{-\int_{\gamma_{mm'}} A_\mu^a(x) T_a dx^\mu\right\}. \tag{4.7}$$

The value of the wave-function at m is $\overline{\boldsymbol{\psi}}(m)$ when expressed in the basis $\bar{e}^d$, where

$$\overline{\boldsymbol{\psi}}(m) = \mathbf{U}^{-1}_{mm'}(C)\boldsymbol{\psi}(m) \tag{4.8}$$

Changing the base point must change the holonomy of all curves: $\mathbf{H}_m(C) \Rightarrow \mathbf{H}_{m'}(C)$ in order that these continue to represent the same non-localized gauge potential properties. For this requires that the transformed holonomy applied to the transformed wave-function at m equal the transformation of the result of applying the original holonomy to the original wave-function at m, i.e.

$$\mathbf{H}_{m'}(C)\overline{\boldsymbol{\psi}}(m) = \overline{\mathbf{H}_m(C)\boldsymbol{\psi}(m)} \tag{4.9}$$

This requirement is met because of the following transformation property of the holonomy under change of base point (see appendix B, equation B.49):

$$\mathbf{H}_{m'}(C) = \mathbf{U}^{-1}_{mm'}(C)\mathbf{H}_m(C)\mathbf{U}_{mm'}(C) \tag{4.10}$$

There is another related problem: in a non-Abelian theory, the holonomy $\mathbf{H}_m(C)$ 4.6 is not gauge invariant, but transforms as follows:

$$\mathbf{H}_m \Rightarrow \overline{\mathbf{H}}_m = \mathbf{U}(m)\mathbf{H}_m\mathbf{U}^{-1}(m) \qquad (4.11)$$

under a non-Abelian gauge transformation $\mathbf{U}(m)$, where $\mathbf{U}(x)$ is a unitary matrix at $x \in M$ corresponding to a smoothly varying rotation of bases in the Lie algebra of the theory's structure group at each point x. (In the Abelian case, $\mathbf{H}$ is not merely covariant but invariant, since $\mathbf{U}$ and $\mathbf{H}$ commute.) This gauge transformation is induced by a smooth alteration in the bundle connection, represented by a change in the vector potential *in the same section* throughout an open subset of M including m:

$$A_\mu^a(x) \Rightarrow \overline{A}_\mu^a(x) = \mathbf{U}(x)A_\mu^a\mathbf{U}(x)^{-1} - \left[\partial_\mu \mathbf{U}(x)\right]\mathbf{U}^{-1}(x) \qquad (4.12)$$

which is a rewriting of equation 3.38. But if holonomies are not gauge invariant, then to claim that they represent non-localized gauge potential properties seems to fly in the face of the motivating idea of the non-localized gauge potential properties view—the idea that the structure of gauge potential properties is given by the gauge-invariant content of a gauge theory.

This problem may be resolved as follows. Since the gauge transformation 3.38 is induced by a change of connection, there is a corresponding change in the covariant derivative on the associated vector bundle, where the wave-function is represented as a section. So the wave-function also transforms under this gauge transformation:

$$\psi(x) \Rightarrow \overline{\psi}(x) = \mathbf{U}(x)\psi(x) \qquad (4.13)$$

with the result that

$$\overline{\psi}(m) = \mathbf{U}(m)\psi(m) \qquad (4.14)$$

We have already noted the universal agreement that the phase of the wave-function at a point has no absolute significance. And on the non-localized gauge potential properties view it is only phase differences around closed curves that represent properties of charged particles. Now we see that even these phase *differences* are gauge dependent: applied at the base point m, a gauge transformation changes all phase differences around closed curves by the same unitary transformation $\mathbf{U}(m)$. In fact this is just what one should expect. These phase differences represent intrinsic properties only in a particular representation of the Lie algebra at m: changing the representation

4.3 THE NON-LOCALIZED GAUGE POTENTIAL PROPERTIES VIEW

will change the phase difference representing given intrinsic properties. Since the holonomies are supposed to represent the non-localized gauge potential properties that are responsible for these intrinsic properties, there must also be a corresponding change in the holonomy of C representing the non-localized gauge potential properties that effect this phase difference around C. This is ensured by requiring that the holonomies transform in accordance with 4.11, for then we have, for all curves C with base point m,

$$\mathbf{H}_m(C)\overline{\boldsymbol{\psi}(m)} = \overline{\mathbf{H}_m(C)\boldsymbol{\psi}(m)} \qquad (4.15)$$

Now there is a quantity related to a holonomy that is both gauge invariant and independent of base point, namely the Wilson loop

$$W(C) = Tr\{\mathbf{H}_m(C)\} \qquad (4.16)$$

This is a complex number rather than a matrix (see chapter 3, section 3.1.2). But it is important to see why it alone does not suffice to represent all the properties postulated by the non-localized gauge potential properties view.[4] For this purpose, consider the example of an SU(2) gauge theory describing the (approximate) isospin symmetry of neutrons and protons. The behavior of a nucleon in a region will clearly depend on whether it is a neutron or a proton. Accordingly, the theory represents their states by orthogonal wave-functions in some bundle section. Vertical bundle automorphisms will implement gauge transformations 3.38 in the connection and the transformation 4.11 in the holonomies. These include an automorphism that transforms a wave-function representing a neutron (in a particular section) into one representing a proton under the same representational conventions, without changing the Wilson loops, since these are gauge invariant. But clearly the gauge potential acts differently on protons and neutrons. So that difference is not captured by the Wilson loops, but only by the holonomies, which (as we saw) are not invariant, but transform by a common similarity transformation under such a gauge transformation.

In applying the theory, it is necessary to choose an arbitrary isospin vector (or rather spinor) in the fiber above some particular space-time point to represent the value of a neutron's rather than a proton's wave-function at that point. This is something we can do because even though isospin is an approximate symmetry of strong nuclear forces, it is completely broken by the electromagnetic interactions. So we can (for example) specify that a particular isospinor is to represent the value of the *positively charged* nucleon's wave-function at p, and thereby set up the needed representational convention. But

---

[4] I am indebted to Tim Maudlin for stressing this point in correspondence.

this simultaneously fixes the representation for the holonomies of neutrons and protons so that it is clear which non-localized gauge potential properties these represent and that they affect a neutron's behavior one way but a proton's behavior a different way. Any theory that uses magnitudes other than scalars to represent intrinsic properties will require a similar procedure for coordinating axes in the theory's representational space with mutually exclusive features (spatial directions, sign of electric charge, etc.) of the physical world.

After dealing with these complications, we may now formulate a non-localized gauge potential properties view of a non-Abelian gauge theory as follows. At, or in arbitrarily small neighborhoods of, each entire loop in space-time there attach qualitative intrinsic non-localized gauge potential properties. What these are, is not determined by any qualitative intrinsic physical properties attached at, or in arbitrarily small neighborhoods of, space-time locations on the loop. These non-localized properties are represented by the holonomy of an oriented, piece-wise smooth, non-self-intersecting, closed curve whose image traces out the loop. Accordingly, they may be referred to as *holonomy properties*. How the holonomy represents these properties depends both on the base point used to define the curve and on the choice of basis at that point for a representation of the Lie algebra of the structure group of the theory.[5] The holonomy properties of a loop act on quantum particles by affecting qualitative intrinsic physical properties that attach to a particle on, or in arbitrarily small neighborhoods of, the entire loop. What these are is not determined by any qualitative intrinsic physical properties attached at, or in arbitrarily small neighborhoods of, space-time locations on the loop. These properties may be represented by the generalized phase difference of the particles' wave-function around the loop. Accordingly, they may be referred to as *phase difference properties*. The form such a phase difference takes also depends both on the base point used to define a curve that traces out the loop and on the choice of basis at that point for a representation of the Lie algebra of the structure group of the theory.

The non-Abelian gauge theories formulated in chapter 3 include not only classical Yang–Mills theories but also general relativity, considered as the theory of a connection and curvature of a principal fiber bundle. For the structure group of general relativity in that formulation is also non-Abelian. One may entertain a non-localized gauge potential properties view of Yang–Mills theories (whether Abelian or non-Abelian), of general relativity, or of both. The previous section provided reasons to reject this view of general relativity in favor of a localized gauge potential properties view. In the rest of this chapter I will further clarify and defend a non-localized gauge potential properties view of classical Yang–Mills theories. I call it a *holonomy interpretation*.

---

[5] The set of holonomies undergoes a common conjugacy transformation with a different choice of base point and/or a transformation of basis at that point.

4.4 A holonomy interpretation

As a first approximation, the holonomy interpretation of a classical Yang–Mills gauge theory maintains that the theory describes or represents qualitative intrinsic holonomy properties of entire regions of space-time, each of which consists of all the points on a loop. But this apparently commits the interpretation to space-time substantivalism, and in any case it is not quite right.

What exactly is a loop? Recall that a loop corresponds to the oriented image of a continuous, and piece-wise smooth, non-self-intersecting, closed curve in the space-time manifold. Although there will be no loops if space-time is not orientable, we have no reason to believe that our space-time is non-orientable, and every reason to believe that it *is* orientable.

On a realist view, space-time exists over and above events and processes occurring in it. On this view one may identify a loop with a set, or fusion, of space-time points plus an orientation: a point lies on the loop if and only if it is a member or part of this. An orientation may be given by a cyclic order relation of the points on the loop. But then holonomy properties are not strictly intrinsic properties of the set or fusion of points on the loop, but only of this together with an order relation. Alternatively, a holonomy property may be considered an *extrinsic* property of the set or fusion of points on the loop, in so far as whether or not it holds depends not just on properties of that set or fusion, but also on a relation among its constituent points.

On a relationist view, neither space-time nor its regions exist independently—there are only spatiotemporal relations among events or processes. In that case, a holonomy property cannot literally be either an intrinsic or an extrinsic property of a space-time region. A relationist who wishes to adopt a holonomy interpretation of a gauge theory owes us an account of what objects (s)he wishes to substitute for space-time regions, and an argument as to how spatiotemporal relations among these can be made to stand in for properties of and relations among points and/or regions of space-time. Relationists have not found it easy to provide such reconstructions of more familiar talk of space-time and its features. Perhaps the added difficulty of accommodating a holonomy interpretation of a Yang–Mills gauge theory will provide a further argument against space-time relationism.

A gauge theory represents holonomy properties of space-time loops in a region whether or not quantum particles are present in that region. If quantum particles are present, then their wave-function (in a particular gauge) takes a value at each point in the region. The value of the wave-function of each type of particle will be an element of a vector space on which acts a particular representation of the structure group. The dimension of that space will depend on the kind of particle, just as the wave-functions of particles of different spin are vectors or spinors in a space whose dimension depends on their spins. So the expression 4.6 for the holonomy

$$\mathbf{H}(C) = \wp \exp\left(-\oint_C A_\mu^a T_a dx^\mu\right) \quad (4.17)$$

is actually multiply ambiguous, since it depends on the dimensionality of the matrices T_a that represent the Lie algebra of the structure group, and this will vary with the type of particle considered. An unambiguous expression representing the holonomy properties of a curve without regard to the type of particles on which these may be thought to act is given by the holonomy map in the principal fiber bundle on which the gauge theory is formulated

$$\mathcal{H}(C) = \wp \exp\left(-\oint_C \mathcal{A}_\mu dx^\mu\right) \quad (4.18)$$

where $\mathcal{A}$ is a Lie-algebra-valued one-form on (an open subset of) the space-time manifold representing the connection ω on the principal fiber bundle.[6]

It is not even necessary to mention any bundle connection to represent holonomy properties. As chapter 3 showed (in section 3.1.2), the holonomy $\mathcal{H}(\gamma)$ may be defined as the image of a homomorphism from the structure group of the gauge theory into the hoop group. This definition permits perhaps the most direct representation of holonomy properties in a gauge theory, and makes it transparent that these attach to loops independently of the presence of quantum particles of one kind or another. But, on the holonomy interpretation, the important point is not how holonomy properties are represented, but that the gauge theory does indeed represent such non-localized properties of loops.

4.4.1 Epistemological considerations

We have now considered three different interpretations of classical Yang–Mills gauge theories: the no gauge potential properties view, the localized gauge potential properties view, and the non-localized gauge potential properties view (in its holonomy properties version). Which, if any, is right?

This question may seem naive or unanswerable. Certainly it cannot be straightforwardly resolved by experiment or observation. I have bracketed the assumption of empirical adequacy in this chapter by simply assuming that all the classical gauge theories under consideration are empirically adequate. Under this assumption, no observation can falsify any interpretation of any of

---

[6] See appendix B, equation B.27. If the bundle is not trivial, $\mathcal{A}_\mu$ may not be defined over the entire base space. A particular curve C may then not be confined to any open set over which a single $\mathcal{A}_\mu$ is definable, in which case this expression for $\mathcal{H}(C)$ would need to be modified. But the holonomy of such a closed curve exists even if the bundle is not trivial.

them. But even without this assumption, there are reasons to believe that the interpretations are empirically equivalent, so no observation could discriminate among them.

For how could observations give access to gauge potential properties, if there are any? It seems that all we are able to observe is the behavior of particles under the influence of gauge interactions. In chapter 2 we saw that all three interpretations of classical electromagnetism predicted the same results for all such observations, whether the particles were classical or quantum. But does this conclusion generalize to all classical Yang–Mills gauge theories?

In the case of non-Abelian theories, one might think that observations of particle behavior could at least establish the existence of gauge potential properties, even if it could not show whether these are localized or non-localized. For recall Wu and Yang's demonstration that, in the case of an SU(2) gauge theory, there may be physically distinct situations in a simply connected region of space-time, represented by inequivalent gauge potentials, even though the gauge field is the same throughout the region in each situation. One reason for taking these situations to be physically distinct was that sources of the field were present in the region in one situation but not in the other. But surely the behavior of particles will be different depending on whether or not there are sources present in the region. And since this difference cannot be accounted for by any difference in the gauge field, it must be because of a different distribution of gauge potential properties in the region.

But even if the behavior of particles in a region with fixed gauge field does depend on the gauge potential, it does not follow that the potential itself represents any properties of the region. For an alternative explanation is available in terms of the sources in the region. In the two situations of Wu and Yang's example, these sources differ everywhere in the region. So an alternative *local* explanation of the different behavior is possible that makes no mention of any gauge potential properties but appeals directly to the different distribution of sources.

Whether two interpretations of a theory are empirically equivalent depends on what can be observed. But it is a familiar point that a theory may be extended so that additional observations may break the tie. For example, the conjunction of Newton's theory of mechanics with a theory of electromagnetism according to which the speed of light *in vacuo* is constant only with respect to a privileged state of absolute rest gives rise to the possibility of discriminating observationally between one interpretation of Newton's theory that commits it to a state of absolute rest and another that does not. Similarly, one cannot rule out a future extension of a Yang–Mills gauge theory that permits observations whose results depend on the existence of a privileged gauge, so that some future Faraday can enter his cage and determine this experimentally. If that were to happen, then his observations would discriminate in favor of an interpretation of the gauge theory that commits it to such a privileged gauge, and against a holonomy interpretation. This has not yet happened. But since we cannot be sure that it never will, it seems that we are in no position to answer the question as to whether a holonomy interpretation is correct.

But the possibility that our evidence may change does not preclude our using the evidence we have as the basis for a reasonable, though fallible, judgment as to how best to interpret a theory. Prior to its extension to encompass electromagnetism, and given that gravitation and other known fundamental forces did not depend on absolute velocity, the evidence justified the belief that, correctly interpreted, Newton's mechanics was not committed to a state of absolute rest.

Even such defeasible reasons in favor of one interpretation rather than another may be dismissed as no more than personal preferences by constructive empiricists like van Fraassen (1980). For a constructive empiricist, the goal of science is not explanatory truth, but empirical adequacy, where a theory is empirically adequate just in case everything it says about observable structures is true; while acceptance of a scientific theory involves as belief only that it is empirically adequate (van Fraassen 1980). Van Fraassen also advocates a voluntarist epistemology, in accordance with which it is up to anyone reflecting on such a theory in science to adopt whatever interpretation he or she likes best. Since it is our working assumption that classical Yang–Mills gauge theories are empirically adequate under all the interpretations presently under consideration, he would dismiss any "reasons" offered in favor of one interpretation rather than another as amounting to no more than expressions of subjective preference, or possibly rhetorical devices designed to mold the preferences of others.

But note that, for a constructive empiricist, there could be no objective reason to adopt an interpretation of a gauge theory that takes it to represent gauge *field* properties, since these are not observable either, except indirectly through observations of their effects on particle behavior. So we can rationally deny the existence of magnetic fields, while remaining careful to remove metal jewelry before having a CAT scan.

One can press harder. The consistent constructive empiricist will also deny that the evidence for the gauge theory of electromagnetism requires a rational person to believe that there *are* electrons whose interference effects allegedly constitute that evidence. And he will deny that the evidence for the gauge theory of chromodynamics requires a rational person to believe that there are colored quarks, or even the hadrons they are said to compose, and whose behavior scientists say they study in the high-energy physics experiments whose results allegedly constitute that evidence. He will not count Rutherford's experiments as observations of electrons and their charge/mass ratio, or Millikan's experiments as observations of their charge. He will not count Dehmelt's experiments as observations of a single barium ion that are at least as direct as observations of the star Sirius. Nor will he be persuaded that since scientists design and build expensive machines specifically in order to produce carefully controlled beams of electrons, protons, antiprotons, and now B-mesons, they (and we) must rationally believe there are such things. He cannot even agree that to withhold belief from the proposition that DNA has a double helix structure is no more rational than to withhold belief from the proposition that Saturn has rings, or that the Earth is roughly spherical.

The point is that a great deal of modern science, including molecular biology and nanotechnology, concerns structures that a constructive empiricist takes to be unobservable. If it is rational to remain agnostic concerning what science says about such structures, then it is rational to dismiss great swaths of modern science as little more than entertaining and impressive speculation that happens to pay off in predicting and controlling events that concern us.

Rather than freeing us from the constraints of scientism, such epistemological license unjustifiably demeans scientific rationality. While remaining grounded on observation, science itself constantly revises the bounds of the observable, both by constructing new instruments to enhance our observational powers and by developing new theories enabling us to use familiar means to observe novel structures. Indeed these advances typically go hand in hand, as new theories suggest new kinds of instruments, while new instruments require theoretical backing to justify their claims to veracity. Revision does not always mean expansion, though. We no longer think that we can observe whether or not someone is a witch, nor that we can tell whether the Earth is moving with our unaided senses.

The constructive empiricist has a different view of observation. He admits that scientific advances affect what we take ourselves to observe, but he denies that this constitutes any change in what is in fact observable—that is fixed by the world, including our own constitution as part of it. We—members of our epistemic community—are humans, and there are natural limits to what a human organism can become aware of via its unaided sense organs. We find out about these limits by doing science, but this does not change them. Only a change in the sensory capacities of members of our epistemic community could change what is observable by that community. This could happen if humans were to develop new sensory abilities, perhaps as a result of natural selection or genetic engineering. Or it could happen if we were to admit non-humans (apes? aliens? androids??) to our epistemic community. Otherwise, the limits of observation are fixed by nature, independent of advances in science and technology. Some things are simply invisible because they are too small to be seen. Other things are unobservable because neither sight nor any other of our senses react to their presence.

The constructive empiricist may have captured *a* notion of observation. But it is not a notion that is of interest to scientists other than those researching into human sensory capabilities. Observational astronomers, for example, speak of observing the universe at all wavelengths of electromagnetic radiation quite without regard to whether these are detectable by unaided human senses. They worry about the impending loss of the Hubble Space Telescope because it will hinder or prevent their observation of certain astronomical events. And they continue to develop new kinds of instrument such as neutrino telescopes and gravitational wave interferometers to enable them to observe the universe by methods that exploit non-electromagnetic processes that are quite inaccessible to human senses. Medical researchers use a microscope to observe what they are doing as they extract individual cells from an embryo for genetic testing. Physicists observe and manipulate individual atoms using scanning tunneling electron microscopes.

There is nothing strained or artificial about such uses of 'observation' and related terms. They all conform to the guiding principle that an observation is a process whereby an epistemic agent gains more or less reliable information about some state of affairs by physically interacting with it. When we find or create such a process we may exploit it to gain information about that state of affairs—i.e. to observe it. The key question for epistemology is not what can we become aware of through our unaided senses, but what can we become reliably informed about by interacting with it through suitable processes. The answers to these questions diverge in both directions because of our sensory limitations. There are many things of which our senses make us aware without reliably informing us about them: and there are many things about which we may become reliably informed by means of suitable interactions though we can never become aware of them through our unaided senses. For scientific epistemology what matters is not sensory awareness but reliable information.

Science itself is our guide to what we can be reliably informed about. This involves us in a kind of epistemic circle, since it is the (supposedly) reliable information provided by observation that justifies our confidence in the guide. But the circle is wide, and it is not vicious. The theories that warrant our confidence in certain observational procedures are typically different from those we use those procedures to test. And even when these are not wholly disjoint sets, the fact that we have relied on a theory when placing our confidence in the reliability of some observational procedure does not guarantee that its results will bear out that theory. We may, for example, trust Euclidean geometry in its application to our rulers and compasses, but the geometry revealed to us through their use could still turn out to be non-Euclidean.

However, it can happen that a theory postulates novel structures while at the same time leaving no way for us to obtain reliable information about their key characteristics. Newton's theory of mechanics provides a notorious example. Newton postulated the existence of a uniquely privileged state of absolute rest. But if all physical processes conformed to his mechanics and no forces depended on absolute velocities, then no physical process could reliably inform us as to which of an infinity of different states of unaccelerated motion that was. If we believe Newton's theory as he interpreted it, then we believe there is a structure—the state of absolute rest—about which there is no way of obtaining reliable information by observation. This is not because of the contingent limitations of human sense organs, but because the theory countenances no physical process that discriminates that state from a host of others.[7]

Faced with such a situation, the prudent epistemologist will recommend caution. No belief identifying absolute rest with a demonstrated state of uniform motion could be warranted. It is not clear that we can even entertain such a belief, as I will explain in a later section (4.4.3). We can certainly entertain the belief that Newton's theory is true as he interpreted it, and so also the belief that there is a unique state of absolute rest. But no matter how much evidence we obtained for Newton's theory, we would have no reason to hold this last belief. Consequently, we would have no reason to believe Newton's theory as he interpreted it. This does not mean we should reject Newton's theory, however. For alternative interpretations are available.

One may adopt an interpretation that resembles a more committed constructive empiricism. This view takes the evidence for Newton's theory to warrant (some degree of) belief that everything it says is (pretty much) true, *except* when this concerns absolute (rather than relative) velocities or absolute rest. About such assertions it remains agnostic. The interpretation requires no reformulation of the theory: it requires only that one take the evidence for it to warrant belief only in parts of it—the kernel of the theory on which such evidence actually bears.

In fact it is now well known that one can reformulate Newton's theory so as to remove the shell. The trick is to set the theory in the framework of a four-dimensional affine space-time, whose time-like lines define inertial motion without singling out any state of rest.[8] The theory does allow processes that would enable us to obtain reliable information as to which states of motion these represent, so the affine structure is observable. One can now interpret the theory realistically, so that evidence for it warrants (some degree of) belief that what it says about the affine structure of space-time is (pretty much) true, along with everything else it says as thus reformulated. Someone may add to this formulation further claims that single out one congruence of time-like curves as those corresponding to a state of absolute rest. But this will lead to no new predictions, and neither the theory nor anything else that we know describes processes we could use to observe which actual trajectories correspond to these privileged curves. Such additional claims do no work in the reformulated theory and we have no more reason to believe them than we do to believe in undetectable ghosts.

---

[7] This may not have seemed such a problem for Newton. He believed that God was in possession of such reliable information through other means. So he may have thought we could come to share this information; not through observation, but through revelation.

[8] See Stein (1967), Sklar (1974), Friedman (1983), etc.

A classical Yang–Mills gauge theory puts us in a very similar situation. On the localized gauge potential properties view, it describes or represents intrinsic properties that attach at space-time points. But if all physical processes conformed to the theory and no other phenomena depended on a particular gauge, then no physical process could reliably inform us as to how among an infinity of different ways these localized gauge properties are actually distributed. If we believe the theory as thus interpreted, then we believe there is a structure—the actual distribution of localized gauge potential properties—about which there is no way of obtaining reliable information by observation. This is not because of the contingent limitations of human sense organs, but because the theory countenances no physical process that discriminates that distribution from a host of others.

Faced with such a situation, our prudent epistemologist will again recommend caution. No belief identifying the distribution of localized gauge potential properties with some demonstrated state of the world could be warranted. It is not clear that we can even entertain such a belief, as I will explain later (in section 4.4.3). We can certainly entertain the belief that the classical Yang–Mills gauge theory truly describes or represents localized gauge potential properties, and so also the belief that these are actually distributed somehow. But no matter how much evidence we obtained for the gauge theory, we would have no reason to hold this last belief. Consequently, we would have no reason to believe the gauge theory on the localized gauge potential properties interpretation. This does not mean we should reject the theory, however. For alternative interpretations are available.

One may adopt an interpretation that resembles a more committed constructive empiricism. This view takes evidence for the classical Yang–Mills gauge theory to warrant (some degree of) belief that everything it says is (pretty much) true, *except* when this concerns the localized gauge potential properties and their distribution. About such assertions it remains agnostic. The interpretation requires no reformulation of the theory: it requires only that one take the evidence for it to warrant belief only in parts of it—the kernel of the theory on which such evidence actually bears.

But one can do better. One can reformulate the theory as a theory of holonomy properties, so that it does not even appear to mention localized gauge potential properties. The cleanest way to do this would be to introduce holonomies as the images of homomorphisms from the hoop group into the structure group of the theory, and to show how these may be used to represent holonomy properties of loops. The theory could then be developed by proving that these holonomies may be represented by means of (any of a gauge-equivalent class of) connections on a principal fiber bundle, and that these in turn can be represented, at least locally, by (a generalization of) a vector potential defined on the space-time manifold. This way of proceeding makes two things clear. First, there is no fundamental object in the theory that is even capable of representing or describing any localized gauge potential properties. Second, gauge symmetry is a formal feature, not of the theory itself, but only of various representations or coordinatizations of its fundamental structures. So unless the world has more structure than the theory can represent or describe, there are no localized gauge potential properties, and gauge symmetry can have no physical consequences.

Someone may add to this formulation further claims that single out one connection and/or one gauge as that corresponding to some actual distribution of localized gauge potential properties. But this will lead to no new predictions, and neither the theory nor anything else that we know describes processes we could use to observe how such properties are actually distributed. So these additional claims do no work in the reformulated theory and we have no more reason to believe them than we do to believe in undetectable ghosts.

4.4.2 Objections considered

But maybe we *should* believe in ghosts! Plenty of people do, and appeal to them as offering an explanation of otherwise inexplicable phenomena. Skeptical scientists reply that we have been unable to detect them, and so have no evidence for their existence. But perhaps such skepticism is too extreme. If ghosts offer the best explanation of otherwise inexplicable phenomena, then does that not provide a good reason to believe in them, even if they are not observable, or even detectable?

The previous paragraph should not be taken too seriously. It was intended only as a provocative introduction to the following consideration of objections to the epistemological defense of the holonomy interpretation of classical Yang–Mills theories offered in the previous section.

Leeds (1999) has argued that only by adopting an interpretation of classical electromagnetism according to which the vector potential is faithfully represented by a particular connection on a fiber bundle can one explain why the canonical momentum operator takes the form $-i\hbar\nabla - e\mathbf{A}$ for a particle with charge e subject to a magnetic vector potential $\mathbf{A}$. He motivates this view by proposing that an adequate interpretation of electromagnetism must take phases seriously. This means taking seriously some "element of reality" that shows up in the various gauges as the phase factor along open as well as closed paths. In the context of the fiber bundle formulation, "taking phases seriously" means believing that the actual distribution of localized EM potential properties may be represented by some particular bundle connection.

There is a three-part reply to Leeds's argument. Note first that a momentum operator is usually *called* canonical just when quantization of a classical theory in its Hamiltonian form converts the classical momentum magnitude in that theory into this operator. But the classical magnitude that plays the role of canonical momentum in a Hamiltonian formulation of classical electromagnetism is just

p − e**A**, which becomes −iℏ**∇** − e**A** on quantization. Lest this be dismissed as a cheap semantic trick, note secondly that only gauge-invariant operators are candidates for representing observables after quantization, and −iℏ**∇** − e**A** is indeed gauge invariant, unlike −iℏ**∇**. Now canonical quantization is not always a well-defined procedure, and is not in any case guaranteed to yield a correct theory. So the fact that −iℏ**∇** − e**A** is a natural gauge-invariant candidate for representing the physical magnitude momentum does not conclusively prove that this must be the correct form for the momentum operator in an empirically adequate theory of electromagnetic interactions of quantum particles. But we have techniques for measuring momentum that yield results in the absence of electromagnetism whose values and statistics conform to quantum mechanical predictions. Applying those same techniques in the presence of electromagnetism yields (gauge-independent) results whose values and statistics conform to quantum mechanical predictions only if momentum is there represented by the operator −iℏ**∇** − e**A**. Ultimately, it is experiment that decides that this operator represents the magnitude whose values our momentum measurements give us.

I have encountered a second objection based on inference to the best explanation against the epistemological defense of the holonomy interpretation of a classical Yang–Mills gauge theory.[9] The objection is that if holonomy properties are indeed basic, then we have no explanation of why they are related to one another in the way that they are. But a very natural explanation is forthcoming if these relations in fact derive from an underlying distribution of localized gauge potential properties. So we have reason to believe there are such localized gauge potential properties, and to reject the holonomy interpretation in favor of the localized gauge potential properties view.

As explained in the previous chapter (section 3.1.2), the Wilson loops of any Yang–Mills theory satisfy a set of identities called the Mandelstam identities. Here is a simple example for the case of electromagnetism:

$$W(C_1)W(C_2) = W(C_1 \circ C_2) \qquad (4.19)$$

where C_1, C_2 are closed curves, $W(C)$ is the Wilson loop (trace of the holonomy) of C, and ∘ indicates composition of curves. In the case of an Abelian theory like electromagnetism, the Wilson loop coincides with the holonomy, and this is independent of both base point and choice of connection, so we have

$$H(C_1)H(C_2) = H(C_1 \circ C_2) \qquad (4.20)$$

Now suppose that C_1, C_2 are closed curves with the same base point that trace out loops l_1, l_2 respectively. The curve $C_1 \circ C_2$ may contain "trees"—segments

---

[9] Here I am indebted to Frank Arntzenius and especially Tim Maudlin in correspondence.

4.4 A HOLONOMY INTERPRETATION

where the curve retraces its steps, forming a constituent curve that encloses no area. From the curve $|C_1 \circ C_2|$ by removing all such trees: then

$$H(C_1 \circ C_2) = H(|C_1 \circ C_2|) \tag{4.21}$$

Provided $|C_1 \circ C_2|$ contains no self-intersections, its image is a loop $l_1 \oplus l_2$ (otherwise its image is a set of intersecting loops). Then the holonomy properties of $l_1 \oplus l_2$ are wholly determined by those of l_1, l_2: they are represented by a holonomy—$H(|C_1 \circ C_2|)$—that is simply the product of the holonomies of (the curves that trace out) the loops that compose it. This is just one example of the general fact that, on the holonomy interpretation of a Yang–Mills gauge theory, the holonomy properties of distinct loops are not independent of one another, but are interrelated in many different ways.

These relations among holonomy properties of distinct loops are explicable if the holonomy properties of each loop are not in fact primitive, but derive from localized gauge potential properties that attach at points on the loop, or their arbitrarily small neighborhoods. Otherwise the relations appear merely as puzzling, brute facts. Hence we should reject the holonomy interpretation in favor of the localized gauge potential properties interpretation of a classical Yang–Mills gauge theory.

It is not clear how good the suggested explanation would be, given the epistemic inaccessibility of the additionally postulated localized gauge potential properties. To see its failings, compare the following Newtonian explanation of why all inertially moving objects move at constant velocities with respect to one another: Inertial motion is just moving through equal intervals of absolute space in equal intervals of absolute time. This would not be a good explanation, precisely because nothing in Newton's theory (or elsewhere) enables experiments to yield information relevant to determining the state of absolute rest it presupposes.

A structure, like absolute rest, that remains radically indeterminate, even given a formulation of a theory that postulates it, does not provide the basis for a credible explanation of a generalization in the domain of that theory. A good theoretical explanation of such a generalization has unifying power—power that accrues when a postulated theoretical structure plays other roles in this or other theories. Typically this is how theoretical structure becomes determinate. (Consider, for example, how the diverse roles played by Avogadro's number permitted multiple independent determinations of its value, while demonstrating the explanatory power of the principle that a mole of gas always contains the same number of molecules.) And it typically leads to predictions of new phenomena—another characteristic of a good theoretical explanation that has often been remarked upon.

Localized gauge potential properties do not unify a classical Yang–Mills gauge theory, because of their severely limited role within the theory. They are unobservable, even given the theory: and they do not lead to predictions of new phenomena. No justifiable principle of inference to the best explanation should saddle us with them, no matter how much evidence supports the theory.

Moreover, the relations among holonomy properties can be given an alternative explanation which appeals to no such additional structure. For these all follow simply from the fundamental postulate of a Yang–Mills gauge theory—that the holonomy properties attached at each loop may be represented by an element of the Lie algebra of its structure group, and specifically the element which is the image of a hoop corresponding to the loop under the homomorphism from the hoop group that defines the theory. In the case of classical electromagnetism, this leads to what might be called the *loop supervenience* of holonomy properties: the holonomy properties of any loop $\oplus_i L_i$ are determined by those of any loops L_i that compose it. Loop supervenience alone suffices to explain all the relations among holonomy properties of loops in classical electromagnetism.

4.4.3 Semantic considerations

The previous subsection argued that evidence supporting a classical Yang–Mills gauge theory may give us reason to believe that there are holonomy properties, but not that there are any localized gauge potential properties underlying them. Of course, some future extension of the theory may warrant a revision in this evaluation. But this could happen only if the extended theory turned out to violate gauge symmetry so as to permit us to obtain reliable information as to the actual distribution of supposed localized gauge potential properties.

There is a close connection between the epistemological objections to the localized gauge potential properties view and the semantic difficulties faced by that view described earlier (in section 4.2.1). But do the semantic difficulties strengthen the epistemological objections or do they, on the contrary, provide an answer to them?[10]

The epistemological problems with the localized gauge potential properties view have a semantic aspect. They are connected to the fact that it leads to unanswerable questions. Neither the theory itself, nor anything we can do when applying it, enables us to give determinate answers to questions about how the supposed localized gauge potential properties are distributed. This is not just a failing of language. We cannot even entertain a thought that they are one way rather than another, from among an infinity of gauge-related distributions. But an advocate of the localized gauge potential properties view may deny that this renders these properties problematic. It may be an epistemological defect in a theory to raise meaningful but unanswerable empirical questions. But the

---

[10] I am indebted to Tim Maudlin for defending the latter possibility in correspondence.

semantic features of the localized gauge potential properties view are such that there are questions it does not even permit one meaningfully to ask. And what appeared as an epistemological vice could be seen rather as a semantic virtue—the virtue of rendering metaphysical questions literally meaningless rather than empirically unanswerable!

But this attempt to rescue the localized gauge potential properties view misfires. The epistemological objection does not presuppose that particular statements as to the distribution of localized gauge potential properties have determinate content. It is directed against the key existence claim of the view—that there *is* such a distribution represented by a gauge potential, irrespective of whether we can know it or say what it is. This claim has determinate (though limited) content, as it must have for the localized gauge potential properties view to be intelligible. But it is a claim that receives no support from the evidence for the gauge theory.

The reason why it receives no support is closely connected to the reason why the localized gauge potential properties view gets into semantic difficulties. In order for the existence claim to play the kind of role in the theory that it would have to play in order to benefit from the theory's confirmation, it would have to be possible to use it in making inferences within the theory—inferences to conclusions, about matters on which experiments or other processes could reliably inform us, that could not be derived without it. Only in this way could the existence claim contribute to the theory's predictive or explanatory success. But a bare existential statement is useless in this regard. It can acquire inferential power only to the extent that it is supplemented by descriptions of what it claims to exist that forge the necessary connections to observable matters. The semantic difficulties of the localized gauge potential properties view show that this is precisely what is lacking in this case. Far from resolving the epistemological problems faced by the localized gauge potential properties view, the semantic difficulties faced by that view should be seen as a symptom of those problems.

4.5 Metaphysical implications: non-separability and holism

The holonomy interpretation of classical Yang–Mills gauge theories has striking metaphysical implications. Some may see this as a reason to reject it. But such rejection would be no more justified than the Cartesian's rejection of Newtonian forces—or so I shall argue. The metaphysician should rather welcome this novel lesson from physics as to how the world might be.

Holonomy properties are non-localized gauge potential properties. Does believing in them involve accepting some kind of failure of locality? No: holonomy properties may act locally even if they are not "locally possessed." It is better to use a distinctive term to capture such descriptive (as opposed to causal) non-locality. Elsewhere (1991, 1994, 2004) I have used the term *non-separability* for this purpose.[11]

Non-separability: Some physical process occupying a region R of space-time is not supervenient upon an assignment of qualitative intrinsic physical properties at space-time points in R.

On the holonomy interpretation, a classical Yang–Mills theory describes or represents non-separable processes. For there are processes in which holonomy properties are assigned at space-time loops that do not supervene upon any assignment of properties (such as localized gauge potential properties) at points on those loops.

This may not seem so novel, since there are reasons to believe that even classical mechanics describes non-separable processes. Take instantaneous velocity, for example: this is usually defined as the limit of average velocities over successively smaller temporal neighborhoods of that point. This provides a reason to deny that the instantaneous velocity of a particle at a point supervenes on qualitative intrinsic properties assigned at that point. Similar skeptical doubts can be raised about the intrinsic character of other "local" magnitudes such as the density of a fluid, the value of an electromagnetic field, or the metric and curvature of space-time (see Butterfield 2006).

But the non-separability of holonomy properties is more radical. According to the holonomy interpretation, a classical Yang–Mills theory describes or represents processes that are strongly non-separable (Healey 2004).

Strong Non-separability: Some physical process occupying a region R of space-time is not supervenient upon an assignment of qualitative intrinsic physical properties at points of R and/or in arbitrarily small neighborhoods of those points.

Any kind of non-separability violates what David Lewis (1986), p. x calls

Humean Supervenience. ... all there is to the world is a vast mosaic of local matters of particular fact, just one little thing and then another. ... We have geometry: a system of external relations of spatiotemporal distance between points. Maybe points of space-time itself, maybe point-sized bits of matter or aether or fields, maybe both. And at those points we have local qualities: perfectly natural intrinsic properties which need nothing bigger than a point at which to be instantiated. For short: we have an

---

[11] This is related to a notion of synchronic locality introduced by Belot (1998, p. 540), which he defines as follows:

The state of the system at a given time can be specified by specifying the states of the subsystems located in each region of space (which may be taken to be arbitrarily small).

In fact, Belot's notion is even closer to what I have elsewhere (1991, 2004) called spatial non-separability. Compare also Aharonov's discussion of what he means by a "non-local" property of a physical system (Aharonov 1984, p. 12).

4.5 METAPHYSICAL IMPLICATIONS: NON-SEPARABILITY AND HOLISM

arrangement of qualities. And that is all. There is no difference without difference in the arrangement of qualities. All else supervenes on that.

But the holonomy property interpretation of a classical Yang–Mills gauge theory radically violates Humean supervenience because it is strongly non-separable. Consider once more what it implies for the magnetic Aharonov–Bohm effect. It says that there are (electro)magnetic holonomy properties of, or at, large loops encircling the central solenoid that do not supervene on an "arrangement of qualities"—not just at points on that loop, but at *any points whatever* outside the solenoid. If the holonomy interpretation is right, then a world in which a Yang–Mills gauge theory is true can be accommodated only by a significant revision in the metaphysics behind Humean supervenience.

Such a world would exhibit an interesting kind of holism. Elsewhere (2004, 2004a) I have analyzed the relevant kind of holism as follows:

Physical Property Holism: There is some set of physical objects from a domain D subject only to type P processes, not all of whose qualitative intrinsic physical properties and relations supervene on qualitative intrinsic physical properties and relations in the supervenience basis of their basic physical parts.

Such holism arises when features of physical wholes fail to be determined by those of their proper parts. The wholes with which we are here concerned are extended regions of space-time (and/or events or processes that occupy them); the parts are regions (and/or their occupants) that in some sense compose these. Whether we take a gauge theory to exhibit holism will depend not only on what we consider to be the relevant features of the parts, but also on how we understand the composition relation between parts and wholes.

Because it is non-separable (under the holonomy interpretation), there is a sense in which a classical Yang–Mills theory exhibits holism. We have a case of physical property holism provided that points on a loop (or events that occur at them) count as basic physical parts of the loop (or of an event that occupies it). For the holonomy properties "at" a loop are not determined by localized gauge potential or any other intrinsic properties of such constituents, even taken together with the spatiotemporal and other intrinsic relations among them. By contrast, classical general relativity is separable, since all the qualitative intrinsic physical properties it ascribes on a loop do supervene on qualitative intrinsic physical properties assigned on (infinitesimal neighborhoods of) space-time points on that loop. Gravity (as described by classical general relativity) does not exhibit this kind of physical property holism.

But there is another way of understanding the part–whole relation in these cases. As we saw in the previous section, two loops l_1, l_2 may be considered to compose a third $l_1 \oplus l_2$ just in case the curves C_1, C_2 that trace them out form the curve $|C_1 \circ C_2|$ that traces out $l_1 \oplus l_2$. On this alternative understanding of the composition relation, one may take any basis of arbitrarily small loops (or events occupying them) as basic parts of (an event occupying) any larger loop they make up. The holonomy properties of this composite object will indeed supervene on those of its basic parts, and so there will be no physical property holism in either a classical Yang–Mills theory or classical general relativity!

Metaphysicians can be a conservative lot. Just think of how much contemporary metaphysics either presupposes a pre-Einsteinian view of time or, having reluctantly given it up, continues the quixotic struggle to square it with some "common sense" view of time (that only the present exists, that the future (unlike the past) is ontologically open, that time travel is a conceptual impossibility...) a century after Einstein's revolution. I expect metaphysical resistance to the thesis that classical Yang–Mills gauge theories are strongly non-separable and manifest a kind of metaphysical holism. What form might this resistance take?

Lewis's Humean supervenience is closely connected to a view of the independence of possibilities that he calls a *patchwork principle* for possibility (1983, p. 77):

if it is possible that X happen intrinsically in a spatiotemporal region, and if it is likewise possible that Y happen in a region, then also it is possible that both X and Y happen in two distinct but adjacent regions. There are no necessary incompatibilities between distinct existences. Anything can follow anything.

As stated, there is not even an appearance of conflict between this principle and the holonomies interpretation of a classical Yang–Mills gauge theory. Lewis has in mind temporal stages of some process (actually, stages in the life of some person) that are "adjacent" in the sense that one immediately follows the other, so that the two together form a connected, four-dimensional spatiotemporal region. But the intuition behind the principle extends beyond that context to apply to processes involving holonomy properties of loops that are adjacent in a different sense: they compose a larger loop in the way l_1, l_2 compose $l_1 \oplus l_2$. Suppose l_1, l_2 are adjacent in this sense, and that it is possible for l_1 to have holonomy properties P_1, and possible for l_2 to have holonomy properties P_2. Is it possible that l_1 have holonomy properties P_1 while l_2 has holonomy properties P_2?

One may deny the possibility of such joint possession on the grounds that this would be incompatible with the actual holonomy properties of $l_1 \oplus l_2$. But this is a severely restricted notion of possibility that is not relevant to the patchwork principle. The gauge theory can certainly represent a situation in which l_1 has holonomy properties P_1 while l_2 has holonomy properties P_2. This will likely be a situation in which $l_1 \oplus l_2$ has holonomy properties that differ from those it actually has (and which are determined by the *actual* holonomy properties of l_1, l_2). But it is possible in the sense that it is represented by a

4.5 METAPHYSICAL IMPLICATIONS: NON-SEPARABILITY AND HOLISM

model of the theory. Even if it were not possible in this sense, that would be no reason to think it would not be *metaphysically* possible—the gauge theory is at most contingently true. Whatever one thinks of the intuition behind the patchwork principle, the holonomy interpretation of a classical Yang–Mills gauge theory gives one no reason to revise that intuition.

Some metaphysicians may recoil from what they perceive to be the non-local nature of the action of holonomy properties on quantum particles. But the constitutive non-locality of non-separability does not imply causal non-locality: holonomy properties can act right where they are. They act on quantum particles by modifying non-localized properties of those particles that are represented by phase differences of their wave-function around closed curves. But they do this not by changing hypothetical localized phase properties at each point on such a curve, but by acting on the whole curve at once. There is still an issue about just what these non-localized phase properties are, and whether its possession of such properties means that a quantum particle is itself in some sense present everywhere on each such curve. But the relation between properties of quantum particles and properties of their wave-functions is a difficult issue that arises outside the context of gauge theories. It is an important issue for any interpretation of quantum mechanics but best left aside here.

But how do holonomy properties propagate, and do they conform to relativistic locality? These questions need to be clarified before they can be addressed. The only holonomy properties that are possessed at a time are those that attach on space-like loops. It is only these properties whose propagation one can question. Answers are most easily given by using a representation in terms of vector potentials. Electromagnetism is the simplest case. Holonomies of space-like curves represented by the four-vector potential A_μ^m that Mattingly co-opted as his current field propagate continuously on the light-cone, as can be readily seen from the definition 2.17

$$A_\mu^m(x) = 4\pi \int_{\text{all space-time}} d^4x' D_r(x - x') J_\mu(x') \qquad (4.22)$$

The holonomy properties these represent therefore conform to relativistic locality. Non-Abelian generalizations of this result would show that the holonomy properties of other Yang–Mills gauge theories also propagate continuously in conformity to relativistic locality.

Cartesian critics of Newtonian forces clung to an outdated metaphysics of contact action of extended bodies until the success of Newton's theories made their criticisms seem irrelevant and led to an advance in metaphysics as well as science; or, better, to an advance in natural philosophy. But the residual problem of how gravity could act at a distance remained until Einstein's revision of Newton's theory provided an account of continuous propagation of gravitational influence at finite speed. The success of Yang–Mills gauge theories reveals the metaphysics of Humean supervenience and the separability of all physical processes to be similarly outdated. Contemporary metaphysicians who object to holonomy properties and their attendant holism and non-separability can appeal to no analogous residual problem of non-local action.

5
Quantized Yang–Mills gauge theories

Electromagnetism yields paradigms of empirically successful gauge theories. Maxwell's classical theory has been extraordinarily successful in its description of the electromagnetic field and its interactions with classically described matter. Historically this success was marked by striking predictions (of radio waves, etc.), explanations (of the Faraday effect, etc.), and the unification of electricity and magnetism with optics. Even today, classical electromagnetic theory is of enormous instrumental value in practical applications to electronics, optics, radio and microwave technology, and so on. But despite these empirical successes, there are good reasons why this theory is no longer believed to be empirically adequate. There are phenomena (the photoelectric effect, laser action, etc.) for which Maxwell's theory cannot account.

But the advent of quantum theory turned such defeats into victories as Maxwell's equations were reinterpreted to apply to a quantum rather than a classical electromagnetic field. Once more, electromagnetism provided a shining example of an empirically successful gauge theory, quantum electrodynamics—the theory of the quantized electromagnetic field and its interaction with electrons and other quantum mechanically described matter. Besides furnishing an understanding of "particulate" properties of light and of laser action, quantum electrodynamics has yielded some of the most accurate predictions in the whole of science (of the magnetic moment of the electron, etc.). The quantum theory of the electromagnetic field provided the first empirically successful account of how a fundamental force of nature operates on subatomic and subnuclear scales. It thereby yielded the paradigm case of an empirically successful quantized gauge theory. This theory has also borne fruit by guiding the construction of quantum theories to describe the other three forces currently considered fundamental—the weak and strong interactions, and gravity. And it has led to a further unification of physics by revealing the electromagnetic and weak interactions to be different manifestations of a single electroweak interaction—once more described by a gauge theory.

Gauge theories of the weak and strong interactions differ from electromagnetic theory in two important respects. By contrast with electromagnetism, the strengths of the weak and strong interactions fall off rapidly, between objects that can be separated from one another far enough for the effects of these interactions to be observed experimentally, as the distance between these objects increases. This means that quantized gauge theories that describe them have no empirical consequences for which a corresponding classical gauge theory could expect to account, as classical electromagnetism accounts (remarkably well) for what are now considered large-scale empirical consequences of quantum electrodynamics. Consequently, there is no empirically successful *classical* gauge theory of either the weak or the strong interaction.

The second difference is that, unlike electromagnetism, weak and strong interactions are described by non-Abelian rather than Abelian gauge theories—i.e. the order in which one applies two distinct elements of an algebraic structure characterizing such a theory may make a difference to the result of their joint application. This theoretical difference turns out to imply two important physical contrasts with electromagnetic interactions. Each of the weak and strong interactions is effected by more than one kind of mediating quantum, whereas electromagnetic interactions are mediated by photons alone. And there are interactions both among the quanta that mediate the strong interaction, and among those that mediate the weak interaction, while photons do not interact with each other directly.

Despite these differences, gauge theories of electromagnetic, weak, and strong interactions share essential similarities of structure, summed up by calling them all Yang–Mills theories (after ground-breaking explorations of this structure by Yang and Mills (1954)).

General relativity is an empirically successful classical field theory, and though (as we saw in chapters 1 and 3) there are reasons why it, too, may be considered a gauge theory of a fundamental interaction, it is not a Yang–Mills gauge theory, but has a rather different structure. Indeed, its structure is sufficiently different that attempts to formulate a quantum theory of the gravitational interaction by starting from general relativity and applying quantization techniques that were successfully applied to classical field theories of the other three interactions have come up against technical and conceptual barriers that have not yet been clearly overcome. Part of the difficulty stems from the fact that general relativity is a theory of space and time themselves, and not just a theory of forces acting within a background of space and time. Another problem is more practical: gravity is so weak that it is extremely difficult to conduct, or even to conceive, experiments capable of revealing any breakdown of classical general relativity in a situation that could provide clues to the construction of a successor, quantum theory of gravity.

This chapter is concerned with quantized Yang–Mills theories. Section 5.1 begins by describing a standard procedure for canonical quantization of a classical field, illustrates this in the case of a real Klein–Gordon field, and points out difficulties that arise when applying this procedure to the Maxwell field of classical electromagnetic theory in a vacuum. Various methods for handling these difficulties have been developed. Sections 5.2–5.6 describe these methods and analyze the relations between the resulting quantum field theories. Sections 5.7–5.8 address problems that arise in generalizing certain of these methods to non-Abelian gauge theories. The empirical success of quantized Yang–Mills theories stems from their use to describe interactions in the Standard Model of high energy physics. It is therefore important to understand how Yang–Mills fields interact with other fields that describe elementary particles like electrons and quarks. Section 5.9 explains how these techniques for quantizing Yang–Mills fields may be extended to include such interactions. Another technique for quantizing a Yang–Mills gauge field leads to the so-called loop representation. This is less familiar than other approaches, and is rarely described in popular works or introductory texts. Since the loop representation will prove important to the interpretative project of this book, it will be described more fully later, in chapter 7.

5.1 How to quantize a classical field

There is a standard procedure for constructing a quantum field theory from a classical field theory. This is called canonical—not to reflect its status, but rather to acknowledge the role played by canonical variables in a Hamiltonian formulation of the classical theory that forms the starting point of the construction.

Canonical quantization begins by deriving equations obeyed by a classical field from the requirement that the action corresponding to a Lagrangian for the field be stationary under variations of the field and its derivatives. For example, the Klein–Gordon equation

$$(\partial_\mu \partial^\mu + m^2)\varphi = 0 \tag{5.1}$$

obeyed by a real scalar field $\varphi(x)$ may be derived as the Euler-Lagrange equation that results from the requirement that the action S corresponding to the Lagrangian density $\mathcal{L}(\varphi, \partial^\mu \varphi)$ be stationary under variations of $\varphi, \partial^\mu \varphi$, where $S = \int \mathcal{L}(\varphi, \partial^\mu \varphi) \mathrm{d}^4 x$, $\mathcal{L}(\varphi, \partial^\mu \varphi) = \frac{1}{2}[(\partial_\mu \varphi)(\partial^\mu \varphi) - m^2 \varphi^2]$. The field canonically conjugate to φ is

$$\pi(x) = \frac{\partial \mathcal{L}}{\partial \dot\varphi(x)} = \dot\varphi(x) \tag{5.2}$$

Replacing the real-valued functions φ and π by self-adjoint operators $\hat{\varphi}, \hat{\pi}$, one now imposes the equal-time canonical commutation relations[1]

$$[\hat{\varphi}(\mathbf{x}, t), \hat{\pi}(\mathbf{x}', t)] = i\delta^3(\mathbf{x} - \mathbf{x}') \tag{5.3}$$
$$[\hat{\varphi}(\mathbf{x}, t), \hat{\varphi}(\mathbf{x}', t)] = [\hat{\pi}(\mathbf{x}, t), \hat{\pi}(\mathbf{x}', t)] = 0 \tag{5.4}$$

Hamilton's equations now entail that the field $\hat{\varphi}$ satisfies the Klein–Gordon equation 5.1. A solution to that equation may be written as a Fourier expansion with coefficients $\hat{a}_p, \hat{a}_p^\dagger$ that satisfy the commutation relations

$$[\hat{a}_p, \hat{a}_{p'}^\dagger] = \delta^3(\mathbf{p} - \mathbf{p}') \tag{5.5}$$
$$[\hat{a}_p, \hat{a}_{p'}] = [\hat{a}_p^\dagger, \hat{a}_{p'}^\dagger] = 0 \tag{5.6}$$

These may be represented as acting as raising and lowering operators on a Fock space—the infinite direct sum of symmetrized "n-particle" Hilbert spaces—so that any state of the free Klein–Gordon field may be expressed as a linear superposition of basis states, each of which represents a state containing n spinless quanta. $\hat{a}_p^\dagger$ is said to be a creation operator, since, when operating on a given state in Fock space, it yields another state with one additional quantum of momentum $\mathbf{p}$: similarly, $\hat{a}_p$ "annihilates" a quantum of momentum $\mathbf{p}$. This is called the *Fock representation* for the Klein–Gordon field; appendix E provides further details, and chapter 8, section 8.2 pursues the interpretative significance of such Fock representations.

Following this procedure in the case of electromagnetism immediately leads to problems. Maxwell's equations in empty space may be written in Lorentz covariant form as

$$\partial^\mu F_{\mu\nu} = 0 \tag{5.7}$$

where the electromagnetic field tensor $F_{\mu\nu}$ is expressible in terms of the four-vector potential A_μ as

$$F_{\mu\nu} = \partial_\mu A_\nu - \partial_\nu A_\mu \tag{5.8}$$

(see appendix A). They may be derived as Euler–Lagrange equations by requiring that the action associated with the following Lagrangian density be stationary

$$\mathcal{L}_{EM}(A_\mu, \partial_\nu A_\mu) = -\frac{1}{4} F_{\mu\nu} F^{\mu\nu} \tag{5.9}$$

---

[1] To keep things simple, I here follow the standard textbook presentation in taking field operators to be defined at a point. A mathematically rigorous presentation would define them on a suitable space of test functions on the manifold of points. Such a presentation will prove necessary later (see chapters 7, 8).

The field canonically conjugate to **A** is $-\mathbf{E}$, where the electric field $\mathbf{E} = -\partial_0 \mathbf{A} - \boldsymbol{\nabla} A_0$. But the field π^0 canonically conjugate to A_0 is identically zero, since $\partial \mathcal{L}/\partial \dot{A}_0 = 0$. Clearly, this rules out imposition of an equal-time commutation relation analogous to equation 5.3. Moreover, A_μ is determined by equation 5.8 only up to a gauge transformation

$$A_\mu \to A'_\mu = A_\mu + \partial_\mu \Lambda \qquad (5.10)$$

where Λ is any smooth function. This gauge freedom means that there are fewer physical degrees of freedom than those represented by any particular A_μ, and the Hamiltonian equations of motion correspondingly underdetermine the time development of A_μ. In order to quantize the electromagnetic field it is therefore necessary either to remove this gauge freedom by simply selecting a gauge, or to represent electromagnetism by some distinct gauge-invariant object. Both approaches have been followed, resulting in the variety of alternative quantizations of the free electromagnetic field that will be described in the next four sections of this chapter and the first section of chapter 7.

5.2 Coulomb gauge quantization

By a transformation of the form given by equation 5.10 one can arrive at a four-vector potential A_μ that satisfies the *Lorenz condition*[2]:

$$\partial^\mu A_\mu = 0 \qquad (5.11)$$

This can be done by choosing a function Λ satisfying $\partial^\mu \partial_\mu \Lambda = -\partial^\mu A_\mu$. By a further transformation of the form 5.10, this time using a function Λ satisfying $\partial_\mu \Lambda = 0$, $\boldsymbol{\nabla}^2 \Lambda = \boldsymbol{\nabla}.\mathbf{A}$ one can arrive at a four-vector potential A_μ that satisfies the *Coulomb condition*

$$A_0 = 0, \boldsymbol{\nabla}.\mathbf{A} = 0 \qquad (5.12)$$

The requirement that A_μ satisfies equation 5.12 eliminates spurious degrees of freedom in the representation, but only by making arbitrary choices that threaten to violate the Lorentz covariance of the resulting theory. Coulomb

---

[2] This is not a misprint. The Lorenz condition that defines the Lorenz gauge is named in honor of the Danish physicist Ludwig Valentine Lorenz (1829–1891) who first applied it. The Lorentz transformations, on the other hand, are named in honor of his better-known near-namesake, the Dutch physicist Hendrik Antoon Lorentz (1853–1928). It is especially tempting to refer to the Lorenz condition as the "Lorentz condition" since (unlike the Coulomb condition) this condition is indeed Lorentz covariant!

5 QUANTIZED YANG–MILLS GAUGE THEORIES

gauge quantization sacrifices manifest Lorentz covariance to achieve such a representation.

The canonical fields in Coulomb gauge are now $\mathbf{A}, -\mathbf{E}$. But the expected equal-time commutation relations

$$[\hat{A}_j(\mathbf{x}, t), \hat{E}^k(\mathbf{x}', t)] = -i\delta_{jk}\delta^3(\mathbf{x} - \mathbf{x}')$$

are inconsistent with 5.12. Consistency is restored by requiring instead that

$$[\hat{A}_j(\mathbf{x}, t), \hat{E}^k(\mathbf{x}', t)] = -i\delta^3_{\perp jk}(\mathbf{x} - \mathbf{x}') \qquad (5.13)$$

$$[\hat{A}_j(\mathbf{x}, t), \hat{A}_k(\mathbf{x}', t)] = [\hat{E}^j(\mathbf{x}, t), \hat{E}^k(\mathbf{x}', t)] = 0 \qquad (5.14)$$

where $\delta^3_{\perp jk}(\mathbf{x} - \mathbf{x}') = (\delta_{jk} - \partial_j \frac{1}{\Delta} \partial_k)\delta^3(\mathbf{x} - \mathbf{x}')$, and $\Delta = \nabla^2$ is the Laplacian operator. Field operators obeying these modified commutation relations 5.13, 5.14 now satisfy the equations

$$\partial_\mu \partial^\mu \hat{A}_j = 0 = \hat{A}_0 : \hat{\mathbf{E}} = -\partial_0 \hat{\mathbf{A}} \qquad (5.15)$$

whose solutions may be written as a Fourier expansion with coefficients $\hat{a}_{\mathbf{k},\lambda}, \hat{a}^\dagger_{\mathbf{k}',\lambda'}(\lambda, \lambda' = 1, 2)$ satisfying commutation relations analogous to 5.5, 5.6

$$[\hat{a}_{\mathbf{k},\lambda}, \hat{a}^\dagger_{\mathbf{k}',\lambda'}] = \delta^3(\mathbf{k} - \mathbf{k}')\delta_{\lambda\lambda'} \qquad (5.16)$$

$$[\hat{a}_{\mathbf{k},\lambda}, \hat{a}_{\mathbf{k}',\lambda'}] = [\hat{a}^\dagger_{\mathbf{k},\lambda}, \hat{a}^\dagger_{\mathbf{k}',\lambda'}] = 0 \qquad (5.17)$$

These coefficients may consequently be thought to represent creation and annihilation operators for photons of momentum $\mathbf{k}, \mathbf{k}'$ and linear polarizations λ, λ' (orthogonal for $\lambda \neq \lambda'$), acting on a Fock space. One can now write general solutions to the field equations 5.15 in terms of these coefficients as follows:

$$\hat{\mathbf{A}}(\mathbf{x}, t) = \int N_k d^3k \sum_{\lambda=1}^{2} \boldsymbol{\epsilon}(\mathbf{k}, \lambda) \left(\hat{a}_{\mathbf{k},\lambda} e^{-ik.x} + \hat{a}^\dagger_{\mathbf{k}',\lambda'} e^{+ik.x} \right) \qquad (5.18)$$

$$\hat{\mathbf{E}}(\mathbf{x}, t) = \int N_k d^3k \sum_{\lambda=1}^{2} i|\mathbf{k}| \boldsymbol{\epsilon}(\mathbf{k}, \lambda) \left(\hat{a}_{\mathbf{k},\lambda} e^{-ik.x} - \hat{a}^\dagger_{\mathbf{k}',\lambda'} e^{+ik.x} \right) \qquad (5.19)$$

where N_k is a normalization constant; $|\mathbf{k}|$ is the magnitude of the vector $\mathbf{k}$; $\boldsymbol{\epsilon}(\mathbf{k}, 1), \boldsymbol{\epsilon}(\mathbf{k}, 2)$ are transverse vectors (at right-angles to one another and to $\mathbf{k}$); and $k.x \equiv |\mathbf{k}|t - \mathbf{k}.\mathbf{x}$. It follows that, in this gauge, both $\hat{\mathbf{A}}$ and $\hat{\mathbf{E}}$ are transverse—as indicated by a sub- or super-scripted T. A term with factor $e^{-ik.x}$ is called *positive frequency*, while a term with factor $e^{+ik.x}$ is called *negative frequency*. So each of $\hat{\mathbf{A}}_T$ and $\hat{\mathbf{E}}_T$ may be decomposed into a positive frequency part made up of terms multiplying annihilation operators, and a negative frequency part made up of terms multiplying creation operators:

$$\hat{\mathbf{A}}_T(\mathbf{x},t) = \hat{\mathbf{A}}_T^{(+)}(\mathbf{x},t) + \hat{\mathbf{A}}_T^{(-)}(\mathbf{x},t) \tag{5.20}$$

$$\hat{\mathbf{E}}_T(\mathbf{x},t) = \hat{\mathbf{E}}_T^{(+)}(\mathbf{x},t) - \hat{\mathbf{E}}_T^{(-)}(\mathbf{x},t) \tag{5.21}$$

Thus one has achieved a Fock representation for the electromagnetic field in Coulomb gauge. But this required the sacrifice of manifest Lorentz covariance, and the form of the modified commutation relation 5.13 was motivated merely to prevent inconsistency. A method of quantization that seeks to avoid this sacrifice is described in the next section, while section 5.5 seeks to supply the missing motivation by setting Coulomb gauge quantization in the framework provided by Dirac's constrained Hamiltonian approach.

5.3 Lorenz gauge quantization

Suppose one were to try to retain manifest Lorentz covariance by adopting the Lorenz gauge condition 5.11, but dropping the restriction to Coulomb gauge, and imposing the Lorentz-covariant equal-time commutation relations

$$[\hat{A}_\mu(\mathbf{x},t), \hat{\pi}_\nu(\mathbf{x}',t)] = i\eta_{\mu\nu}\delta^3(\mathbf{x}-\mathbf{x}')$$
$$[\hat{A}_\mu(\mathbf{x},t), \hat{A}_\nu(\mathbf{x}',t)] = [\hat{\pi}^\mu(\mathbf{x},t), \hat{\pi}^\nu(\mathbf{x}',t)] = 0$$

where $\eta_{\mu\nu}$ represents the Minkowski metric.

This brings us back to the original problem that when Maxwell's equations are derived from the Lagrangian density $\mathcal{L}_{\text{EM}}(A_\mu, \partial_\nu A_\mu) = -\frac{1}{4}F_{\mu\nu}F^{\mu\nu}$, the field π^0 canonically conjugate to A_0 is identically zero, which is incompatible with these covariant commutation relations. That problem may be finessed, if not solved, by adding a Lorentz-covariant "gauge-fixing" term proportional to $(\partial^\mu A_\mu)^2$ to $\mathcal{L}_{\text{EM}}$. The resulting classical Euler–Lagrange equations are then equivalent to Maxwell's equations *in Lorenz gauge*, and the new π_0 is no longer zero.

But quantum mechanics now raises a new problem: The above covariant commutation relations turn out to be inconsistent with the Lorenz gauge condition 5.11 applied to $\hat{A}_\mu(\mathbf{x},t)$, as one can show by using them to evaluate the (non-zero) commutator $[\partial^\mu \hat{A}_\mu(\mathbf{x},t), \hat{A}_\nu(\mathbf{x}',t)]$. The inconsistency may be removed by restricting the vectors Ψ in the Hilbert space on which the field operators act to those for which the *expectation values* of the Lorenz condition are zero: $(\Psi, \partial^\mu \hat{A}_\mu(\mathbf{x},t)\Psi) = 0$. This can conveniently be arranged by imposing an even stronger condition—that the Hilbert space vectors be annihilated, not by $\partial^\mu \hat{A}_\mu(\mathbf{x}, t)$ itself, but by $\partial^\mu \hat{A}_\mu^{(+)}(\mathbf{x}, t)$, where $\hat{A}_\mu^{(+)}(\mathbf{x}, t)$ is the positive frequency part of $\hat{A}_\mu(\mathbf{x}, t) = \hat{A}_\mu^{(+)}(\mathbf{x}, t) + \hat{A}_\mu^{(-)}(\mathbf{x}, t)$. (This effectively restricts $\hat{A}_\mu^{(+)}(\mathbf{x}, t)$ to annihilation operators: even the vacuum state Ψ_0 violates $\partial^\mu \hat{A}_\mu^{(-)}(\mathbf{x}, t) \Psi_0 = 0$.) All these more or less ad hoc moves do turn out to yield a manifestly Lorentz covariant quantum theory of the free electromagnetic field which, when extended to an interacting field theory and subjected to suitable approximation techniques (coded by Feynman diagrams) yields remarkably accurate predictions. But the way in which the theory has been developed leaves a lot to be desired from a foundational perspective, and it remains unclear how this theory relates to the theory of the previous section.

5.4 Classical electromagnetism as a constrained Hamiltonian system

Seminal work by Dirac (1964) resulted in a technique for quantizing a classical gauge theory, regarded as describing a certain kind of Hamiltonian system. This corresponds to a perspective on gauge theories that is rather different from that arising from the fiber bundle formulation favored by Trautman and introduced in chapter 1. Since this perspective will prove important to the analysis to follow, it is appropriate to begin by reviewing the theory of classical electromagnetism, whose fiber bundle formulation was given in chapter 1, section 1.2, from this new perspective. Further details are provided in appendix C.

The new perspective views classical electromagnetism as (describing) a type of constrained Hamiltonian system. To describe such a system, one uses a set of *configuration* variables that generalize the position coordinates of a system of particles, and a corresponding set of generalized *momentum* variables. In a field theory, the former represent components ϕ_i of a field, while the latter correspond to corresponding field momenta π^i. The usual Hamiltonian formulation of a theory treats all these variables as independent. But when certain theories are formulated this way, the independence assumption fails—there turn out to be functional relations among the variables that hold independently of the dynamical evolution of systems to which the theory applies. Such identities are called *constraints*, and a system described by a theory whose Hamiltonian formulation involves constraints is called a *constrained Hamiltonian system*.

In arriving at a formulation of classical electromagnetic theory as a theory of constrained Hamiltonian systems, the first step is to regard the equations of motion for the free electromagnetic field (Maxwell's equations A.2) as Euler–Lagrange equations deriving from Hamilton's principle as applied to an appropriate Lagrangian. The Lagrangian for electromagnetism is the spatial

5.4 ELECTROMAGNETISM AS A CONSTRAINED HAMILTONIAN SYSTEM

integral of the following Lagrangian density, expressed in terms of the electric and magnetic field as

$$\mathcal{L}_{EM} = -\frac{1}{4}F_{\mu\nu}F^{\mu\nu} = 1/2(\mathbf{E}^2 - \mathbf{B}^2) \tag{5.22}$$

Using 1.4, 1.1 this may be rewritten in terms of the electric and magnetic potentials as

$$\mathcal{L}_{EM} = 1/2(\boldsymbol{\nabla}\varphi+\dot{\mathbf{A}})^2 - (\boldsymbol{\nabla}\times\mathbf{A})^2) \tag{5.23}$$

The next step is to switch from a Lagrangian to a Hamiltonian formulation by first replacing generalized configuration and velocity coordinates $(\phi, \dot\phi)$ by corresponding configuration and momentum coordinates (ϕ, π) and then using these to express the Hamiltonian density in terms of the Lagrangian density as

$$\mathcal{H} = \sum_i \phi_i \pi^i - \mathcal{L} \quad (i = 0, 1, 2, 3) \tag{5.24}$$

Here the generalized coordinates ϕ_i are taken to be the electric and magnetic potentials $\varphi, \mathbf{A}$, and the generalized momenta are given by $\pi^i = \partial \mathcal{L}/\partial \dot\phi_i$. It follows that

$$\pi^0 = 0 \tag{5.25}$$
$$\pi^i = -E^i \tag{5.26}$$

The vanishing of π_0 indicates that the Hamiltonian formulation of electromagnetism is constrained; equation 5.25 is called a *primary constraint*. A so-called *secondary constraint*

$$\boldsymbol{\nabla}.\mathbf{E} = 0 \tag{5.27}$$

arises from the need to ensure that this primary constraint is preserved by the equations of motion, i.e. that $\dot\pi^0 = 0$. (More correctly, both primary and secondary constraints here form an infinite family, each member of which corresponds to a particular value of **x** at which the relevant field is evaluated.)

In a Hamiltonian formulation of a theory, the basic equations of motion are Hamilton's equations. These specify the rates of change of the variables ϕ, π with respect to time, in terms of the Hamiltonian function H. In a field theory H is the spatial integral of $\mathcal{H}$ over the domain of the field. For classical electromagnetism as presently formulated the Hamiltonian density derived from the Lagrangian density 5.23 is

$$\mathcal{H} = 1/2(\mathbf{E}^2 + \mathbf{B}^2) + \dot\varphi\pi^0 - \varphi(\boldsymbol{\nabla}.\mathbf{E}) \tag{5.28}$$

138   5 QUANTIZED YANG–MILLS GAUGE THEORIES

The derivation is given in appendix C, where 5.28 appears as equation C.45. Hamilton's equations then yield

$$\dot{\mathbf{E}} = \mathbf{\nabla} \times (\mathbf{\nabla} \times \mathbf{A}) \tag{5.29}$$
$$\dot{\mathbf{A}} = -\mathbf{E} - \mathbf{\nabla}\varphi \tag{5.30}$$

(see appendix C). These equations determine how an initial state $(\mathbf{A}, \mathbf{E})$ will change with time. Now because of the primary and secondary constraints, the value of the Hamiltonian density 5.28 (and consequently also the Hamiltonian) depends on $\mathbf{E}$ and $\mathbf{A}$ but not on $\varphi, \dot{\varphi}$, and so one might expect the state to develop in a way that is uniquely determined by this initial state. But equation 5.30 shows that how $\mathbf{A}$ evolves depends not only on the value of $\mathbf{E}$ but also on that of φ. Consequently, the Hamiltonian equation of motion for $\mathbf{A}$ is radically indeterministic: if the initial state is $(\mathbf{A}_0, \mathbf{E}_0)$, then $(\mathbf{A}_t + \mathbf{V}_t, \mathbf{E}_t)$ represents a state at $t \neq 0$ that is equally consistent with this equation, irrespective of the value of $\mathbf{V}_t$. But both $\mathbf{E}$ and $\mathbf{B}$ are uniquely defined at every moment, and each evolves deterministically: an initial state $(\mathbf{B}_0, \mathbf{E}_0)$ uniquely determines the state $(\mathbf{B}_t, \mathbf{E}_t)$ at any time $t \neq 0$. This squares with the classical intuition that it is $\mathbf{E}$ and $\mathbf{B}$ that represent the true physical state of the electromagnetic field, while their generating potentials $\varphi, \mathbf{A}$ are mere mathematical tools.

The evolution of an arbitrary dynamical variable $F(\phi, \pi)$ in the Hamiltonian formalism is given by the equation of motion

$$dF/dt = \partial F/\partial t + \{F, H\} \tag{5.31}$$

where $\{F, H\}$ symbolizes the so-called Poisson bracket. Here this may be rewritten as

$$dF/dt = \partial F/\partial t + \{F, H_0\} + \{F, \int (\dot{\varphi}\pi^0 - \varphi \mathbf{\nabla}.\mathbf{E}) d^3x\} \tag{5.32}$$

where $H_0 = \int \frac{1}{2}(\mathbf{E}^2 + \mathbf{B}^2) d^3x$. In this form the time evolution of F may be thought to compose a physical part, generated by the Hamiltonian, and a non-physical part generated by the constraints 5.25 and 5.27, though this decomposition is not unique because of the ambiguity in the Hamiltonian 5.28. It is the non-physical part that corresponds to a gauge transformation in the constrained Hamiltonian formulation of classical electromagnetism. More generally, Dirac proposed that the gauge transformations of a theory in the constrained Hamiltonian formulation be identified as just those transformations which are generated in this way by what he called a *first-class* constraint (see appendix C): the first-class constraints in the case of classical electromagnetism turn out to be precisely 5.25 and 5.27. Applied to the constrained Hamiltonian formulation of the theory of the free classical electromagnetic field, Dirac's proposal classifies the transformations 5.10 as gauge transformations of A_μ, as expected.

5.5 The free Maxwell field as a Hamiltonian system

Dirac (1964) established a method for quantizing a classical theory by first formulating it as a Hamiltonian system, and then applying commutation relations to operators that have taken the place of canonical variables of that system. A key advance was to apply the method to Hamiltonian systems in which the canonical coordinates are subject to constraints, such as the constraint $\pi^0 = 0$ we have encountered repeatedly in the theory of electromagnetism.

Gambini and Pullin (1996) break Dirac's procedure into a sequence of steps. First, select an algebra of quantities of the classical theory (typically, the algebra generated by taking Poisson brackets of the canonical variables). Then represent this algebra as a set of operators acting on a space of functionals, typically of the form $\Psi(q)$, where Poisson brackets of classical variables are replaced by commutators of operators, and q is short for all the canonical configuration variables. If constraints are present, they may be of two kinds (see appendix C). Any constraint restricts the available phase space to a certain region: it may be expressed in the form $\varphi_i(q,p) = 0$, where q,p are short for all the canonical variables. A constraint is first class if the Poisson bracket of $\varphi_i(q,p)$ with all the other constraint functions $\varphi_j(q,p)$ is itself a linear combination of constraint functions: otherwise it is second class. Focus, for now, on theories subject only to first-class constraints. The first-class constraints define a region of the phase space called the constraint surface, on which they are all satisfied together. Each classical constraint function $\varphi_i(q,p)$ has a quantum analog $\hat{\varphi}_i(\hat{q},\hat{p})$, in which variables have been replaced by corresponding operators. The functional space is now restricted to those functionals that are annihilated by the operator $\hat{\varphi}_i(\hat{q},\hat{p})$ corresponding to each constraint equation $\varphi_i(q,p) = 0$: i.e., $\hat{\varphi}_i(\hat{q},\hat{p})\Psi(q) = 0$. Now one introduces an inner product on the functional space, making it into a Hilbert space on which self-adjoint operators of the form $\hat{\varphi}(\hat{q},\hat{p})$ are taken to represent observable quantities. Finally, the state of a system is taken to be represented by a functional Ψ that obeys a Schrödinger equation $i\partial\Psi/\partial t = \hat{H}\Psi$.

One can now apply Dirac's method to a classical electromagnetic field obeying Maxwell's equations in empty space. The canonical configuration variables are the four components of A_μ: the electric field $\mathbf{E}$ gives three of the momentum variables as $-\mathbf{E}$, while the vanishing of the fourth momentum variable π^0 represents a constraint. A second constraint follows from Hamilton's equation of motion for π^0: it is just the Gauss law $\nabla\cdot\mathbf{E} = 0$. Since the constraint functions commute, these are first-class constraints. Accordingly, the method requires one to impose commutation relations on the canonical variables as follows:

$$[\hat{A}_j(\mathbf{x}, t), \hat{E}^k(\mathbf{x}', t)] = -i\delta_{jk}\delta^3(\mathbf{x} - \mathbf{x}') \tag{5.33}$$

$$[\hat{A}_0(\mathbf{x}, t), \hat{\pi}^0(\mathbf{x}', t)] = i\delta^3(\mathbf{x} - \mathbf{x}') \tag{5.34}$$

$$[\hat{A}_j(\mathbf{x}, t), \hat{A}_k(\mathbf{x}', t)] = [\hat{E}^j(\mathbf{x}, t), \hat{E}^k(\mathbf{x}', t)] = 0 \tag{5.35}$$

Equation 5.33 no longer leads to an inconsistency, since the Gauss law is interpreted not as an operator equation, but as a restriction on the wave-functionals in the Hilbert space on which these operators act. Equation 5.34 looks strange in light of the classical constraint $\pi^0 = 0$, but again the corresponding operator equation is reinterpreted after quantization not as an identity, but as imposing a restriction on physical wave-functionals. These commutation relations may be realized by representing the action of the operators $\hat{A}_j, \hat{E}^k$ on the space of wave-functionals $\Psi[A]$ as follows:[3]

$$\hat{E}^k \Psi[A] = i\frac{\delta}{\delta A_k}\Psi[A] \tag{5.36}$$

$$\hat{A}_j \Psi[A] = A_j \Psi[A] \tag{5.37}$$

The Gauss' law is now represented by the equation

$$\partial_k \hat{E}^k \Psi[A] \equiv i\partial_k \frac{\delta}{\delta A_k}\Psi[A] = 0 \tag{5.38}$$

The Hamiltonian may be written in terms of the operators $\hat{A}_j, \hat{E}^k$, giving the Schrödinger equation $\hat{H}(\hat{A}_j, \hat{E}^k)\Psi[A] = i\partial \Psi[A]/\partial t$.

Armed with the distinction between first- and second-class constraints, we may now take up the motivation for the modified commutation relations 5.13 imposed in the course of quantizing the free Maxwell field in Coulomb gauge. The Coulomb gauge condition 5.12 now acts as a constraint, and the Gauss law provides another constraint. The Poisson bracket relations satisfied by $\nabla.\mathbf{E}, \nabla\mathbf{A}$ imply that these constraints are second class. There is a

---

[3] The operator $\frac{\delta}{\delta A_j}$ represents the functional derivative of the expression to which it is applied. In a functional derivative, instead of differentiating a function with respect to a variable, one differentiates a functional with respect to a function. If $F[f(x)]$ is a functional, then its functional derivative with respect to the function $f(y)$ is defined by

$$\frac{\delta F[f(x)]}{\delta f(y)} \equiv \lim_{\epsilon \to 0} \frac{F[f(x) + \epsilon\delta(x - y)] - F[f(x)]}{\epsilon}$$

general result that such second-class constraints may be converted into first-class constraints by systematically altering the original algebra so that Poisson brackets are replaced by so-called *Dirac brackets*. The commutation relations 5.13 emerge automatically as the result of replacing the relevant Dirac brackets by commutators.

5.6 Path-integral quantization

The techniques described in previous sections for developing a quantum theory of the electromagnetic field were all based on canonical formulations of classical electromagnetism. The basic idea was to replace number-valued functions representing classical fields by operator-valued functions representing quantum fields, where the field operators act on a Hilbert space of states. The probability for a transition from an initial to a final state is calculated by determining how the relevant operators evolve as a result of interactions between the corresponding fields.

There is an alternative technique for quantizing a classical theory which treats such interactions in a different way. It associates an amplitude—a complex number—with *every* continuous sequence of states of a classical field leading from initial to final state, not just those permitted by the classical field equations. Each such sequence of classical states is called a path. The amplitude for the process is the result of superimposing all the individual path amplitudes, and the transition probability is the square (modulus) of this total amplitude. Mathematically, the superposition is not just a sum, but a special kind of integral—not over a single variable or even over a finite-dimensional space, but over an infinite-dimensional, continuous space of functions, each defined by a different path from initial to final classical state. Path-integral quantization is often preferred in its application to non-Abelian gauge theories since it automatically handles many of the complexities that attend their canonical quantization. It is not easy to give rigorous definitions of these path integrals, and it is common for physicists to deploy them without concern for such mathematical niceties. For the limited purposes of this book it will prove neither necessary nor possible to pursue these technical questions, nor indeed to give more than a superficial introduction to the ideas behind path-integral quantization.[4]

Much of the testable empirical content of the quantum theory of a particular set of interacting fields is arrived at by means of an approximation

---

[4] The classic introduction to path integrals is by Feynman and Hibbs (1965). While many contemporary physics texts present the path-integral quantization of gauge field theories, and the mathematics of this technique have been intensively studied, I know of no sustained critical discussions of its conceptual foundations.

technique called perturbation theory, which is applicable when the interaction is sufficiently weak. It is largely through applying the results of perturbation theory calculations to experimental data that quantized gauge theories have established their empirical credentials. A key input to perturbative calculations of scattering cross-sections and lifetimes of bound states for matter fields interacting through a particular gauge field is the set of *n-point functions* of the theory—vacuum expectation values of time-ordered products of field operators. In the case of a scalar field $\hat{\varphi}(\mathbf{x}, t) \equiv \hat{\varphi}(x)$ such an n-point function may be written as $T(x_1, \ldots, x_n) \equiv \langle 0 | T(\hat{\varphi}(x_1) \ldots \hat{\varphi}(x_n)) | 0 \rangle$, where $|0\rangle$ represents the vacuum state, and the field operators are ordered so that $\hat{\varphi}(x_i)$ is placed to the left of $\hat{\varphi}(x_j)$ if $t_i > t_j$. In the path-integral approach this is derived by repeated functional differentiation of the so-called *generating functional* $W[J]$—the vacuum–vacuum transition amplitude in the presence of an external source J:

$$T(x_1, \ldots x_n) = (-1)^n \frac{\delta^n W[J]}{\delta J(x_1) \ldots \delta J(x_n)}\bigg|_{J=0}$$

The generating functional itself is expressed as a path integral in the form

$$W[J] = N \int D\varphi \exp\left[i \int d^4x \left(\mathcal{L}(\varphi, \dot{\varphi}) + J\varphi\right)\right]$$

where N is a normalization constant chosen so that $W[0] = 1$. The path integration is taken over all classical field configurations from $\varphi(\mathbf{x}, -\infty)$ before the source is switched on until $\varphi(\mathbf{x}, +\infty)$ after it is switched off again. These tricky mathematical ideas will only be needed once after section 5.8 (in section 6.4).

Path-integral quantization is equivalent to canonical quantization in the following sense: When correctly applied, it yields exactly the same n-point functions as a standard canonical quantization method such as those described in earlier sections. Hence that canonical and path-integral formulations of a gauge theory yield the same n-point functions implies that they are in one natural sense empirically equivalent. In fact they are equivalent in a stronger sense. As Wightman (1956) showed, any quantum field theory of a neutral scalar field is wholly specified by its set of vacuum expectation values for products of field operators. His methods have generalizations to any quantum field theory. Now path-integral and canonical formulations of a quantized gauge theory yield the same vacuum expectation values for time-ordered products of field operators. Wightman's result does not by itself prove that they are strongly equivalent, in that they yield the very same quantum field theory. But it does indicate that the path-integral and canonical quantizations of the free Maxwell field and other gauge theories are not merely empirically equivalent.

5.7 Canonical quantization of non-Abelian fields

Attempts to quantize a non-Abelian gauge theory using canonical techniques involving an initial choice of gauge bring new problems. The first problem manifests itself if one tries to impose a natural generalization of the Coulomb gauge condition considered in section 5.2. Gribov (1977) pointed out that this involves an essential ambiguity. If we denote a non-Abelian Yang–Mills gauge potential by $A_\mu^a = (A_0^a, -\vec{A}^a)$, where the a superscript indicates a component in some basis for the Lie algebra of the theory's structure group, and the arrow symbolizes the spatial component of a four-vector field, then the generalized Coulomb condition would be $A_0^a = \vec{\nabla}.\vec{A}^a = 0$. But even if we impose the further condition that $\vec{A}^a$ vanish sufficiently fast at infinity, this still permits distinct A_μ^a that differ by a finite gauge transformation, and so the Coulomb gauge is not well defined.

Moreover, this is a symptom of a deeper problem. As Singer (1978) showed, for a Yang–Mills theory based on a compact non-Abelian group such as SU(N) ($N = 2, 3, \ldots$), *no* choice of gauge (corresponding to a choice of a continuous, single-valued A_μ^a) is consistent with the condition that $\vec{A}^a$ always vanish, or even tend to a constant value, at infinity. Not only is there no analog to the Coulomb gauge in such theories, but a wide variety of other apparent gauge choices are also ruled out on topological grounds.[5]

Still, other choices of gauge are available that do not suffer from this problem—for example, the choice of axial gauge $A_3^a = 0$. But these sacrifice other symmetries besides Lorentz invariance—for example, rotational invariance. Moreover, any canonical quantization technique that begins by imposing a choice of gauge involves significant algebraic complications.

An alternative approach to canonical quantization of a (possibly non-Abelian) Yang–Mills field generalizes the method of section 5.5. One starts with a constrained Hamiltonian formulation of the corresponding classical gauge field, whose Lagrangian density is given by

$$\mathcal{L}_{\text{YM}}(A_\mu^a, \partial_\nu A_\mu^a) = -\frac{1}{4} F_{\mu\nu}^a F^{a\mu\nu} \tag{5.39}$$

where $\mathbf{F}_{\mu\nu} \equiv \partial_\mu \mathbf{A}_\nu - \partial_\nu \mathbf{A}_\mu + [\mathbf{A}_\mu, \mathbf{A}_\nu]$. Here bold-face type indicates a vector in the "internal space" of the structure group. Its expansion in terms of a basis of generators for the group's Lie algebra is then

$$\mathbf{A}_\mu \equiv \sum_a A_\mu^a T_a \tag{5.40}$$

---

[5] See chapter 3, section 3.1.3.

5 QUANTIZED YANG–MILLS GAUGE THEORIES

These generators satisfy the commutation relations

$$[T_a, T_b] = f_{abc} T_c \tag{5.41}$$

The structure group is Abelian if and only if all f_{abc} are zero. Note that the commutator here occurs in a *classical* non-Abelian gauge theory.

The canonical coordinates are the A^a_μ, and the canonical momenta $\pi^{a\mu}$ are given by $\pi^{a0} = 0$; $\pi^{aj} = -E^{aj} \equiv -F^{a0j} = \partial^0 A^{aj} - \partial^j A^{a0} + f_{abc} A^{b0} A^{cj}$. The vanishing of π^{a0} is a first-class constraint, as is the "Gauss law" that follows from its preservation under time evolution

$$\dot{\boldsymbol{\pi}}^0 = \partial_j \mathbf{E}^j + [\mathbf{A}_j, \mathbf{E}^j] \equiv D_j \mathbf{E}^j = 0 \tag{5.42}$$

Following the Dirac quantization prescription, one converts the classical Poisson bracket relations of the canonical variables into commutation relations of corresponding abstract operators, and seeks a representation of these by operators acting on a space of functionals of the configuration variables A^a_μ. The only non-vanishing classical canonical Poisson bracket relations are

$$\{A^a_j(\mathbf{x}), E^{bk}(\mathbf{x}')\} = -\delta_{ab}\delta_{jk}\delta^3(\mathbf{x}-\mathbf{x}') \tag{5.43}$$

$$\{A^a_0(\mathbf{x}), \pi^{b0}(\mathbf{x}')\} = \delta_{ab}\delta^3(\mathbf{x}-\mathbf{x}') \tag{5.44}$$

The corresponding quantum algebra of operators may be represented as follows:

$$\hat{A}^a_j \Psi(\overrightarrow{A^a}, A^a_0) = A^a_j \Psi(\overrightarrow{A^a}, A^a_0) \tag{5.45}$$

$$\hat{E}^{ak} \Psi(\overrightarrow{A^a}, A^a_0) = i\delta/\delta A^a_k \Psi(\overrightarrow{A^a}, A^a_0) \tag{5.46}$$

$$\hat{A}^a_0 \Psi(\overrightarrow{A^a}, A^a_0) = A^a_0 \Psi(\overrightarrow{A^a}, A^a_0) \tag{5.47}$$

$$\hat{\pi}^{a0} \Psi(\overrightarrow{A^a}, A^a_0) = -i\delta/\delta A^a_0 \Psi(\overrightarrow{A^a}, A^a_0) \tag{5.48}$$

Operators corresponding to the classical constraints are now taken to annihilate the physical wave-functionals. The classical constraint $\pi^{a0} = 0$ implies that the physical wave-functionals $\Psi(\overrightarrow{A^a}, A^a_0)$ do not depend on A^a_0, and so the generalized Gauss constraint has the form

$$D_k E^{ak} \Psi(\overrightarrow{A^a}) = 0 \tag{5.49}$$

To complete the quantization program, one selects an appropriate inner product for the Hilbert space of wave-functionals, constructs a Hamiltonian operator from the canonical operators, and formulates the corresponding Schrödinger equation.

5.8 Path-integral quantization of non-Abelian fields

The difficulties associated with canonical quantization of non-Abelian gauge fields have motivated the adoption instead of path-integral techniques that generalize path-integral quantization of the free Maxwell field. Indeed, this method for quantizing a gauge theory is now preferred by those who wish to extend the quantization program to the case of interacting fields in order to apply perturbation techniques to derive Feynman rules that can be conveniently applied in calculating experimentally accessible quantities like scattering cross-sections for (electro)weak or strong interactions. Not only is path-integral quantization in some ways more elegant, but also it makes manifest the gauge-invariant nature of the resulting theory. Even though the generating functionals and n-point functions to which it gives rise still depend on a choice of gauge, the arbitrariness of that choice is manifested by the fact that all such choices lead to the same amplitudes for processes in all orders of perturbation theory, and consequently the same empirical predictions.

But path-integral quantization of gauge fields does involve a difficulty associated with gauge invariance. The path integrals are functional integrals over the space of potentials A_μ^a, and since many such potentials are gauge equivalent to one another, this involves massive "over-counting." Not surprisingly, the resulting functional integrals diverge! Techniques have been developed for getting around this problem, by Fadeev and Popov and others, but they purchase calculational efficiency and predictive success only by introducing further apparent conceptual difficulties to be discussed in chapter 6. Specifically, these techniques involve modifying the original Lagrangian of the theory by the addition of terms that appear to correspond to additional "ghost" fields, associated with "non-physical" particles—fermions with integral spin that never appear in incoming or outgoing states of any interaction. This comes about as follows.

Suppose that one were to write down the following generating functional for a non-Abelian Yang–Mills gauge field—specifically an SU(2) field:

$$W[J] = N \int D\mathbf{A}_\mu \exp\left[i \int d^4x(\mathcal{L}_{YM} + \mathbf{J}_\mu \mathbf{A}^\mu)\right] \quad (5.50)$$

This involves an integral over all possible paths $\mathbf{A}_\mu$ from initial to final gauge potential configuration. Clearly each physically distinct path is counted infinitely many times in this integral, because each particular $\mathbf{A}_\mu$ is gauge equivalent to infinitely many others. It is therefore not surprising that equation 5.50 fails to yield the required n-point functions. Some way is needed to restrict the range of the integral to physically distinct configurations. It turns out that there is a neat trick for doing this which involves inserting two additional terms in the path integral, yielding the following form:

$$W_f[J] = N \int D\mathbf{A}_\mu (det \ M_f) \delta(f_a(\mathbf{A}_\mu)) \exp\left[i \int d^4x (\mathcal{L}_{YM} + \mathbf{J}_\mu \mathbf{A}^\mu)\right] \quad (5.51)$$

This modification of 5.50 is a result of imposing a general gauge-fixing condition of the form $f_a(\mathbf{A}_\mu) = 0$. It is not necessary for our purposes to understand in detail the meanings of the two gauge-dependent additional factors in the path integral. The important point is that these may be evaluated in such a way that equation 5.51 assumes the form

$$W_f[J] = N' \int D\mathbf{A}_\mu D\mathbf{c} D\mathbf{c}^\dagger \exp\left(iS_{eff}[J]\right) \quad (5.52)$$

A similar transformation may be effected for a general Yang–Mills theory. Here N' is a new normalization factor, and two terms have been added to the action $S[J] = \int d^4x (\mathcal{L}_{YM} + \mathbf{J}_\mu \mathbf{A}^\mu)$ to give a new *effective action*

$$S_{eff} = S[J] + S_{gf} + S_{FPG} \quad (5.53)$$

These are a gauge-fixing term S_{gf} that depends on f_a, and the *Fadeev-Popov ghost term* S_{FPG} which has the form of an action associated with two new fields **c**, **c**† that now appear in the path integral 5.52. We shall have to evaluate the status of these so-called "ghost" fields in chapter 6. For now, notice only that in the case of an Abelian gauge theory like quantized electromagnetism, no such terms appear. For an Abelian theory, the additional determinant term in the analog of equation 5.51 is independent of A_μ and so may be absorbed in the normalization factor; while the δ term gives rise only to a gauge-fixing term S_{gf}. Consequently, no ghost fields with accompanying ghost action occur for the case of an Abelian Yang–Mills theory. That is why it was not necessary to mention this possibility in section 5.6.

5.9 Interacting fields in the Lagrangian formulation

So far, this chapter has considered how to quantize only a free Yang–Mills gauge theory. But Yang–Mills fields are never in fact free; moreover they are characteristically manifested only when they interact with matter fields. Indeed, it is common to mark a distinction between matter fields (whose quanta are fermions such as leptons and quarks) and force fields, so that the role of Yang–Mills theories is just to describe how matter interacts, by "exchanging" the photons, weak bosons, and gluons that are the quanta associated with the electromagnetic, weak, and strong forces, respectively. But the situation cannot be quite so simple. For in the case of non-Abelian gauge theories,

5.9 INTERACTING FIELDS IN THE LAGRANGIAN FORMULATION

the field equations derived as Euler–Lagrange equations from the free non-Abelian Yang–Mills Lagrangian 5.39 are non-linear. This implies that the gauge potential A_μ^a is itself a source of the gauge field $F_{\mu\nu}^a$. It follows that while photons (the quanta associated with the Abelian theory of electromagnetism) carry no electric charge, and so experience no direct electromagnetic interactions with one another, the quanta of a quantized non-Abelian gauge theory are themselves subject to the same interaction that they mediate.

Consider first how the theory of quantum electrodynamics describes electromagnetic interactions that may involve electrons and positrons (their positively charged anti-particles). The Lagrangian (density) 5.9 for the free Maxwell field is modified as follows:

$$\mathcal{L}_{\text{QED}} = \bar{\psi}(i\gamma^\mu \partial_\mu - m)\psi - \frac{1}{4}F_{\mu\nu}F^{\mu\nu} - e\bar{\psi}\gamma^\mu\psi A_\mu \qquad (5.54)$$

The second term in this expression is recognizable as the Lagrangian of the free Maxwell field; the first term is the Lagrangian of the free Dirac field that describes electrons and positrons. It is the third term that describes the electromagnetic interaction among these two otherwise free fields. It is noteworthy that the Lagrangian 5.54 is invariant under the simultaneous gauge transformations

$$\psi \to \exp\{-ie\Lambda(x^\mu)\}\psi \qquad (5.55)$$
$$\bar{\psi} \to \exp\{+ie\Lambda(x^\mu)\}\bar{\psi} \qquad (5.56)$$
$$A_\mu \to A_\mu + \partial_\mu \Lambda(x^\mu) \qquad (5.57)$$

This will be important when we come to consider the so-called gauge argument in chapter 6. It is the small size of the coupling constant e (or, more precisely, of the dimensionless number $\alpha \equiv \frac{e^2}{\hbar c} \simeq \frac{1}{137}$) that corresponds to the relatively small strength of the electromagnetic interaction and thereby facilitates the use of perturbation theory in extracting remarkably accurate predictions from the theory of quantum electrodynamics based on this Lagrangian. It will prove convenient to rewrite 5.54 in the following form:

$$\mathcal{L}_{\text{QED}} = \bar{\psi}(i\gamma^\mu D_\mu - m)\psi - \frac{1}{4}F_{\mu\nu}F^{\mu\nu} \qquad (5.58)$$

where $D_\mu \equiv (\partial_\mu + ieA_\mu)$ gives the covariant derivative of ψ.

One can now generalize this treatment of interactions in the Abelian theory of quantum electrodynamics to write a Lagrangian for the interaction of a non-Abelian Yang–Mills theory. For example, the Lagrangian for quantum chromodynamics may be written as

$$\mathcal{L}_{\text{QCD}} = \bar{\psi}_a(i\gamma^\mu D_\mu - m)\psi^a - \frac{1}{4}F_{a\mu\nu}F^{a\mu\nu} \qquad (5.59)$$

where the Lagrangian for free quarks is

$$\mathcal{L}_{\text{quark}} = \overline{\psi}_a(i\gamma^\mu \partial_\mu - m)\psi^a \tag{5.60}$$

in which the quark field ψ^a ($a = 1, 2, 3$) is a vector in a representation of the non-Abelian group SU(3),

$$D_\mu \psi^a \equiv \partial_\mu \psi^a - ig f_{abc} A_\mu^b \psi^c \tag{5.61}$$

is the covariant derivative, and

$$F_{\mu\nu}^a = \partial_\mu A_\nu^a - \partial_\nu A_\mu^a - ig f_{abc} A_\mu^b A_\nu^c \tag{5.62}$$

(cf. 3.25), where all repeated indices are summed over, and the f_{abc} are structure constants of the Lie algebra of the group SU(3), whose generators T_a have commutation relations

$$[T_a, T_b] = f_{abc} T_c \tag{5.63}$$

A dimensionless parameter α_s related to the coupling "constant" g gives the strength of the interaction as a function of interaction energy. At low energies, it is large compared to α. But it decreases monotonically with increasing energy ("asymptotic freedom"). At the energies characteristic of quarks confined within hadrons, it is small enough to facilitate the application of perturbation theory in quantum chromodynamics.

6

The empirical import of gauge symmetry

If gauge symmetry is just a formal feature of the way a gauge theory represents its subject matter, then that a gauge theory does or does not have this feature can by itself have no empirical consequences. But several lines of thought seem to lead to a contrary conclusion. Gauge symmetry would have direct empirical consequences if it could be manifested by physically realizing gauge transformations. And there is a so-called "gauge argument" that purports to show that requiring a "globally" gauge-symmetric matter field also to be "locally" gauge symmetric implies the existence of an associated gauge field with definite properties.

If, on the contrary, there are physical phenomena that are manifest only in a particular gauge, then *failure* of gauge symmetry has empirical consequences. "Ghost" fields, or the Higgs boson, come to mind in this context. Ghost fields arise only in certain gauges, while the Higgs mechanism may seem to operate precisely when gauge symmetry is spontaneously broken.

Some explanations of the so-called θ-vacuum in a non-Abelian gauge theory apparently presuppose that "large" (as opposed to "small") "local" gauge transformations relate physically distinct states. A recent text (Bertlmann 1996) might be taken to attribute physical significance to violations of gauge symmetry in its discussion of anomalies in a quantum field theory. For it claims that anomalies are responsible for physical phenomena including the two-photon decay of neutral pions, while maintaining also that anomalies "signal the breakdown of gauge symmetry and, in consequence, the ruin of the consistency of the theory" (p. 1).

This chapter is devoted to tracing these lines of thought in order to show that none of them impugns the thesis that gauge symmetry is just a formal feature of the way a gauge theory represents its subject matter. But first it is necessary to clarify this thesis by distinguishing two kinds of symmetry that may be associated with a theory and its subject matter.

6.1 Two kinds of symmetry

Abstractly, a symmetry of a structure is an automorphism—a transformation that maps the elements of an object back onto themselves so as to preserve the structure of that object. Many different structures may be distinguished in a given object. A physical object may have a certain size, shape, pattern of colors, etc., and an abstract object may also exhibit a variety of structures—SU(2) is a group, it is non-Abelian, it is a Lie group (and so also a differentiable manifold), it is compact, it is simple, etc. We are concerned here with physical theories and the situations to which they may be applied, so we need to say what kinds of objects these involve.

A physical theory specifies a set of models—mathematical structures—that may be used to represent various different situations, actual as well as merely possible, and to make claims about them. Any application of a physical theory is to a situation involving some system, actual or merely possible. Only rarely is that system the entire universe: typically, one applies a theory to some subsystem, regarded as a relatively isolated part of its world. The application proceeds by using the theory to model the situation of that subsystem in a way that abstracts from and idealizes the subsystem's own features, and also neglects or idealizes its interactions with the rest of the world.

For example, general relativity may be applied to the motion of Mercury around the Sun by using the Schwarzschild model to describe the space-time geometry. This application idealizes the source of the gravitational field experienced by Mercury as a single, spherically symmetric massive body and Mercury itself as an infinitesimal, structureless test body; and it treats the rest of the universe merely by imposing the boundary condition that space-time tend to Minkowski form at large distances. In this application, the situation modeled would consist of Mercury's changing orbit around the sun. The Schwarzschild model may be perfectly adequate for this purpose, even though it fails to adequately represent the space-time geometry far from the solar system. It may be thought to represent a world that general relativity countenances as possible, but which resembles the actual world only in respect of the behavior of Mercury's orbit.

A set of physical situations may possess symmetries for which we can acquire evidence, whether or not we have developed a sophisticated theory to apply to them. The geometric symmetries of a crystal can be studied without a theory to explain why they obtain. A piece of music preserves its structure when transposed up or down in pitch, displaying another kind of symmetry in its performances that can be recognized and studied with no theoretical understanding of the physics of sound. Such symmetries are *empirical*: they can be recognized even without a physical theory to account for them. But they do not cease to be empirical if and when such a theory becomes available. A theory may entail an empirical symmetry.

Sometimes one can recognize an empirical symmetry among a set of situations even though one can discriminate among them by recognizing features they do not share. Transposing a piece of music up or down in pitch does not preserve the sound of each note, but only how one note sounds relative to another: someone with perfect pitch can immediately tell the transposed versions apart, while the rest of us may need instruments. But other empirical symmetries are more complete: it may be difficult or even impossible to discriminate among situations related by an empirical symmetry by any observations or measurements confined to each situation. Snowflakes have hexagonal symmetry. If this were perfect, then it would be impossible to tell whether a snowflake had been rotated through some non-zero multiple of 60 degrees while one was out of the room by observing it after returning.

Galileo (1632/1967, pp. 186–7) illustrated his relativity principle by describing a famous empirical symmetry of this kind:

Shut yourself up with some friend in the main cabin below decks on some large ship, and have with you there some flies, butterflies and other small flying animals. Have a large bowl of water with some fish in it; hang up a bottle that empties drop by drop into a wide vessel beneath it. With the ship standing still, observe carefully how the little animals fly with equal speed to all sides of the cabin. The fish swim indifferently in all directions; the drops fall into the vessel beneath; and, in throwing something to your friend, you need throw it no more strongly in one direction than another, the distances being equal; jumping with your feet together, you pass equal spaces in every direction. When you have observed all these things carefully (though there is no doubt that when the ship is standing still everything must happen in this way), have the ship proceed with any speed you like, so long as the motion is uniform and not fluctuating this way and that. You will discover not the least change in all the effects named, nor could you tell from any of them whether the ship was moving or standing still.

His implicit claim is that a situation inside the cabin when the ship is in motion is indistinguishable from another situation inside the cabin when the ship is at rest by observations confined to those situations. The claim follows from a principle of the relativity of all uniform horizontal motion. We know today that an unqualified form of Galileo's claim is false, since the motion of the ship is not inertial, whether it is at rest or moving with constant speed. Extremely delicate measurements made within the cabin could discriminate between two such states of motion of the ship, if they were not rendered unfeasible by the non-uniformities of its actual motion. But no such measurements were possible in Galileo's day, and the sorts of observations to which he appeals do provide some support for the claim that no measurement solely of the behavior of a set of objects is capable of distinguishing one state of collective uniform horizontal motion of those objects from another. A weaker form of that claim would be that situations related by a uniform collective horizontal motion are empirically symmetrical with respect to observations and measurements of the type Galileo describes, where

A 1-1 mapping $\varphi: \mathcal{S} \to \mathcal{S}$ of a set of situations onto itself is an *empirical symmetry with respect to C-type measurements* if and only if no two situations related by φ can be distinguished by means of measurements of type C.

We can take a stronger form of Galileo's implicit claim to be that situations related by a uniform collective horizontal motion are strongly empirically symmetrical, where a strong empirical symmetry is a limiting case of an empirical symmetry in which C-type measurements are just those that are confined to each situation. Specifically

A 1-1 mapping $\varphi: \mathcal{S} \to \mathcal{S}$ of a set of situations onto itself is a *strong empirical symmetry* if and only if no two situations related by φ can be distinguished by means of measurements confined to each situation.

A measurement is confined to a situation just in case it is a measurement of intrinsic properties of (one or more objects in) that situation. Note that the reference to measurement is not superfluous here, in so far as a situation may feature unmeasurable intrinsic properties. We shall see an example of this soon.

If every function $\varphi \in \Phi$ is an empirical symmetry of $\mathcal{S}$ with respect to C, then $\mathcal{S}$ is symmetric under Φ-transformations. These form a $\in$ group: the group identity is the identity mapping on $\mathcal{S}$—a trivial empirical symmetry. Spatial translations and rotations provide familiar examples of strong empirical symmetries of situations involving geometrical figures in Euclidean space. If s is any figure in Euclidean space, then a translation and/or rotation φ yields a congruent figure $\varphi(s)$. Note that situations in $\mathcal{S}$ related by a transformation φ may be in the same or different possible worlds: if φ is a strong empirical symmetry, then $\varphi(s)$ may be in the same world w as s, but only if w is itself sufficiently symmetric.

Uniform velocity boosts are strong empirical symmetries of a set of situations accurately modeled either by Newtonian or by special relativistic mechanics (excluding electromagnetic phenomena), since a Galilean (respectively Lorentz) boost by velocity v applied to the situation s of a mechanical system in a world w yields a situation $\varphi_v(s)$ that is indistinguishable from s with respect to all intrinsic properties modeled. Special relativity extends this strong empirical symmetry to include electromagnetic phenomena. Even when a situation s actually obtains, $\varphi_v(s)$ will rarely do so. In some cases, careful laboratory manipulations may actually bring it about, but the situation $\varphi_v(s)$ will more typically obtain only in some "merely" (i.e. non-actual) possible world.

One may distinguish symmetries of the set of situations to which a theory may be applied from symmetries of the set of the theory's models.

A mapping $f: \mathcal{M} \to \mathcal{M}$ of the set of models of a theory Θ onto itself is a *theoretical symmetry of* Θ if and only if the following condition obtains: For every model m of Θ that may be used to represent (a situation s in) a possible world w, $f(m)$ may also be used to represent (s in) w.

If every function $f \in F$ is a symmetry of Θ, then Θ is symmetric under F-transformations. These form a group. Coordinate transformations provide familiar examples of theoretical symmetries. Consider, for example, a theory of Euclidean geometry whose models specify the locations of various objects in a plane in terms of systems of rectangular Cartesian coordinates. If the vertices of a triangle may be represented by coordinates (x_1, y_1), (x_2, y_2), (x_3, y_3) in one model, then they may also be represented in a different model by coordinates $(\bar{x}_1, \bar{y}_1)$, $(\bar{x}_2, \bar{y}_2)$, $(\bar{x}_3, \bar{y}_3)$, where $\bar{x}, \bar{y}$ are related to x, y by a linear transformation (corresponding to a spatial translation of the origin of coordinates) and/or an orthogonal transformation (corresponding to a rotation of coordinate axes). General relativity provides another example of a theoretical symmetry, namely diffeomorphism invariance. For if $\langle M, g, O_1, \ldots, O_n \rangle$ is a model of general relativity that may be used to represent a situation s in a possible world w, then so is $\langle M, d^*g, d^*O_1, \ldots, d^*O_n \rangle$, where d^* is the carry-along of the diffeomorphism d. A theoretical symmetry of a theory establishes an equivalence relation among its models: two models are equivalent just in case either may be used to represent (the same situation in) the very same possible worlds. But models related by a theoretical symmetry may sometimes be used to represent *different* situations.

Theoretical symmetries may be purely formal features of a theory, in so far as they relate different but equivalent ways the theory has of representing one and the same empirical situation. One model may be more conveniently applied to a given situation than another model related to it by a theoretical symmetry, but the theoretical as well as empirical content of any claim made about that situation will be the same no matter which model is applied. But a theoretical symmetry of a theory may entail an empirical symmetry, in which case it is not a purely formal feature of the theory.

To the extent that the Schwarzschild model is adequate to represent the actual behavior of Mercury's orbit around the Sun, so also is it adequate to model a distinct possible situation corresponding to a time translation of this behavior, or a time-independent spatial rotation of it about the center of the Sun. So general relativity implies that time translations and these spatial rotations are empirical symmetries *of these situations*. But they are not empirical symmetries of all situations to which general relativity applies, since that theory also models situations on which no such transformations are defined.

But Galilean relativity is a strong empirical symmetry of every situation in a Newtonian world, provided that all masses and forces in that situation are independent of time. For under these assumptions, Newton's theory implies that in any situation involving purely mechanical systems, any aspect of their behavior that is measurable within that situation is independent of the uniform absolute velocity of their center of mass, which velocity is not itself detectable by any measurement. So Galilean relativity is not a purely formal feature of Newton's theory. Indeed, it is not even a consequence of a theoretical symmetry of Newton's theory, since that theory itself distinguishes among strongly empirically symmetric situations with respect to the absolute velocities of their constituents.

Contrast this with the strong empirical symmetry associated with uniform velocity boosts in special relativity. This strong empirical symmetry *is* a consequence of a theoretical symmetry of special relativity: the Lorentz boost of a model of a situation is also a model of that situation. The Lorentz boost of any model may be used to represent the same situation as the original model, but (unlike the Newtonian case) it may also be used to represent a boosted *duplicate* of that situation. (Here a duplicate of a situation is a situation that shares all its intrinsic properties.) The special theory of relativity entails the strong empirical symmetry associated with Lorentz invariance by implying that these strongly empirically equivalent situations are not merely empirically indistinguishable by means of measurements confined to those situations, but indistinguishable by reference to any intrinsic properties or relations of entities each involves.

Even if they are entailed by a theory, non-trivial empirical symmetries are not purely formal features of that theory, since they relate *distinct* situations. As already noted, one may acquire evidence for an empirical symmetry prior to developing any theory. If one were to assume that the representation relation between models of a theory and situations modeled is 1–1, one could offer an alternative characterization of an empirical symmetry φ of the situations to which a theory may be applied in terms of a map f_φ from model m of situation s to model $f_\varphi(m)$ of situation $\varphi(s)$. That might mislead one into thinking that theoretical symmetries are never just formal features of a theory, in so far as they characterize empirical symmetries among distinct situations represented by its models. But that would be a mistake: an empirical symmetry cannot be characterized in this way without a theory to model the situations it relates; and it can be characterized in this way only if the symmetry is *not* a consequence of a theoretical symmetry of the theory.

Relativity principles assert empirical symmetries. If "local" gauge transformations reflect some similar empirical symmetry, then they also represent distinct but indistinguishable situations. But this chapter defends the thesis that the successful employment of Yang–Mills theories warrants the conclusion that "local" gauge transformations are only theoretical symmetries of these theories that reflect no corresponding non-trivial empirical symmetries among the situations they represent. "Local" gauge symmetry is a purely formal feature of these theories.

6.2 Observing gauge symmetry?

It would be a powerful objection to the view that "local" gauge symmetry is a purely formal feature of a theory Θ if one could exhibit pairs of situations, indistinguishable by means of measurements confined to each, that Θ distinguishes by modeling one situation in each pair as related to the other situation by a non-trivial "local" gauge transformation. For this would establish that "local" gauge transformations reflect empirical symmetries among situations that Θ represents as different. Now even though a situation may obtain either in the actual world or in some merely possible world, a situation can be exhibited only in the actual world. To show that a "local" gauge transformation reflects a non-trivial empirical symmetry φ, one would therefore need to exhibit such distinct but internally empirically indistinguishable situations s, s' in the actual world related by a mapping φ corresponding to that "local" gauge transformation. Observing such situations would count as a direct observation of "local" gauge symmetry. But such observation is impossible, according to contemporary Yang–Mills gauge theories.

This claim may seem surprising in the light of Faraday's experiments discussed in the introduction. Recall that he constructed a hollow cube with sides 12 feet long, covered it with good conducting materials but insulated it carefully from the ground, and electrified it so that it was at a large potential difference from the rest of his laboratory, "went into this cube and lived in it, but though [he] used lighted candles, electrometers, and all other tests of electrical states, [he] could not find the least influence on them." This provides a close analogy to Galileo's ship scenario that showed how observations provide evidence supporting the empirical symmetry of uniform velocity boosts. The situation inside Faraday's cube when electrified is indistinguishable by measurements confined to the cube from its situation when not electrified. But one can observe that the two situations are related by a transformation apparently corresponding to the constant difference in potential between them, by observing phenomena associated with the electric field produced between the cube when electrified and the ground (such as passage of a current between them). Similarly, the situation inside Galileo's ship's cabin when moving uniformly is (he supposed) indistinguishable by measurements confined to the cabin from its situation when stationary. But one can observe that the two situations are related by a transformation corresponding to the constant velocity difference between them, by observing phenomena associated with the ship when it is moving relative to the harbor and ocean (such as production of a wake). If charging Faraday's cube induces a "local" gauge transformation, then his observations inside it before and after charging constitute direct observations of at least one kind of empirical "local" gauge symmetry, in which case "local" gauge transformations cannot be purely formal features of the gauge theory of electromagnetism.

However, a more careful examination reveals that, while the situation inside Faraday's cube when charged differs from its situation when uncharged, this is an extrinsic rather than an intrinsic difference. More importantly, the different situation inside the cube does not reflect a "local" gauge transformation there. For to establish that this difference reflects a "local" gauge transformation applied to a model of the contents of the cube, it would be necessary to show that the theory of electromagnetism must itself model the difference in the situations as one involving such a "local" gauge transformation. But this is not so. That theory also contains a model of the combined situation incorporating both subsituations (cube uncharged, cube charged), in which the electromagnetic potentials inside the cube in *both* subsituations are everywhere zero. In this combined model, there are electric (but not magnetic) fields outside the cube in each subsituation; but these arise not only from the difference in the charges on the outside of the cube, but also from the existence of a (spatio) temporally varying magnetic vector potential outside the cube that gives rise to no magnetic field.

Classical electromagnetism can model the situation inside Faraday's cube when charged in just the same way it models its situation when uncharged, so that it counts these situations as duplicates, assigning them each the same intrinsic properties. The two clearly differ in their extrinsic properties—in one situation the cube is surrounded by unbalanced electric charges, while in the other it is not. But the difference between these situations does not correspond to a physical "local" gauge transformation. The theory of electromagnetism contains a model of the *joint* situation involving both conditions of the cube that represents the state of its interior in exactly the same way in each condition, whether or not its exterior is charged. So classical electromagnetism does not imply that the situation of Faraday's cube when charged is related to its situation when uncharged by a "local" gauge transformation. And we have no other reason to believe that these situations are so related. Consequently, the theoretical symmetry of classical electromagnetism associated with "local" gauge transformations implies no corresponding empirical symmetry. "Local" gauge symmetry is a purely formal feature of classical electromagnetism.

Contrast this with special relativity's explanation of the empirical symmetry associated with uniform velocity boosts. Special relativity models situations involving a system (such as the contents of Galileo's cabin) in different states of uniform motion. Because Lorentz boosts are a theoretical symmetry of the theory, it can model each separate state of uniform motion in the same way, thereby certifying such different situations as duplicates in so far as the systems involved have all the same intrinsic properties. But, in special relativity, *every* joint model of duplicate situations in which a system is in different states of uniform motion is a model in which these duplicate situations are related to each other by a non-zero velocity transformation. That is why the theoretical Lorentz boost symmetry implies the empirical symmetry associated with the relativity of uniform motion.

Note that, in the context of a theory of pure electrostatics, observations inside Faraday's cube *would* constitute evidence that a constant electric potential transformation represents an empirical symmetry. Such a theory entails that the situation involving the charged cube differs from that of the uncharged cube, since the only way it can *jointly* model the relevant duplicate situations is with a non-zero electric potential difference between them. Hence, in the context of such a theory, charging the cube does change the situation inside it—not intrinsically, but by altering its relative electric potential—a change reflected by a constant electric potential transformation. This is the context in which Galileo's ship provides a good analogy to Faraday's cube. But it is not the appropriate context for assessing the empirical status of the "local" gauge symmetry of the full theory of classical electromagnetism.

As we saw in chapter 2, this full theory of classical electromagnetism has interesting new empirical consequences in conjunction with the quantum mechanics of particles. One might think that one such consequence, the electric Aharonov–Bohm effect, could count as a manifestation of an empirical "local" gauge symmetry, despite the above argument. In that effect a beam of electrons is split into two coherent beams, each of which passes through a shielded cylinder that functions like a Faraday cage, and the two beams interfere after they are recombined. One cylinder is raised to a constant electric potential with respect to the other *only while the electrons are passing through the cylinders*. The resulting interference pattern then depends on this difference in potential of the two cylinders, even though there is no electromagnetic field inside either cylinder while the electrons pass through them.[1] Observation of the interference pattern in this experiment appears to provide direct evidence that the two situations are related by a "local" gauge transformation corresponding to a difference in electric potential. In principle, this transformation could be shown to be a symmetry by further experiments establishing that the situations inside the two cylinders could not be distinguished by observations (of electron trajectories, interference patterns, or anything else) confined to the interior of each cylinder in their relevantly different charge states.

---

[1] In fact, the experiment is very hard to perform in this set-up. Instead, a different experiment was proposed involving a ring geometry interrupted by tunnel barriers, with a bias voltage V relating the potentials of the two halves of the ring. This situation results in an Aharonov–Bohm phase shift as above, and was observed experimentally in 1998.

But even if the result of such experiments had their expected results, this would not count as an empirical manifestation of a "local" gauge symmetry, for two reasons. The first reason is familiar. By a suitable gauge transformation, one can model the joint situation involving a split electron beam passing through both cylinders in a gauge in which the electromagnetic potentials inside the two cylinders are *identical* while the beam passes. In that gauge, the phase difference responsible for the resulting interference between the component beams would then be attributed solely to differences in the electromagnetic potential in different regions *outside* the cylinders while the different components traversed those regions. Classical electromagnetism represents the electromagnetic situation inside each cylinder as identical, reflecting no physical transformation from one situation to the other.

In this (quantum) context, a "local" gauge transformation also includes a variable phase transformation applied to the electrons' wave-function. Suppose one were to apply such a variable phase transformation just to the component of the electron beam that passes through the charged cylinder, assuming that the phase variation of that component wave-function is a direct physical manifestation of the fact that this cylinder is charged while the component beam passes through it. Then one might conclude that the resulting phase difference between the two component wave-functions reflects a physical "local" gauge transformation applied to the electrons' wave-function. But this would be a mistake, for reasons Brown and Brading (2004) clearly explained.

Charging the second cylinder in the electric Aharonov–Bohm effect modifies the *total* wave-function of the electrons passing through the cylinders in a way that does *not* correspond to a variable phase transformation of the total wave-function, but rather as follows:

$$\Psi = a\Psi_1 + b\Psi_2 \rightarrow \Psi' = a\Psi_1 + be^{i\Lambda(x,t)}\Psi_2 \qquad (6.1)$$

The phase relation between the modified and unmodified wave-function is undefined, for the whole beam or for either of its components. So the physical transformation of charging one cylinder cannot correspond to a variable phase transformation applied either to a component or to the whole beam. Only the phase relations between the two components of a single wave-function are well defined, and then only at the same space-time point. As one part of a "local" gauge transformation, one can formally apply a variable phase transformation to either Ψ or Ψ' that has the effect of varying only the phase of its second component in a region where the first component vanishes (i.e. inside the charged cylinder, in an idealization). But that does not correspond to charging the cylinder, or to any other physical transformation. However, by appropriate choice of such transformations one can arrive at a wave-function for that component that represents its phase *as it passes through the cylinder* as unaffected by charging the cylinder. This shows that quantum mechanics together with classical electromagnetism can model the situations inside the two cylinders of the electric Aharonov–Bohm effect in exactly the same way, independent of the charge state of the cylinders while the electrons pass. There is no intrinsic difference between these situations.

These considerations may be generalized to the context of quantized Yang–Mills gauge theories, where classical electromagnetism is replaced by a quantized gauge field, and wave-functions of quantum particles are replaced by quantized matter fields. There is no sound basis for comparing the phase of a quantum wave-function or field representing a matter system in one situation with the phase of that or a related system in a different situation. Either the quantum theory contains *no* joint models of the combined situation, or it contains many models, each permitting a different comparison. Consequently, there can be no justification for regarding two such situations as physically related in a way that reflects a "local" gauge transformation. Neither purely classical phenomena, nor quantum mechanical phenomena, provide any reason to suppose that a "local" gauge transformation can be physically implemented. "Local" gauge symmetry is not an empirical symmetry: it is merely a theoretical symmetry of Yang–Mills gauge theories.

Brown and Brading (2004) express a similar conclusion as follows:

there can be no analogue of the Galilean ship experiment for local gauge transformations, and therefore local gauge symmetry has only *indirect* empirical significance (being a property of the equations of motion).

There is no analog to Galileo's ship experiment that would enable one to observe empirical manifestations of the "local" gauge symmetry of a theory of classical electromagnetism, or of any quantized Yang–Mills gauge theory.

6.3 The gauge argument

A principle of "local" gauge symmetry, understood now as a symmetry of a theory of matter associated with a variable generalized phase transformation such as 1.21 in the wave-function or quantum field that represents it, is often believed to require the existence of a gauge field with prescribed properties that interacts with this matter in a definite way. One of the chief architects of the gauge theories underlying the Standard Model of high-energy physics (Weinberg 1992, p. 142) said

Symmetry principles have moved to a new level of importance in this [i.e. the 20th!] century and especially in the last few decades: there are symmetry principles that dictate the very existence of all the known forces of nature.

The literature contains frequent references to a "gauge principle," supposed to entail, or at least explain, the existence and properties of gauge fields including not just electromagnetism, but also fields responsible for the entire unified electroweak interaction, as well as a strong interaction field and even gravity. Weyl (1929) was perhaps the first to deploy a "gauge principle" to (re)derive the existence and (known) properties of electromagnetism. As he put it,

It seems to me that this new principle of gauge invariance, which follows not from speculation but from experiment, tells us that the electromagnetic field is a necessary accompanying phenomenon ... of the material wave-field represented by ψ. (quoted from O'Raifeartaigh, 1997, p. 122)

Clearly, it is an empirical claim of the highest importance that there is a field with definite properties responsible for a fundamental interaction, and establishing these claims is a major goal of physics. If a principle of "local" gauge symmetry entails such a claim, it cannot be a purely formal requirement on a theory. Indeed, in the quoted passage, Weyl maintains that one such principle follows from experiment. Does an empirical principle of "local" gauge symmetry entail the existence and features of any fundamental interaction?

Teller (2000) has referred to attempts to motivate the introduction of specific gauge fields along these lines collectively as "the gauge argument." He, Brown (1999), Martin (2002), and others have analyzed the structure of attempts to deduce the existence of a gauge field with definite properties from an empirically based assumption of "local" gauge symmetry. They have concluded that such attempts are at most of heuristic or pedagogical value, and careful authors of textbooks such as Aitchison and Hey (2003) have effectively admitted as much. They write on pages 72–3

... we must emphasize that there is ultimately no compelling logic for the vital leap to a local phase invariance from a global one. The latter is, by itself, both necessary and sufficient in quantum field theory to guarantee local charge conservation. Nevertheless, the gauge principle—deriving interactions from the requirement of local phase invariance—provides a satisfying conceptual unification of the interactions present in the Standard Model.

It is worthwhile examining the gauge argument more closely to show in detail why it fails to refute the thesis that "local" gauge symmetry is a purely formal feature of a theory. There is also a secondary motivation. One's appreciation of a good conjuring trick may well be enhanced rather than diminished by learning exactly how it was executed. I begin by taking as stalking horse a version of the gauge argument given by Ryder (1996, p. 90–7) in a widely used introductory text.[2]

---

[2] I claim no originality for this section's critique of the gauge argument, which is largely based on the work of Teller (2000), Brown (1999), Martin (2002) and Aitchison and Hey (2003). It would be useful

6.3 THE GAUGE ARGUMENT

Consider once more a complex scalar field ϕ satisfying the Klein–Gordon equation and its conjugate

$$\partial_\mu \partial^\mu \phi + m^2 \phi = 0 \qquad (6.2)$$

$$\partial_\mu \partial^\mu \phi^* + m^2 \phi^* = 0 \qquad (6.3)$$

As noted in chapter 1, these field equations may be derived as Euler–Lagrange equations by requiring that the action associated with the Lagrangian density

$$\mathcal{L}_0 = (\partial_\mu \phi)(\partial^\mu \phi^*) - m^2 \phi^* \phi \qquad (6.4)$$

be stationary under independent variations in ϕ, ϕ^*. Since this action is invariant under the transformation (cf. equation 1.16)

$$\phi \to \exp(i\Lambda)\phi \qquad (6.5)$$

with constant Λ, Noether's first theorem implies that there will be a conserved current

$$J^\mu = i(\phi^* \partial^\mu \phi - \phi \partial^\mu \phi^*) \ : \ \partial_\mu J^\mu = 0 \qquad (6.6)$$

and associated conserved Noether charge N

$$N = \int J^0 d^3x \qquad (6.7)$$

Now suppose that one has empirical reasons to believe that ϕ represents charged matter, and that its charge Q is conserved. One can associate this conservation of charge with conservation of the Noether charge by taking Q to equal eN, where e may be identified with the charge of the quanta of the field's quantized counterpart. In that case, experiments confirming conservation of charge yield empirical evidence for the constant phase invariance of $\mathcal{L}_0$.

The next critical step of the argument is to extend this "global" invariance of the matter Lagrangian (density) $\mathcal{L}_0$ to "local" invariance under transformations of the form (cf. equation 1.21)

$$\phi \to \exp(i\Lambda(x))\phi \qquad (6.8)$$

to deepen our understanding by analyzing various versions of "the" gauge principle in the literature to see whether any could form the basis for a more convincing gauge argument. I commend this task to the interested reader.

in which Λ is permitted to be an arbitrary smooth function of the space-time coordinates $x^\mu (\equiv x)$. Ryder (1996, p. 93) attempts to justify this step by appeal to "the letter and spirit of relativity," which allegedly rules out the transformation 6.5 by preventing us from performing the same transformation everywhere at once. He continues

To get round this problem we simply abandon the requirement that Λ is a constant, and write it as an arbitrary function of space-time, $\Lambda(x)$.

But this cannot get round any problem, in so far as a constant phase transformation (such as 6.5) is a special case of a variable phase transformation (such as 6.8). And relativity raises no such problem, since constant phase invariance does not require that anyone be able to perform the transformation 6.5. Since the representation of the matter field is determined only up to a constant overall phase, it is unclear how any physical operation could correspond to this transformation. Even if constant phase invariance did reflect an empirical symmetry, this might be observed by comparing pre-existing situations rather than by physically transforming one situation into another.

But in fact constant phase invariance does not reflect an empirical symmetry (observable or not), since the phase relations between two distinct matter fields are not defined—at most relative phases within a single field could turn out to be empirically significant. The "global" invariance of the matter Lagrangian density has empirical content, but this is only indirectly manifested in observations of conservation of the associated Noether charge.

There may be better ways than Ryder's of justifying the extension from constant to variable generalized phase invariance of the Lagrangian (density). Auyang (1995) follows Weyl (1929) (see O'Raifeartaigh 1997 and Yang and Mills 1954) in trying to justify the extension from constant to variable phase invariance as an abandonment of the assumption that meaningful comparisons even of relative phases may be made without adoption of some prior convention as to what is to count as the same phase at different space-time points. The fiber bundle formulation of a gauge theory described earlier gives content to this line of argument. Within that formulation, one can regard a choice of section in the bundle of phases as implementing the required convention. Alternatively, the matter field may itself be represented by a section in a vector bundle associated to a principal bundle with the structure group of the theory (U(1) in the case of electromagnetism), in which case the convention is implemented by choosing one section rather than another of the principal bundle, thereby specifying the phase of the field the vector bundle section represents (see figure 1.6).[3] But understood in this way, a variable generalized

---

[3] The gauge argument has not to this point offered any reason for why the connection or curvature of the auxiliary principal bundle in this second representation should itself be taken to represent any physical field.

6.3 THE GAUGE ARGUMENT

phase transformation corresponds merely to a change in the way a single situation of the matter field is represented. Contrary to Weyl's assertion, any corresponding "local" gauge invariance in a theory of this field follows neither from speculation nor from experiment, but simply from a conventional system of representation.

This becomes clearer when we consider the next step in the gauge argument, which supposedly first introduces a gauge field in order to preserve "local" gauge invariance. Under the variable phase transformation 6.8, the Lagrangian density $\mathcal{L}_0$ is transformed into $\mathcal{L}'_0$, where

$$\mathcal{L}'_0 = \mathcal{L}_0 + (\partial_\mu \Lambda)[-i\phi \partial^\mu \phi^* + i\phi^* \partial^\mu \phi] \quad (6.9)$$
$$= \mathcal{L}_0 + (\partial_\mu \Lambda) J^\mu$$

and so variable phase invariance has failed. One can avoid the unwanted term by introducing an additional term $\mathcal{L}_1$ in the Lagrangian density involving a new magnitude C_μ to give

$$\mathcal{L} = \mathcal{L}_0 + \mathcal{L}_1 \quad (6.10)$$

where

$$\mathcal{L}_1 = -J^\mu C_\mu \quad (6.11)$$

This will cancel the unwanted term $(\partial_\mu \Lambda) J^\mu$ in $\mathcal{L}'_0$ if the variable phase transformation in φ is accompanied by the following transformation in C_μ:

$$C_\mu \rightarrow C'_\mu = C_\mu + \partial_\mu \Lambda \quad (6.12)$$

which is of exactly the same form as the gauge transformation 1.6. But $\mathcal{L}_0 + \mathcal{L}_1$ is still not invariant under this joint transformation: it is necessary to add one further term $\mathcal{L}_2$, where

$$\mathcal{L}_2 = C_\mu C^\mu \phi^* \phi \quad (6.13)$$

Now $\mathcal{L}_0 + \mathcal{L}_1 + \mathcal{L}_2$ is invariant under the combination of a variable phase transformation 6.8 and an equation 6.12 that closely resembles a gauge transformation in the electromagnetic potential A_μ. It is convenient to write this Lagrangian as

$$\mathcal{L}_0 + \mathcal{L}_1 + \mathcal{L}_2 = (D_\mu \phi)(D^\mu \phi^*) - m^2 \phi^* \phi \quad (6.14)$$

where $D_\mu \phi \equiv (\partial_\mu + iC_\mu)\phi$, $D_\mu \phi^* \equiv (\partial_\mu - iC_\mu)\phi^*$. D_μ is the covariant derivative operator (see appendix B), so called since $D_\mu \phi$ transforms in the same way as ϕ, and $D_\mu\phi^*$ transforms in the same way as ϕ^*, under 6.8, 6.12. It is just such a replacement in a matter field Lagrangian of an ordinary derivative operator ∂_μ by a covariant derivative operator like D_μ that O'Raifeartaigh (1997, pp. 6 and 118) calls the gauge principle.

It is important to appreciate that thus far no reason has been offered beyond a suggestive similarity of form between 6.12 and 1.6 for relating the magnitude C_μ to the gauge potential A_μ of electromagnetism. As Teller (2000), Brown (1999), and Aitchison and Hey (2003) all note, there is as yet no reason to suppose that C_μ corresponds to the potential for any new physical field: it may simply be an artefact of extending the models of the theory of the matter field alone to those corresponding to arbitrary choices of section—of the bundle of phases, or of the principal bundle to which the vector bundle is associated. O'Raifeartaigh (1997, p. 118) argues that C_μ must correspond to some field, "since otherwise [it] is reduced to a Lagrange multiplier, and the theory is constrained." More precisely, he offers this as a reason to add a further kinetic term to the Lagrangian, representing the energy of the field associated with C_μ. Such a term must itself be invariant under 6.12 to preserve overall invariance, and this suggests (but certainly does not entail) the addition to the Lagrangian density of a term of the form

$$\mathcal{L}_3 = \lambda E^{\mu\nu} E_{\mu\nu} \tag{6.15}$$

where

$$E_{\mu\nu} \equiv \partial_\mu C_\nu - \partial_\nu C_\mu \tag{6.16}$$

and λ is some constant, for $\mathcal{L}_3$ is Lorentz invariant as well as invariant under 6.12. Again, it is tempting to relate $E_{\mu\nu}$ to the electromagnetic tensor $F_{\mu\nu}$ of 1.7: this temptation is further increased if one notes that variation of the action associated with $\mathcal{L}_3$ with respect to C_μ yields the following analog to the source-free Maxwell equations governing the electromagnetic field:

$$\partial_\nu E^{\mu\nu} = 0 \tag{6.17}$$

But note that nothing in the argument so far shows that either $E_{\mu\nu}$ or C_μ represents any new physical field. For it is consistent with the foregoing that $E_{\mu\nu} \equiv 0$, so that any such hypothetical field would carry no energy or momentum, and would not physically interact in any way with the matter field.

Nevertheless, if one had not already known of the existence of the electromagnetic field, these considerations might have provided heuristic reasons to consider the possibility that there is indeed a new physical field corresponding

6.3 THE GAUGE ARGUMENT

to $E_{\mu\nu}$ and C_μ, so that in certain circumstances $E_{\mu\nu} \neq 0$. And of course we do know that electromagnetism *is* such a field, with

$$\partial_\nu F^{\mu\nu} = -J^\mu \ : \ J^\mu \equiv (\rho, \mathbf{j}) \tag{6.18}$$

Now variation of the action associated with 6.14 with respect to ϕ, ϕ^* yields the following Euler–Lagrange equations:

$$D^\mu D_\mu \phi + m^2 \phi = 0 \tag{6.19}$$
$$D^\mu D_\mu \phi^* + m^2 \phi^* = 0$$

which are just the equations of motion for a charged scalar field with quanta of charge e subjected to an electromagnetic potential A_μ, provided one sets $C_\mu = eA_\mu$. Moreover, variation of the action associated with $\mathcal{L}_{tot} \equiv \mathcal{L}_0 + \mathcal{L}_1 + \mathcal{L}_2 + \mathcal{L}_3$ with respect to A_μ now yields the following Euler–Lagrange equations:

$$\partial_\nu F^{\mu\nu} = -i(\phi^* D^\mu \phi - \phi D^\mu \phi^*) \tag{6.20}$$

if one sets $E_{\mu\nu} = eF_{\mu\nu} : \lambda = -\frac{1}{4e^2}$. These may be identified as Maxwell equations for a conserved Noether current $\mathcal{J}^\mu$ of $\mathcal{L}_{tot}$ that is the natural covariant generalization of 6.6

$$\mathcal{J}^\mu = i(\phi^* D^\mu \phi - \phi D^\mu \phi^*) \ : \ \partial_\mu \mathcal{J}^\mu = 0 \tag{6.21}$$

It is the comparison of already known equations of motion for electromagnetism with the Euler–Lagrange equations derived from 6.14 and $\mathcal{L}_{tot}$, together with the observed conservation of the electric charge $Q = eN$ associated with the Noether charge N of the matter field ϕ that finally connects magnitudes C_μ, $E_{\mu\nu}$ to electromagnetism via the identifications

$$C_\mu = eA_\mu \ : \ E_{\mu\nu} = eF_{\mu\nu} : \lambda = -\frac{1}{4e^2} \tag{6.22}$$

so that the final Lagrangian density for the matter field ϕ interacting with electromagnetism is

$$\mathcal{L}_{tot} = (D_\mu \phi)(D^\mu \phi^*) - m^2 \phi^* \phi - \frac{1}{4} F^{\mu\nu} F_{\mu\nu} \tag{6.23}$$

where the covariant derivatives are now $D_\mu \phi \equiv (\partial_\mu + ieA_\mu)\phi$, $D_\mu \phi^* \equiv (\partial_\mu - ieA_\mu)\phi^*$. An entirely parallel line of reasoning starting from a different matter field, the Dirac field ψ, leads to the Lagrangian density 5.54 for quantum electrodynamics

$$\mathcal{L}_{\text{QED}} = \overline{\psi}(i\gamma^\mu D_\mu - m)\psi - \frac{1}{4}F_{\mu\nu}F^{\mu\nu} \qquad (6.24)$$

There is no sound argument here from an experimentally based principle of gauge invariance to the existence of an electromagnetic interaction whose properties are consequences of Lagrangian densities like $\mathcal{L}_{\text{QED}}$ and $\mathcal{L}_{\text{tot}}$. The argument as examined contains several inadequately defended premises as well as a number of dubious inferences. But there is, perhaps, a weak sense in which the extension of a principle of constant generalized phase symmetry of matter fields, indirectly supported by observations of charge conservation, explains these properties of electromagnetism, *assuming* the existence of an interaction field for which this extension makes room. For if there is such a field, then one achieves a dramatic simplification and unification of theory by identifying it with electromagnetism in accordance with 6.22. Moreover, it turns out that further explanatory unification may be achieved by following similar heuristic lines of argument in analogous cases, starting from observed conservation of magnitudes that may be associated with Noether charges, derived from variation of the action corresponding to matter field Lagrangian densities under constant generalized phase transformations.

For example, the properties of the strong interaction derived from the quantum chromodynamics Lagrangian density 5.59 may be explained by starting from a symmetry of the Lagrangian density for free quarks 5.60 under "global" SU(3) transformations that may be associated with conservation of color charge (though it is much less clear how well experiments support this conservation). Again, this "global" symmetry may be rendered "local" (or "gauged") without entailing the existence of any accompanying interaction: but if there is such an interaction, then many of its properties follow from the requirement that the combined matter/force field Lagrangian be "locally" gauge-invariant. But this example highlights a further lacuna in the gauge argument, for 5.59 is not the only "locally" gauge-invariant Lagrangian density that one can arrive at by gauging the "global" SU(3) symmetry of the quark field.

In the case of electromagnetism, addition of the kinetic term $-\frac{1}{4}F_{\mu\nu}F^{\mu\nu}$ to arrive at $\mathcal{L}_{\text{tot}}$ could be justified on several grounds. By itself, this term leads to the familiar source-free Maxwell equations. But one could add further terms to $\mathcal{L}_{\text{tot}}$ without sacrificing "local" gauge invariance. However, various possibilities are ruled out by additional constraints, importantly including the requirement that the total Lagrangian lead to a renormalizable theory. O'Raifeartaigh (1979) showed that the minimal addition of $-\frac{1}{4}F_{\mu\nu}F^{\mu\nu}$ yields the simplest, renormalizable, Lorentz- and "locally" gauge-invariant Lagrangian yielding second-order equations of motion for the coupled system. So while the presence of this lacuna further undermines the soundness of the gauge argument, it does little to weaken the associated explanation of the properties of electromagnetism.

But in the case of quantum chromodynamics, it has proved necessary to consider addition of a further term to 5.59 proportional to

$$\epsilon_{\mu\nu\rho\sigma} F^{a\mu\nu} F^{a\rho\sigma} \tag{6.25}$$

where $\epsilon_{\mu\nu\rho\sigma} = +1$ [for $\mu\nu\rho\sigma$ an even permutation of 0123], $\epsilon_{\mu\nu\rho\sigma} = -1$ [for $\mu\nu\rho\sigma$ an odd permutation of 0123], $\epsilon_{\mu\nu\rho\sigma} = 0$ [otherwise].[4] This term preserves "local" gauge invariance, Lorentz invariance, and renormalizability; but nothing in the "logic" of the gauge argument dictates its presence. If it proves necessary (or even possible) to throw it in as an afterthought, this weakens an explanation of the properties of the strong interaction which takes these to be required by gauge symmetry.

To sum up, while the gauge argument effects a significant explanatory unification among the properties of diverse fundamental interactions, it certainly does not dictate their very existence. And while observations of charge conservation may yield indirect support for an empirical constant phase symmetry of matter fields, the gauge argument neither rests on nor entails a principle of "local" gauge symmetry with any empirical import, direct or indirect. "Local" gauge symmetry is a theoretical, not an empirical, symmetry. It is merely a feature of the way gauge theories of electromagnetic, electroweak, and strong interactions are conventionally formulated.

6.4 Ghost fields

As noted in the previous chapter (section 5.8), a standard procedure for conveniently quantizing a non-Abelian Yang–Mills gauge field—say, with structure group SU(2)—involves the addition of two terms to the action $S[J]$ corresponding to the Yang–Mills Lagrangian density $\mathcal{L}_{YM}$ in the presence of a source J:

$$S_{\text{eff}} = S[J] + S_{\text{gf}} + S_{\text{FPG}} \tag{6.26}$$

The first term, S_{gf}, is introduced conventionally as a way of fixing the gauge. It is a function only of the Yang–Mills field $\mathbf{A}_\mu$ and the gauge-fixing function

---

[4] The term arises in case the parameter θ characterizing the eponymous θ-vacuum is non-zero (see section 6.6). The fact that strong interactions appear to be symmetric under the joint transformations C (charge conjugation) and P (parity), even though including such an additional term violates CP symmetry, is known as the strong-CP problem for QCD.

f_a: it is therefore not only gauge dependent, but also involves no new fields, so there is no temptation to endow its presence in the effective action with more than formal significance. But in the path integral 5.52

$$W_f[J] = N' \int D\mathbf{A}_\mu D\mathbf{c} D\mathbf{c}^\dagger \exp\left(iS_{\text{eff}}[J]\right) \tag{6.27}$$

the new fields $\mathbf{c}, \mathbf{c}^\dagger$ both contribute to the definitions of the paths over which the integration is taken and also apparently enter into a new term in the Lagrangian density for the action S_{FPG} in a way that physical fields characteristically do. This makes them look like new physical fields, whose quanta might be expected to be experimentally detectable. Moreover, it is often convenient to apply a (covariant) gauge-fixing condition, after which $\mathbf{c}, \mathbf{c}^\dagger$ are treated as fields on a par with the Yang–Mills field $\mathbf{A}_\mu$ when performing perturbative calculations necessary to make empirical predictions. But there are also other gauges, including the so-called axial gauge, in which the term M_f in 5.51 is independent of $\mathbf{A}_\mu$, and the path integral of S_{FPG} over $\mathbf{c}, \mathbf{c}^\dagger$ gives a constant which merely readjusts the normalization of $W_f[J]$. Consequently, $\mathbf{c}, \mathbf{c}^\dagger$ appear as fields in some gauges but not others.

If these are indeed new physical fields that are manifest in some gauges but not in others, then choice of gauge would have empirical consequences, and "local" gauge symmetry could not be a purely formal feature of a Yang–Mills gauge theory. One could hope to determine experimentally whether a system was in the axial gauge rather than a covariant gauge by attempting to observe quanta of these fields.

But that would be a mistake: it is not for nothing that they are called ghost fields! For even in a gauge in which $\mathbf{c}, \mathbf{c}^\dagger$ figure in calculations of observable effects, those effects do not include any that would either directly or indirectly reveal the presence or properties of these fields. In particular, one can show that ghost quanta—quanta of ghost fields—have no properties that would permit one to observe their presence. Moreover, the existence of fields like $\mathbf{c}, \mathbf{c}^\dagger$ is inconsistent with the spin statistics theorem, a fundamental theorem of relativistic quantum field theory. For the $\mathbf{c}, \mathbf{c}^\dagger$ fields would have to describe fermions with zero spin, while the theorem implies that fermions cannot have integral spin. Taking the requirements of relativistic quantum field theory seriously, ghost fields are not even candidates for physical reality.

Weingard (1988) makes an illuminating comparison between ghost fields and the electromagnetic potential A_μ in a simply connected region outside a perfectly shielded solenoid in the Aharonov–Bohm effect. In each case, one can choose to adopt a gauge in which the magnitude appears, or a gauge in which it does not appear. And in each case, this shows that the magnitude represents nothing physically real. There remains a crucial difference between the two phenomena, however. The ghost fields may be made to disappear everywhere by a suitable choice of gauge, but there is no choice of gauge that makes (a single-valued) A_μ zero *everywhere* outside the solenoid when a current flows through it. Ghost fields, like Coriolis "forces," are merely an artefact of a conventional choice of a system of theoretical representation: really, there are no such things, even if the theory is true. A_μ does represent something real, but in a potentially misleading way. For what it represents is non-separable, and present outside a perfectly shielded solenoid in the Aharonov–Bohm effect only in multiply connected regions, while the representation is in terms of a magnitude that takes a value at every point.

6.5 Spontaneous symmetry-breaking

The gauge argument may help to explain various features of the electromagnetic, weak, and strong interactions. But one thing it does not do is to explain why only the electromagnetic interaction gives rise to long-range forces like the $\frac{1}{r^2}$ Coulomb force. This is typically taken to be a consequence of the fact that this force is mediated by a field whose quanta are *massless* bosons, namely photons. On the other hand, the weak and strong interactions are known to give rise to short-range forces that are observed to fall off rapidly with increasing distance, and heuristic considerations take this to imply that they must be mediated by massive particles. However, Lagrangian (densities) such as 5.54 and 5.59 that are motivated by the gauge argument do not contain any term associating a mass with the quanta of the gauge fields they describe that would be analogous to the term $-m^2\phi^*\phi$ in the Lagrangian 6.4 or 6.23 that represents the (bare) mass m of the quanta associated with the matter field ϕ. Moreover, if one were to add an analogous term such as $-m^2 A_\mu^a A^{a\mu}$ to 5.59 in the attempt to allow the A_μ^a to represent a massive gauge field, the resulting Lagrangian would *fail* to be "locally" gauge invariant (and also non-renormalizable).

In the case of the strong interaction, it proved unnecessary to introduce a massive gauge field, since the short range of the strong force was in the end explained differently by quantum chromodynamics. In this theory, gluons—the "colored" quanta of the strong interaction gauge field—are indeed taken to be massless, and the strong force acts directly not on hadrons like protons, neutrons, and pi mesons, but rather on their bound quark constituents. It is therefore the resultant force between such hadrons that falls off rapidly with distance, even though the underlying force between quarks is long range, and even increases with increasing distance, explaining why observed hadrons consist only of color-neutral combinations of their confined quark constituents.

But to understand the behavior of the weak interaction it did turn out to be necessary to acknowledge the existence of massive vector bosons as mediators of the weak force, apparently in the face of the gauge argument.

Attempts to add a term like $-m^2 A^a_\mu A^{a\mu}$ to a weak interaction Lagrangian failed: the resulting theory proved to be non-renormalizable, and so failed to yield sensible predictions within perturbation theory.

Analogies with various thermodynamic phenomena in condensed matter physics suggested another way for a massless field to acquire mass—namely, spontaneous symmetry breaking. The basic idea is simple enough. The laws of a theory may describe the behavior of a system that is not itself symmetric under a transformation that is a symmetry of those laws. For example, a model of a time-symmetric theory may fail to be time symmetric. While a symmetry of the laws must map each model of the theory into a model of the theory, it need not map each model into itself. In particular, the lowest-energy, or ground, state of a system may not be symmetric under a symmetry of the theory that describes it. A more detailed analysis of such systems as ferromagnets, superconductors, and superfluids revealed situations in which the spontaneous breaking of a symmetry is associated with the following phenomenon: If that system is perturbed from its ground state, then it manifests a kind of inertia that is analogous to that associated with mass. This suggested that gauge fields might acquire mass through spontaneous breaking of some symmetry, and more specifically that a massless Yang–Mills gauge field might acquire mass as a consequence of a spontaneous breaking of gauge symmetry.

Initial attempts to implement this suggestion ran into a serious difficulty, namely Goldstone's theorem. This states (roughly) that the spontaneous breaking of any continuous "global" symmetry of a quantum field theory is associated with the introduction of a number of *massless* quanta of associated quantum fields—so-called Goldstone bosons. Since no corresponding particles associated with weak interactions were observed, this seemed to rule out spontaneous symmetry breaking as a way of introducing mass into a gauge theory of weak interactions.

But then a way around Goldstone's theorem presented itself in what became known as the Higgs mechanism after one of its discoverers. As an example of the Higgs mechanism, consider the theory one arrives at by gauging the following theory of a self-interacting complex scalar field ϕ whose Lagrangian density is symmetric under "global" U(1) phase transformations:

$$\mathcal{L}_0 = (\partial_\mu \phi)(\partial^\mu \phi^*) - m_0^2 \phi^* \phi - \lambda_0 \left(\phi^* \phi\right)^2 \tag{6.28}$$

where m_0, λ_0 are constants, and $\phi = \frac{1}{\sqrt{2}}(\phi_1 + i\phi_2)$. Considered as a function of classical fields, if $m_0^2 < 0$, the potential term $m_0^2 \phi^* \phi + \lambda_0 (\phi^* \phi)^2$ in $\mathcal{L}_0$ has a minimum at $|\phi|^2 = -\frac{m_0^2}{2\lambda_0}$. This corresponds to a non-zero vacuum expectation value $\langle \hat{\phi} \rangle^2 = -\frac{m_0^2}{2\lambda_0}$ of the corresponding quantum field $\hat{\phi}$ in any of a continuous infinity of degenerate ground states in which the continuous "global" U(1) symmetry of $\mathcal{L}_0$ has been spontaneously broken. Goldstone's theorem implies the existence of a single massless scalar boson field $\phi_2' = \phi_2$, and this is accompanied by a massive scalar boson field $\phi_1' \equiv \phi_1 - v$ of mass $\sqrt{-2m_0^2}$, where $v \equiv \sqrt{-\frac{m_0^2}{\lambda_0}}$.

The gauge argument motivates extending the constant phase invariance of $\mathcal{L}$ to "local" gauge invariance by requiring symmetry under variable as well as constant phase transformations, yielding

$$\mathcal{L} = (D_\mu \phi)(D^\mu \phi^*) - m_0^2 \phi^* \phi - \lambda_0 \left(\phi^* \phi\right)^2 - \frac{1}{4} F^{\mu\nu} F_{\mu\nu} \quad (6.29)$$

where D_μ is the covariant derivative. If one makes the substitution

$$\phi = \exp\left(\frac{i\xi}{v}\right) \frac{v + \eta}{\sqrt{2}} \quad (6.30)$$

and neglects third- and higher-order terms, this Lagrangian becomes

$$\mathcal{L} \approx -\frac{1}{4} F^{\mu\nu} F_{\mu\nu} + \frac{1}{2} \partial_\mu \eta \partial^\mu \eta + \frac{1}{2} \partial_\mu \xi \partial^\mu \xi + m_0^2 \eta^2$$
$$+ \frac{1}{2} e^2 v^2 A_\mu A^\mu + e v A_\mu \partial^\mu \xi - \frac{1}{4} v^2 m_0^2 \quad (6.31)$$

Because of the cross-term $evA_\mu \partial^\mu \xi$, this does not yet have any obvious interpretation in terms of separate fields and the masses of their quanta. But if one makes the following transformation:

$$A_\mu' = A_\mu + \frac{1}{ev} \partial_\mu \xi \quad (6.32)$$
$$\phi' = \frac{v + \eta}{\sqrt{2}}$$

then 6.29 becomes

$$\mathcal{L} = -\frac{1}{4} F^{\mu\nu} F_{\mu\nu} + \frac{1}{2}(\partial_\mu \eta \partial^\mu \eta + 2m_0^2 \eta^2) + \frac{1}{2} e^2 v^2 A_\mu' A'^\mu$$
$$- \lambda_0 v \eta^3 - \frac{1}{4} \lambda_0 \eta^4 + \frac{1}{2} e^2 A_\mu' A'^\mu \eta (2v + \eta) - \frac{1}{4} v^2 m_0^2 \quad (6.33)$$

Neglecting terms of third and higher order, this gives

$$\mathcal{L} \approx -\frac{1}{4} F^{\mu\nu} F_{\mu\nu} + \frac{1}{2}(\partial_\mu \eta \partial^\mu \eta + 2m_0^2 \eta^2) + \frac{1}{2} e^2 v^2 A_\mu' A'^\mu \quad (6.34)$$

which is just the Lagrangian density for a scalar field η of mass $\sqrt{-2m_0^2}$ together with a *massive* vector field A_μ' of mass $ev = \sqrt{-\frac{m_0^2 e^2}{\lambda_0}}$. Hence, in the presence of the scalar field ϕ, the massless vector field A_μ has acquired a mass. This is a simple example of the Higgs mechanism: textbook discussions often describe it by saying that A_μ has acquired a mass by "swallowing" the Goldstone boson ϕ'_2 field, which no longer figures in the final Lagrangian. Note that as a massive vector field, A'_μ has one extra degree of freedom than the massless field A_μ. So it is as if A_μ has not only put on weight, but also increased its abilities as a result of this "consumption"!

The transformation 6.32 has the form of a "local" gauge transformation. So it may seem that the acquisition of mass by the vector field A_μ occurs only in a particular gauge, and that this is the gauge resulting from spontaneous breaking of "local" gauge symmetry. An analogous, but more complex, instance of the Higgs mechanism figures in the unified electroweak theory incorporated in the Standard Model of elementary particles, where it is taken to give mass to three otherwise massless vector fields ($W^\pm$ and Z^0) as a result of "swallowing" three scalar Goldstone bosons arising from spontaneous symmetry breaking of gauge symmetry by the Higgs mechanism. In this case, the masses are manifested in what is called the unitary gauge. So it seems that by spontaneously breaking "local" gauge symmetry, nature picks out the unitary gauge as physically significant: the $W^\pm$ and Z^0 particles appear only in this gauge, and experiments reveal them to be real, so (failure of) "local" gauge symmetry has detectable empirical consequences!

But this argument is confused. A first step toward removing the confusion is to analyze more carefully how spontaneous symmetry breaking arises in a quantum field theory, so as better to appreciate the significance, but also the limitations, of Goldstone's theorem. As Earman has argued, spontaneous symmetry breaking in this context may be analyzed in terms of the algebraic approach to quantum field theory.[5]

Rather than starting from a set of quantum field operators acting on a fixed Hilbert space of states, one thinks of a quantum field theory associated with a Lagrangian density like 6.28 as defined by an abstract Weyl algebra $\mathcal{W}$ of "observables" with abstract states defined on it, where a state is just a (normed, positive) linear functional $\omega : \mathcal{W} \to \mathbb{C}$. One can retrieve a formulation in terms of self-adjoint operators acting on a Hilbert space of concrete states by finding a representation of $\mathcal{W}$—a structure-preserving map $\pi : \mathcal{W} \to \mathcal{B}(\mathsf{H})$ into the set of bounded operators on a Hilbert space H. But there are many such representations, including many that are not unitarily equivalent.[6] It follows from the Gelfand–Naimark–Segal (GNS) theorem (see appendix E) that each abstract state ω itself defines a so-called GNS representation $(\mathsf{H}_\omega, \pi_\omega)$; but distinct abstract states may define representations that are not unitarily equivalent.

---

[5] Appendix E provides a brief introduction to algebraic quantum field theory. Earman has championed this approach in (Earman 2003, 2004).

[6] A Weyl algebra is a C^* algebra. Two representations π, π' of an abstract C^* algebra $\mathcal{A}$ are *unitarily equivalent* if and only if there is a unitary map $U : \mathcal{B}(\mathsf{H}_\pi) \to \mathcal{B}(\mathsf{H}_{\pi'})$ such that $\pi'(\hat{A}) = U\pi(\hat{A})U^{-1}$ for all $\hat{A} \in \mathcal{A}$. For details, see appendix E.

6.5 SPONTANEOUS SYMMETRY-BREAKING

A symmetry of a Lagrangian density such as 6.28 corresponds in the algebraic approach to an automorphism θ of the Weyl algebra $\mathcal{W}$ associated with that Lagrangian density—a structure-preserving mapping of $\mathcal{W}$ onto itself. The symmetry θ is said to be *unitarily implementable* in state ω if and only if there is a unitary operator $\hat{U}$ on the Hilbert space H_ω determined by the GNS representation determined by ω such that $\pi_\omega(\theta(W)) = \hat{U}\pi_\omega(W)\hat{U}^{-1}$ for all $W \in \mathcal{W}$. It may turn out that a symmetry of the Lagrangian density is not unitarily implementable in some state ω: if it does, then we have a case of spontaneous symmetry breaking. Moreover, as Earman (2004) shows, an automorphism θ is unitarily implementable in state ω if and only if states ω and $\widehat{\theta\omega} \equiv \omega \circ \theta$ define unitarily equivalent GNS representations. So if there are vacuum state vectors in both Hilbert spaces of the GNS representations determined by ω, $\widehat{\theta\omega}$, then there is a degeneracy of the vacuum in the sense that these vacuum state vectors belong to unitarily inequivalent representations of $\mathcal{W}$, and so appear in distinct Hilbert spaces.

Physicists often express these results without the language of the algebraic approach. They say, for example, that even though there are infinitely many distinct lowest-energy states of the field related by "global" gauge transformations, these cannot be connected by a unitary transformation: the states do not lie in the same (superselection sector of the) Hilbert space, and consequently may not be superposed. "Global" gauge symmetry is then broken by the field's being in one of these vacuum states rather than another—or, indeed, in a low-lying excitation of just one of these vacuum states.

The Goldstone theorem applies only to "global" gauge transformations. More precisely, it applies to continuous symmetries of the Lagrangian density generated by elements of a finite-parameter Lie group. Noether's first theorem shows that, in this case, it is a consequence of the laws of motion that there will be a conserved current, and corresponding conserved Noether charge. But when such a "global" gauge symmetry is made "local," as in moving from 6.28 to 6.29, Noether's second theorem applies, for the symmetries of the Lagrangian are now generated by a Lie group parametrized by a finite set of functions, rather than scalar parameters. This theorem shows that in such a case there is no additional conserved charge, since the conservation of the current expresses a trivial mathematical identity, but that the equations of motion are not independent of each other, and so initial data do not determine a unique solution. Such a situation is conveniently handled by treating the system as a constrained Hamiltonian system (see appendix C), since this provides an elegant way to exhibit and neutralize the resulting indeterminism.

In the constrained Hamiltonian formulation, the constraints to which the system is subject define a subspace of the total phase space called the constraint surface. Motion, on this surface, of the phase point representing the state of the system mixes genuine dynamical evolution with mere change of representation. Motion generated by the so-called first-class constraints is taken to be of the

174   6 THE EMPIRICAL IMPORT OF GAUGE SYMMETRY

latter kind, and is associated with gauge freedom. One can try to get rid of gauge redundancy by moving to the so-called reduced phase space, a point of which corresponds to an entire "gauge orbit" through a point on the constraint surface, generated by the first-class constraints.

Earman (2004, p. 190) asks

> What is the upshot of applying this reduction procedure to the Higgs model and them quantizing the resulting unconstrained Hamiltonian system? In particular, what is the fate of spontaneous symmetry breaking? To my knowledge, the reduction has not been carried out.

He takes the fate of spontaneous symmetry breaking in the Higgs model to hinge on the results of this task, noting (ibid. p. 191) that

> While there are too many what-ifs in this exercise to allow any firm conclusions to be drawn, it does suffice to plant the suspicion that when the veil of gauge is lifted, what is revealed is that the Higgs mechanism has worked its magic of suppressing zero mass modes and giving particles their masses by quashing spontaneous symmetry breaking. However, confirming the suspicion or putting it to rest require detailed calculations, not philosophizing.

It would be nice to have the results of such detailed calculations, but we should be astonished if they did *not* confirm Earman's "suspicion." For while physicists often speak loosely of spontaneous symmetry breaking in the Higgs model, more careful authors have noted (t'Hooft 2005, p. 63) that it is misleading to speak of spontaneous breaking of a *"local"* gauge symmetry (cf. Elitzur 1975). As we have seen, in the case of spontaneous breaking of a *"global"* symmetry, there is a genuine degeneracy in the vacuum state of the quantum field. But in the case of a "locally" gauge-symmetric Lagrangian density like 6.29, there is simply no reason to suspect that the vacuum state is degenerate; moreover, the Hilbert space vacuum state is "locally" gauge symmetric, and so there is no spontaneous breaking of "local" gauge symmetry. It is remarkable enough that the conditions for the applicability of Goldstone's theorem are met prior to application of the Higgs mechanism; it would be astonishing if the reduction procedure showed that they are also met *after* its application. Even if these conditions did turn out to be met by the reduced theory, this would not establish that "local" gauge symmetry reflects any corresponding empirical symmetry, since the reduction procedure itself would already have removed this as a theoretical symmetry of the reduced theory.

For the purposes of this chapter, perhaps the most important lesson to draw from this discussion of spontaneous symmetry breaking is this. The manipulations involved in passing from 6.29 to 6.34 are just ways of revealing the empirical content of the original "locally" gauge-symmetric Lagrangian density. They show, in particular, that excitations of the (unique) vacuum state associated with 6.29 will include quanta of a massive vector field as well as those of a massive scalar field. The equations 6.32 do not represent a choice of gauge for fixed fields, but rather definitions of new fields in terms of existing fields. Hence nowhere in the passage from 6.29 to 6.34 was any particular gauge chosen, or "local" gauge symmetry violated.

The Higgs mechanism is a vital part of contemporary physics's explanation of the distinctive properties of the weak interaction. Indeed, it explains not only why the carriers of this interaction are *massive* vector bosons, but also why weakly interacting fermions including the electron have mass. But while that mechanism is closely related to the phenomenon of spontaneous symmetry breaking of "global" gauge symmetry, it does not in fact require any violation of gauge symmetry, either "global" or "local." The profound empirical consequences of the Higgs mechanism do nothing to show that the "local" gauge symmetry of the Yang–Mills gauge theories to which it applies itself has any empirical import. Their "local" gauge symmetry remains a purely theoretical symmetry, with no corresponding empirical symmetry.

6.6 The θ-vacuum

The ground state of a quantized non-Abelian Yang–Mills gauge theory is usually described by a real-valued parameter θ—a fundamental new constant of nature. The structure of this vacuum state is often said to arise from a degeneracy of the vacuum of the corresponding classical theory. The degeneracy allegedly follows from the fact that "large" (but not "small") local gauge transformations connect physically distinct states of zero field energy. In a classical non-Abelian Yang–Mills gauge theory, "large" gauge transformations supposedly connect models of distinct but indistinguishable situations. If this is so, it shows that at least "large" local gauge symmetry is an empirical symmetry.

In clarifying the distinction between "large" and "small" gauge transformations we will be driven to a deeper analysis of the significance of gauge symmetry. But understanding the θ-vacuum will require refining, not abandoning, the thesis that "local" gauge symmetry is a purely theoretical symmetry.

Consider a classical SU(2) Yang–Mills gauge theory with action

$$S = \frac{1}{2g^2} \int \mathrm{Tr}(\mathbf{F}_{\mu\nu}\mathbf{F}^{\mu\nu})\mathrm{d}^4 x \tag{6.35}$$

where

$$\mathbf{F}_{\mu\nu} = \partial_\mu \mathbf{A}_\nu - \partial_\nu \mathbf{A}_\mu + [\mathbf{A}_\mu, \mathbf{A}_\nu] \tag{6.36}$$

and $\mathbf{A}_\mu = A_\mu^j \frac{\sigma_j}{2\mathrm{i}}$ (where σ_j ($j = 1, 2, 3$) are Pauli spin matrices) transform as

$$\mathbf{A}_\mu \to \mathbf{A}'_\mu = \mathbf{U}\mathbf{A}_\mu \mathbf{U}^\dagger + (\partial_\mu \mathbf{U})\mathbf{U}^\dagger, \quad \mathbf{F}_{\mu\nu} \to \mathbf{U}\mathbf{F}_{\mu\nu}\mathbf{U}^\dagger \tag{6.37}$$

under a "local" gauge transformation $\mathbf{U}(\mathbf{x}, t)$.[7] The field energy is zero if $\mathbf{F}_{\mu\nu} = 0$: this is consistent with $\mathbf{A}_\mu = 0$ and gauge transforms of this. Now restrict attention to those gauge transformations for which $\mathbf{A}'_0 = 0$, $\partial_0 \mathbf{A}'_j = 0$, i.e.

$$\mathbf{A}_\mu = 0 \rightarrow \mathbf{A}'_j(\mathbf{x}) = \{\partial_j \mathbf{U}(\mathbf{x})\}\mathbf{U}^\dagger(\mathbf{x}), \quad \mathbf{A}'_0 = 0 \quad (6.38)$$

These are generated by functions $\mathbf{U} : \mathbb{R}^3 \rightarrow \mathrm{SU}(2)$. Those functions that satisfy $\mathbf{U}(\mathbf{x}) \rightarrow \mathbf{1}$ for $|\mathbf{x}| \rightarrow \infty$ constitute smooth maps $\mathbf{U} : S^3 \rightarrow \mathrm{SU}(2)$, where S^3 is the three-sphere. Some of these may be continuously deformed into the identity map $\mathbf{U}(\mathbf{x}) = \mathbf{1}$. But others cannot be: suppose, for example, $\mathbf{A}'_\mu$ is related to $\mathbf{A}_\mu$ by

$$\mathbf{U}(\mathbf{x}) = \exp i\pi\{(\boldsymbol{\sigma}\cdot\hat{\mathbf{x}})F(|\mathbf{x}|) + \sigma_3\}, \quad \text{where } F(|\mathbf{x}|) = \frac{|\mathbf{x}|}{\sqrt{|\mathbf{x}|^2 + \rho^2}} \quad (6.39)$$

for arbitrary positive ρ. The maps divide into a countable set of equivalence classes, each characterized by an element of the homotopy group $\pi_3(\mathrm{SU}(2)) = \mathbb{Z}$ called the *winding number*. Maps in the same equivalence class as the identity map are said to generate "small" "local" gauge transformations; these are taken to relate alternative representations of the same classical vacuum. But $\mathbf{A}'_\mu$, $\mathbf{A}''_\mu$ generated from $\mathbf{A}_\mu = 0$ by maps $\mathbf{U}(\mathbf{x})$ from different equivalence classes are often said to represent *distinct* classical vacua, and $\mathbf{A}'_\mu$, $\mathbf{A}''_\mu$ are said to be related by "large" gauge transformations. (It is important to distinguish this claim from the quite different proposition considered in the previous section, according to which degenerate *quantum* vacua may be related by a *"global"* gauge transformation in cases of spontaneous symmetry breaking. We are concerned at this point with a possible degeneracy in the *classical* vacuum of a non-Abelian Yang–Mills gauge theory.)

But if "local" gauge symmetry is a purely formal feature of a theory, then a gauge transformation cannot connect representations of physically distinct situations, even if it is "large"! And yet, textbook discussions of the *quantum* θ-vacuum typically represent this by a superposition of states, each element of which is said to correspond to a distinct state from the degenerate classical vacuum.

Such discussions frequently appeal to a simple analogy from elementary quantum mechanics. Consider a particle moving in a one-dimensional periodic potential of finite height, like a sine wave. Classically, the lowest-energy state is infinitely degenerate: the particle just sits at the bottom of one or other of

---

[7] Compare 3.23, 3.14, 3.27; or 3.42, 3.38, 3.47 respectively. All differences in signs and numerical factors are solely the result of differing arbitrary choices in how $\mathbf{A}_\mu$, $\mathbf{F}_{\mu\nu}$, and $\mathbf{U}$ have been defined in each case.

the identical wells in the potential. But quantum mechanics permits tunneling between neighboring wells, which removes the degeneracy. In the absence of tunneling, there would be a countably infinite set of degenerate ground states of the form $\psi_n(x) = \psi_0(x - na)$, where a is the period of the potential. These are related by the translation operator $\hat{T}_a$: $\hat{T}_a\psi(x) = \psi(x - a)$. $\hat{T}_a$ is unitary and commutes with the Hamiltonian $\hat{H}$. Hence there are joint eigenstates $|\theta\rangle$ of $\hat{H}$ and $\hat{T}_a$ satisfying $\hat{T}_a |\theta\rangle = \exp(i\theta) |\theta\rangle$.

Such a state has the form

$$|\theta\rangle = \sum_{n=-\infty}^{+\infty} \exp\{-in\theta\} |n\rangle \qquad (6.40)$$

where $\psi_n(x)$ is the wave-function of state $|n\rangle$. When tunneling is allowed for, the energy of these states depends on the parameter $\theta \in [0, 2\pi)$. It is as if quantum tunneling between the distinct classical ground states has removed the degeneracy, resulting in a spectrum of states of different energies parametrized by θ, each corresponding to a different superposition of classical ground states.

An alternative analogy is provided by a charged pendulum swinging from a long, thin solenoid whose flux Φ is generating a static Aharonov–Bohm potential **A**. The Hamiltonian is

$$\hat{H} = \frac{1}{2m}[-i(\nabla - ie\mathbf{A})]^2 + V \qquad (6.41)$$

With a natural "tangential" choice of gauge for **A** this becomes

$$\hat{H} = -\frac{1}{2ml^2}\left(\frac{d}{d\omega} - ielA\right)^2 + V(\omega) \qquad (6.42)$$

where the pendulum has mass m, charge e, length l, and angle coordinate ω. If the wave-function is transformed according to

$$\psi(\omega) = \exp\left[ie\int_0^\omega lA d\omega'\right] \varphi(\omega) \qquad (6.43)$$

then the transformed wave-function satisfies the Schrödinger equation with simplified Hamiltonian

$$\hat{H}_\varphi = -\frac{1}{2m}\frac{d^2}{d\omega^2} + V(\omega) \qquad (6.44)$$

The boundary condition $\psi(\omega + 2\pi) = \psi(\omega)$ now becomes

$$\varphi(\omega + 2\pi) = \exp\{-ie\Phi\}\varphi(\omega) \qquad (6.45)$$

which is of the same form as in the first analogy: $\hat{T}_{2\pi}\varphi = \exp\{i\theta\}\varphi$, with $\theta = -e\Phi$.

Unlike the periodic potential, the charged pendulum features a *unique* classical ground state. The potential barrier that would have to be overcome to "flip" the pendulum over its support can be tunneled through quantum mechanically, but the tunnel ends up back where it started from! This produces a θ-dependent ground-state energy as in the analogy of the periodic potential. But in this case there is a *single* state corresponding to an *external* parameter θ rather than a spectrum of states labeled by an internal parameter θ.

But which is the better analogy? Is the θ-vacuum in a quantized non-Abelian gauge theory more like a quantum state of the periodic potential, or a state of the charged quantum pendulum? After describing both the periodic potential and the charged quantum pendulum in his chapter 11, Rubakov (2002) distinguishes vacua of a classical Yang–Mills gauge theory (such as the above $\mathbf{A}_\mu, \mathbf{A}'_\mu$) in chapter 13, and provides a clear description of transitions between them in terms of an instanton.[8] These vacua are topologically inequivalent, since their so-called Chern–Simons numbers are different. The Chern–Simons number n_{CS} associated with potential $\mathbf{A}_\mu$ is defined as follows:

$$n_{CS}\left(\mathbf{A}_\mu\right) \equiv \frac{1}{16\pi^2} \int d^3x \epsilon^{ijk} \left(A_i^a \partial_j A_k^a + \frac{1}{3}\epsilon^{abc} A_i^a A_j^b A_k^c \right) \quad (6.46)$$

and $n_{CS}\left(\mathbf{A}'_\mu\right) = n_{CS}\left(\mathbf{A}_\mu\right) + 1$. But in a semi-classical treatment, quantum tunneling between them is possible through quantum tunneling. Moreover, gauge transformations of the form 6.38 result in changes in Chern–Simons number just in case they are "large"—i.e. have non-zero winding number. This suggests that the classical vacua are indeed distinct, and that a "large" gauge transformation represents a change from one physical situation to another. If so, symmetry under "large" gauge transformations is not just a theoretical symmetry but reflects an empirical symmetry of a non-Abelian Yang–Mills gauge theory.

But Rubakov (2002) goes on to offer an alternative perspective when he says (on page 277)

We note that the instanton can be given a slightly different interpretation (Manton 1983) which, in fact, is equivalent to the above. From the point of view of gauge-invariant quantities, topologically distinct classical vacua are equivalent, since they differ only by a gauge transformation. Let us identify these vacua. Then the situation becomes analogous to the quantum-mechanical model of the pendulum.

---

[8] An instanton is a localized solution to the Yang–Mills equation *in four-dimensional Euclidean space* rather than Minkowski space-time. Mathematical analysis of the tunneling process is facilitated by such a Euclidean representation.

On this interpretation, even "large" gauge transformations lead from a single classical vacuum state back into an alternative representation of that same state! The reference to (Manton, 1983) is somewhat misleading, since Manton is discussing a theory in which instantons do not occur. But Manton (1983) does contain passages suggestive of Rubakov's "slightly different interpretation." The question remains as to whether this interpretation is legitimate, and, if it is, whether it can be claimed to be equivalent to an interpretation according to which a "large" gauge transformation represents an empirical transformation between distinct states of a non-Abelian Yang–Mills gauge theory.

Consider first a purely classical non-Abelian Yang–Mills gauge theory. If it has models that represent distinct degenerate classical vacua, what is the physical difference between these vacua? Models related by a "large" gauge transformation are characterized by different Chern–Simons numbers, and one might take these to exhibit a difference in the intrinsic properties of situations they represent. But it is questionable whether the Chern–Simons number of a gauge configuration represents an intrinsic property of that configuration, even if a *difference* in Chern–Simons number represents an intrinsic *difference* between gauge configurations. Perhaps Chern-Simons numbers are like velocities in models of special relativity. As we saw in section 6.2, the velocity assigned to an object in a model of special relativity does not represent an intrinsic property of that object, even though that theory does distinguish in its models between situations involving objects moving with different *relative* velocities. It was this latter distinction that proved critical to establishing that Lorentz boosts are empirical symmetries of situations in a special relativistic world. So does a *difference* in Chern–Simons number represent an intrinsic *difference* between classical vacua in a purely classical non-Abelian Yang–Mills gauge theory? There is no reason to believe that it does. For it to do so, the theory would have to have *joint* models that incorporate *more than one* vacuum state, where the distinct vacua are represented by different Chern–Simons numbers in *every* such model. Such distinct vacua extend over all space, and so could be represented in the same model only if they are represented as occurring at different times. But topologically distinct vacua are separated by an energy barrier, and in the purely classical theory this cannot be overcome. So no single model of the purely classical theory represents vacua with different Chern–Simons numbers. There is no reason to believe that a "large" gauge transformation represents an empirical transformation between distinct vacuum states of a purely classical non-Abelian Yang–Mills gauge theory.

According to a semi-classical theory, vacua with different Chern–Simons numbers *can* be connected by tunneling through the potential barrier that separates them. So such a theory can model a single situation involving more than one such vacuum state, each obtaining at a different time. Moreover, no model of this theory represents these states as having the *same* Chern–Simons numbers. Perhaps this justifies the conclusion that in a world truly described by such a theory a "large" gauge transformation *would* represent an empirical transformation between distinct vacuum states. There is a parallel here with the status of constant electric potential transformations in a world truly described by a theory of electrostatics, as considered in section 6.2. In both cases, had the world been different, the relevant kind of gauge transformation would have represented an empirical symmetry (though this would presumably have been *observable* only in the latter case.) But in each case, the theory is at best a step along the way to an empirically adequate theory of our world, and taking a further step (to a fully quantized non-Abelian Yang–Mills gauge theory, or to a full theory of classical electromagnetism, respectively) reopens the issue of the status of gauge transformations of that kind.

The θ-vacuum of a fully quantized non-Abelian Yang–Mills gauge theory is non-degenerate and symmetric under "large" as well as "small" gauge transformations. Analogies with the periodic potential and quantum pendulum suggest that it be expressed in the form

$$|\theta\rangle = \sum_{n=-\infty}^{+\infty} \exp\{-in\theta\} |n\rangle \qquad (6.47)$$

where state $|n\rangle$ corresponds to a classical state with Chern–Simons number n. But not only the θ-vacuum but the whole theory is symmetric under "large" gauge transformations. So a generator $\hat{U}$ of "large" gauge transformations commutes not only with the Hamiltonian but with all observables. It acts as a so-called "superselection operator" that separates the large Hilbert space of states into distinct superselection sectors, between which no superpositions are possible. Physical states are therefore restricted to those lying in a single superselection sector of the entire Hilbert space. Hence every physical state of the theory, including $|\theta\rangle$, is an eigenstate of $\hat{U}$. Now there is an operator $\hat{U}_1$ corresponding to a "large" gauge transformation with winding number 1,

$$\hat{U}_1 |n\rangle = |n+1\rangle \qquad (6.48)$$

from which it follows that none of the states $|n\rangle$ is a physical state of the theory! This theory cannot model situations involving *any* state corresponding to a classical vacuum with definite Chern–Simons number, still less a situation involving two or more states corresponding to classical vacua with *different* Chern–Simons numbers. Consequently, "large" gauge transformations in a fully quantized non-Abelian Yang–Mills gauge theory do *not* represent physical transformations, and symmetry under "large" gauge transformations is not an empirical symmetry. There is no difference in this respect between "large" and "small" gauge transformations.

There are several reasons why it remains important to better understand the difference between "large" and "small" gauge transformations. One reason is

6.6 THE θ-VACUUM

that doing so will help to resolve the following apparent paradox that may have struck an acute reader of section 6.2. That section argued that "local" gauge transformations implement no empirical symmetry and therefore have no direct empirical consequences, while acknowledging that *"global"* gauge transformations have *indirect* empirical consequences via Noether's theorem, including the conservation of electric charge. The paradox arises when one notes that a "global" gauge transformation appears as a special case of a "local" gauge transformation. If "local" gauge symmetry is a purely formal symmetry, how can (just) this special case of it have even *indirect* empirical consequences? Another motive is to appreciate why some (e.g. Giulini 2003, p. 289) have proposed that we make

a clear and unambiguous distinction between proper physical symmetries on one hand, and gauge symmetries or mere automorphisms of the mathematical scheme on the other.

The proposed distinction would classify invariance under "small" gauge symmetries as a gauge symmetry, but invariance under "large" gauge transformations as a proper physical symmetry. It is founded on an analysis of gauge in the framework of constrained Hamiltonian systems as described in chapter 5 and appendix C. The guiding principle is to follow Dirac's proposal by identifying gauge symmetries as just those transformations on the classical phase-space representation of the state of such a system that are generated by its first-class constraint functions. In a classical Yang–Mills gauge theory, these are precisely those generated by the so-called Gauss constraint functions, such as the function on the left-hand side of equation 5.27 in the case of pure electromagnetism.

Giulini (2003) applies this principle to a quantized Hamiltonian system representing an isolated charge distribution in an electromagnetic field, and concludes that the gauge symmetries of this system consist of all and only asymptotically trivial gauge transformations—i.e. those "local" gauge transformations on the quantized fields that leave unchanged both the asymptotic electromagnetic gauge potential $\hat{A}_\mu$ and the distant charged matter field. A "global" gauge transformation corresponding to a constant phase rotation in the matter field does *not* count as a gauge symmetry since it is not generated by the Gauss constraint (or any other first-class constraint) function. Rather, "global" U(1) phase transformations would correspond to what Giulini calls *physical* symmetries. According to Giulini (2003, p. 308),

This is the basic and crucial difference between local and global gauge transformations.

The formalism represents the charge of the system dynamically by an operator $\hat{Q}$ that generates translations in a coordinate corresponding to an additional degree of freedom on the boundary in the dynamical description. A charge superselection rule, stating that all observables commute with the charge operator, is equivalent to the impossibility of localizing the system in this new coordinate. Consequently, conservation of charge implies that translations in this additional degree of freedom count as physical symmetries for Giulini. So conservation of charge is equivalent both to the existence of these symmetries, and (by Noether's first theorem) to the "global" gauge symmetry of the Langrangian. But these physical symmetries do not correspond to gauge symmetries, either "global" or "local", since they affect neither the gauge potential nor the phase of the matter field.

It is hard to argue that these novel physical symmetries are empirical. No operational procedures are specified to permit measurement of the additional degrees of freedom, and these attach on a boundary which is eventually removed arbitrarily far away. But even if such a new physical symmetry were empirical, it would not correspond to any constant phase change. A "global" gauge symmetry would still not entail any corresponding empirical symmetry. This delicate relation between "global" gauge transformations and some other physical symmetry helps to resolve the apparent paradox outlined above. A "global" gauge transformation is not simply a special case of a "local" gauge transformation. Indeed, the constrained Hamiltonian approach provides a valuable perspective from which it may not even appear to *be* a gauge transformation.

This perspective illuminates the distinction between "large" and "small" gauge transformations more generally. As Giulini (1995) puts it, in Yang–Mills theories

it is the Gauss constraint that declares some of the formally present degrees of freedom to be physically nonexistent. But it only generates the identity component of asymptotically trivial transformations, leaving out the long ranging ones which preserve the asymptotic structure imposed by boundary conditions as well as those not in the identity component of the asymptotically trivial ones. These should be considered as proper physical symmetries which act on physically existing degrees of freedom.

Whether the constrained Hamiltonian approach to gauge symmetry establishes that "large" gauge transformations correspond to empirical symmetries is more sensitive to theoretical context than Giulini's last sentence seems to allow. But the approach certainly shows that not only a "global" gauge transformation but any "large" gauge transformation not generated by a Gauss constraint is very different from the "local" gauge symmetries that it does generate.

6.7 Anomalies

An anomaly is said to arise in a quantum field theory when the quantum analog of a classically conserved current is no longer conserved. Anomalies are interesting for all sorts of reasons—physical, mathematical, and conceptual. At first sight they constitute counterexamples to Noether's theorem—loosely, that a continuous symmetry of a theory gives rise to a corresponding conserved current. Since Noether proved a *theorem* (actually several) this cannot be right; but it is highly educational to see *why* it is wrong.

A theme of a recent book (Bertlmann 1996) is that while you cannot live without anomalies, you cannot live with them either. They are needed to account for experimental facts like the two-photon decay of the π^0-meson: but they also allegedly "signal the breakdown of gauge symmetry and, in consequence, the ruin of the consistency of the theory."It is tempting to conclude that a study of anomalies will reveal that "local" gauge symmetry is not a purely formal requirement, since its breakdown has profound consequences—some good, some bad.

Tempting, but wrong! There are different kinds of anomalies with different implications.

One kind may be viewed as arising from the failure of gauge symmetry in a theory in which massless "handed" fermion fields interact with non-Abelian gauge fields. This failure renders the theory non-renormalizable and useless. Although the action for the theory is invariant under "local" gauge transformations, the theory itself is not—the path-integral measure, and hence the "quantum action", is *not* invariant under a "local" gauge transformation. One way to remove (or rather cancel) such non-Abelian anomalies is to modify the theory by adding terms to the original Lagrangian. Such a modified Lagrangian in the Weinberg–Salam unified electroweak theory implies the existence of the top quark and that quarks come in three colors in QCD. But these are not implications of "local" gauge symmetry itself, but only of a specific theory heuristically motivated as a replacement for a theory beset by internal difficulties associated with its violation.

The ABJ anomaly that permits the two-photon decay of the π^0-meson involves a violation of gauge symmetry in the "quantum action" $W[A_\mu]$ associated with an external Schwinger source field A_μ. But this source field merely figures as a calculational device for evaluating quantities like vacuum-to-vacuum transition probabilities that are gauge invariant even though neither their amplitudes nor their generating function $Z[A_\mu] = \exp iW[A_\mu]$ are invariant under "local" gauge transformations in A_μ.

In neither case does a violation of "local" gauge symmetry associated with anomalies reflect any asymmetry in nature. Where it occurs, violation of "local" gauge symmetry is an artefact of theory. Sometimes its theoretical role is benign (as in the ABJ anomaly): sometimes its removal requires radical theoretical surgery (as in the non-Abelian anomaly). One can acknowledge the occurrence of anomalies while maintaining that "local" gauge symmetry is a purely formal feature of a theory.

7

Loop representations

Following the pioneering work of Mandelstam (1962), there have been a number of proposals for canonically quantizing gauge fields by starting with variables that are gauge-invariant objects associated with paths or loops, rather than with gauge-dependent quantities like $A_\mu^a(x)$. This has the comparative advantage of making the gauge symmetry of the theory manifest throughout, though it does lead to certain technical complications. In recent formulations, the algebra of these variables is represented after quantization by operators acting on a space of wave-function(al)s of what are known as loops, or holonomy loops (hoops, for short). This contrasts with the quantization methods described in chapter 5 (sections 5.5, 5.7), in which the domain of wave-functionals was the set of (gauge-dependent) quantities like $A_\mu^a(x)$: such quantities are often referred to as connections, because of their role in connecting fibers above different points in a fiber bundle formulation of the gauge theory.[1]

Since quantum states now become function(al)s of "loops" rather than connections, this has come to be called a loop representation—an unfortunate name, given the multiple ambiguity of the term 'loop'! In chapter 3, section 3.1.2 I gave my reasons for preferring the term 'hoop' in this context. But to attempt to promote a superior convention by referring instead to "hoop representations" would be to risk incomprehension, and so I will reluctantly continue to conform to the established usage. The existence of loop representations is of considerable importance for the interpretative project of this book, as I will explain in the first section of this chapter. The next section describes loop representations of the free Maxwell field, while loop representations of non-Abelian gauge fields are considered in the section after that. The chapter concludes by explaining how loop representations of

---

[1] A connection on a vector bundle associates a covariant derivative with each vector field. This defines parallel transport of vectors along curves in the base space, and so permits comparison of vectors at different points (relative to a curve linking those points). The covariant derivative on a vector bundle associated to a principal bundle is determined by the connection on the principal bundle—a Lie-algebra-valued one-form that permits comparison of elements in fibers above different points in the principal bundle (relative to a curve linking those points) and in this way connects those fibers. For further details, see appendix B.

quantized non-Abelian Yang–Mills theories provide a novel perspective on the θ-vacuum that offers a resolution to the so-called strong CP problem.

7.1 The significance of loop representations

The first part of this book (chapters 1–4) argued for an interpretation of classical Yang–Mills theories according to which gauge potentials directly represent no localized gauge properties, but rather indirectly represent non-localized holonomy properties. On this interpretation, the gauge symmetry of these theories is a purely formal symmetry of certain representations that may figure in their formulation, reflecting no corresponding empirical symmetry. A more intrinsic formulation of a classical Yang–Mills theory would not even mention gauge, and so the issue of its gauge symmetry would not arise. In a purely classical context, such a formulation may be given in terms of field strengths, but in the context of a quantum particle theory, holonomies are also needed. Coordinating field strengths or holonomies with the intrinsic properties they represent requires a representational convention since these are vectorial magnitudes (in ordinary space and, for a non-Abelian theory, also in an internal space). But this does *not* amount to a choice of gauge, and so both the formulation of the theory and its application remain independent of that notion.

A loop representation of a quantized Yang–Mills field theory offers the prospect of a similar elimination of gauge from the theory. Chapter 6 defended this possibility against various objections that sought to establish that the "local" gauge symmetry of a quantized Yang–Mills field theory has empirical content. If "local" gauge symmetry is just a formal feature of the way a conventional formulation of a quantized Yang–Mills field theory represents its subject matter, then the way is open for a reformulation whose representations involve no mention of gauge, so a "local" gauge transformation cannot even be applied to them. Such a reformulation would neither be, nor fail to be gauge symmetric. It could represent no gauge properties. Following the line of thought from the first part of the book, one might suspect that it rather represents non-localized holonomy properties.

Chapter 8 begins to pursue this suspicion to see how far it may be confirmed. That pursuit quickly enters a thicket of problems faced when one attempts to say what *any* quantum field theory represents and how it represents it. While I cannot resolve these problems in the present book, I do hope to convince the reader that loop representations of a quantized Yang–Mills field theory at least offer a novel perspective on them. Certainly the task of interpreting such theories is of central importance if one is to understand what the world is like according to our best contemporary physics, and any adequate interpretation must address the significance of loop representations.

7.2 Loop representations of the free Maxwell field

As we saw in chapter 5, an essential component of a quantum field theory arrived at by the procedure of canonical quantization is a set of so-called equal-time commutation relations like equations 5.33–5.35. In the Minkowski space-time of the special theory of relativity, these are imposed on each space-like hypersurface Σ—a simultaneity slice corresponding to a moment in time defined by $t = x^0 =$ constant in some global frame with coordinates x^μ ($\mu = 0, 1, 2, 3$). Basic concepts of hoops (holonomy equivalent loops) introduced in chapter 3 will now be reviewed as applied to curves in Σ.

Consider a continuous, piecewise smooth, closed curve C generated by a continuous mapping of the circle S^1 into a three-dimensional spacelike hyperplane Σ.[2] Any curve that results from C by an orientation-preserving reparametrization is equivalent to C by virtue of tracing out the same image in Σ; the equivalence class constitutes a corresponding unparametrized curve. Now consider any unparametrized curve that traces out an image that differs at most by including a finite number of "trees"—i.e. images of closed curves that enclose no area. The set of all their constituent closed curves forms an equivalence class $[C]$ I call the *hoop* associated with C.

Here the *holonomy* of C ($H_A(C)$) is the quantity $\exp i \oint_C A_j(\mathbf{x}).dx^j$, where $\mathbf{A}(\mathbf{x})$ is the magnetic vector potential of the Maxwell field at point $\mathbf{x}$ on Σ.[3] This is an instance of the more general notion of holonomy we first encountered in chapter 3, which will be reviewed in the next section. The term 'hoop' is appropriate as a contraction of "holonomy loop," since all curves in the hoop $[C]$ have the same holonomy, which may therefore be considered a property of the hoop $\gamma = [C]$ itself, and written $H_A(\gamma)$. Since the structure group U(1) of Maxwellian electromagnetism is Abelian, the holonomy $H_A(\gamma)$ is invariant under gauge transformations of the form $\mathbf{A} \to \mathbf{A}' = \mathbf{A} - \nabla\Lambda$, i.e. variable potential transformations. Indeed, the quantity $\oint_C A_j(\mathbf{x}).dx^j$ is also invariant under such gauge transformations and may be written as $A(\gamma)$ since its value is the same for every element of γ. Either $A(\gamma)$ or $H_A(\gamma)$ may be chosen as a basic gauge-invariant variable suitable for canonical quantization, along with the gauge-invariant quantity $\mathbf{E}$. Quantization schemes based on both choices will be considered, after a brief introduction to the more rigorous mathematical framework in which they are set. Further details are given in appendices D and E.

The fundamental Heisenberg commutation relations satisfied by operators $\hat{\mathbf{x}}, \hat{\mathbf{p}}$ representing the canonical dynamical variables $\mathbf{x}, \mathbf{p}$ of a particle in quantum

---

[2] This restriction to *spacelike* curves implies a corresponding restriction on the kinds of hoops that figure in loop representations.

[3] The electromagnetic four-vector potential $A_\mu(x)$ has components $(\varphi, -\mathbf{A})$ in an inertial coordinate system for which Σ is a simultaneity slice $x_0 =$ constant. $\mathbf{A}$ has components $A_j(\mathbf{x})$ ($j = 1, 2, 3$) at the point in Σ with coordinates $\mathbf{x} \equiv (x_1, x_2, x_3)$ in such a system.

mechanics are

$$[\hat{x}_j, \hat{p}_k] = i\hbar \delta_{jk} \hat{I} \qquad (7.1)$$
$$[\hat{x}_j, \hat{x}_k] = [\hat{p}_j, \hat{p}_k] = 0$$

These generalize to equal-time commutation relations (ETCRs) for field systems such as the following for operators corresponding to a real classical scalar field $\varphi(\mathbf{x}, t)$:

$$[\hat{\varphi}(\mathbf{x}, t), \hat{\pi}(\mathbf{x}', t)] = i\hbar \delta^3(\mathbf{x} - \mathbf{x}') \qquad (7.2)$$
$$[\hat{\varphi}(\mathbf{x}, t), \hat{\varphi}(\mathbf{x}', t)] = [\hat{\pi}(\mathbf{x}, t), \hat{\pi}(\mathbf{x}', t)] = 0$$

as well as anti-commutation relations for field operators acting on states of fermionic systems such as electrons and quarks. The following commutation relations for the free Maxwell field were given in chapter 5, section 5.5

$$[\hat{A}_j(\mathbf{x}, t), \hat{E}^k(\mathbf{x}', t)] = -i\delta_{jk}\delta^3(\mathbf{x} - \mathbf{x}') \qquad (7.3)$$
$$[\hat{A}_0(\mathbf{x}, t), \hat{\pi}^0(\mathbf{x}', t)] = i\delta^3(\mathbf{x} - \mathbf{x}') \qquad (7.4)$$
$$[\hat{A}_j(\mathbf{x}, t), \hat{A}_k(\mathbf{x}', t)] = [\hat{E}^j(\mathbf{x}, t), \hat{E}^k(\mathbf{x}', t)] = 0 \qquad (7.5)$$

While some such commutation relations are basic to a quantum field theory, they do not by themselves constitute the theory. To further develop the theory, it is necessary further to specify its dynamical variables (traditionally known as "observables" in any quantum theory) and states. States in a quantum theory are usually represented by vectors in a Hilbert space, and observables by self-adjoint operators on that space (see appendix D). So to develop a theory based on a set of commutation relations like 7.2 it is necessary to characterize the representations of those relations by means of self-adjoint operators acting on an appropriate Hilbert space. For a theory with only a finite number of kinematically independent dynamical variables ("degrees of freedom" in its Lagrangian formulation), the characterization problem was substantially solved by a theorem stated by Stone and proved by von Neumann.

The Stone–von Neumann theorem basically states that all irreducible representations of the Weyl form of the canonical commutation relations expressing the quantization of a finite-dimensional classical theory are unitarily equivalent.[4] What is the Weyl form, and why is it relevant here? The Heisenberg commutation relations 7.1 apply to unbounded operators. But not every vector in the space can lie in the domain of an unbounded operator. The relations 7.1 are therefore not well defined unless and until one specifies

---

[4] A precise statement is given in appendix D, which points out the potential significance of this equivalence for the interpretation of a quantum theory to which the theorem applies.

a domain of definition for all unbounded operators they involve. Doing this involves technical complications that may be avoided by moving to an alternative form of commutation relations proposed by Weyl, namely

$$\hat{U}(\mathbf{a})\hat{V}(\mathbf{b}) = \exp(-i\mathbf{a}.\mathbf{b})\,\hat{V}(\mathbf{b})\hat{U}(\mathbf{a}) \tag{7.6}$$

where $\mathbf{a}, \mathbf{b}$ are vectors in the $3n$-dimensional configuration space of an n-particle system, and one thinks of $\hat{U}(\mathbf{a})$, $\hat{V}(\mathbf{b})$ as related to $\hat{x}_j, \hat{p}_k$ ($j, k = 1, .., 3n$) by $\hat{U}(\mathbf{a}) = \exp(i\mathbf{a}.\hat{\mathbf{x}})$, $\hat{V}(\mathbf{b}) = \exp(i\mathbf{b}.\hat{\mathbf{p}})$. $\hat{U}(\mathbf{a}), \hat{V}(\mathbf{b})$ are unitary operators, and are therefore bounded and everywhere defined, so it is not necessary to attend to their domains of definition. Although the Weyl commutation relations are not equivalent to the Heisenberg CCRs, the latter are essentially the infinitesimal form of the former.

Just as the Heisenberg CCRs for a particle theory generalize to ETCRs for a field theory such as 7.2, so also the Weyl relations 7.6 generalize.[5] To state the generalization, first define new *Weyl operators* by

$$\hat{W}(\mathbf{a},\mathbf{b}) \equiv \exp(i(\mathbf{a}.\mathbf{b})/2)\,\hat{U}(\mathbf{a})\hat{V}(\mathbf{b}) \sim \exp i(\mathbf{a}.\hat{\mathbf{x}} + \mathbf{b}.\hat{\mathbf{p}}) \tag{7.7}$$

which therefore obey the multiplication rule

$$\hat{W}(\mathbf{a},\mathbf{b})\hat{W}(\mathbf{c},\mathbf{d}) = \hat{W}(\mathbf{a}+\mathbf{c},\mathbf{b}+\mathbf{d})\exp(-i(\mathbf{a}.\mathbf{d}-\mathbf{b}.\mathbf{c})/2) \tag{7.8}$$

Equation 7.8 is equivalent to the Weyl relations 7.6. We seek a generalization of 7.8 that will yield a more rigorous form of the ETCRs 7.2 of a scalar field theory.

The first step toward the generalization is to note that a field operator like $\hat{\varphi}(\mathbf{x}, t)$ is not well defined at space-time point $x \equiv (\mathbf{x}, t)$, and so in a rigorous formulation must be replaced by a so-called "smeared field operator" that is defined instead over an appropriate class of "test" functions on Σ, including functions sharply peaked at $\mathbf{x}$. Smeared field operators $\hat{\varphi}, \hat{\pi}$ now act on test functions g, f from the appropriate classes, so we write $\hat{\varphi}(g), \hat{\pi}(f)$. Just as the pair of vectors $(\mathbf{a}, \mathbf{b})$ serves to pick out a point in the finite-dimensional phase space of a particle system, so also a pair of test functions (g, f) picks out a point in the infinite-dimensional phase space of a field system. Weyl operators $\hat{W}(\mathbf{a}, \mathbf{b})$ therefore generalize to *Weyl operators* $\hat{W}(g, f)$.

Now on the classical phase space for a field theory like that of the real scalar field there is a so-called symplectic form $\sigma(f, g)$ that generalizes the form $(\mathbf{a}.\mathbf{d} - \mathbf{b}.\mathbf{c})$ on the phase space of a classical particle system. The multiplication rule 7.8 accordingly generalizes to

$$\hat{W}(g_1,f_1)\hat{W}(g_2,f_2) = \hat{W}(g_1+g_2,f_1+f_2)\exp(-i\sigma(f,g)/2) \tag{7.9}$$

[5] See appendix E for further details of this generalization.

which specifies the so-called abstract *Weyl algebra* for the real scalar field and provides the required rigorous form of the ETCRs 7.2. The explicit expression for the symplectic form in this case is given by the following integral over a spacelike "equal time" hyperplane Σ:

$$\sigma(f,g) = \int_\Sigma d^3x(g_1 f_2 - g_2 f_1) \qquad (7.10)$$

The Stone–von Neumann theorem does not generalize to representations of field Weyl algebras like those specified by 7.9. While such an algebra does possess Hilbert space representations, these are not all unitarily equivalent to one another. Indeed, there is a continuous infinity of inequivalent representations of equation 7.9's algebra. These include both Fock representations inequivalent to one another, and representations equivalent to no Fock representation. Fock representations are important, since (as appendix E explains) the basis states in such a representation have often been interpreted as exhibiting the particle content of the field, thereby justifying talk of photons and other gauge particles at least as emergent phenomena in situations modeled by a Fock representation of a quantized gauge field. Chapter 8, section 8.2 assesses the significance of the existence of all these inequivalent representations, both Fock and non-Fock.

Ashtekar and Isham (1992) consider a variety of different methods for canonically quantizing the free Maxwell field, some involving connection variables, others "loop" variables. They begin with a set of Weyl operators of the general form $\hat{W}[\bullet,f] = \exp\{i[\hat{A}(\bullet) + \hat{E}(f)]\}$, where $A(\bullet)$ is a classical "loop" or connection variable, and $E(f)$ is a classical electric field variable corresponding to "smearing" the quantized field $\hat{E}^k(\mathbf{x})$ with the (co)vector field f_k. The classical variables define a so-called Poisson algebra through the Poisson bracket operation.[6] Ashtekar and Isham show that the choice $A(\bullet) = A(g)$ (where g^j is a vector field used to "smear" the quantized magnetic vector potential operator $\hat{A}_j(\mathbf{x})$) yields a Poisson algebra corresponding to the conventional Weyl algebra generated by 7.9. To arrive at a theory of quantized free Maxwell fields, one considers unitary representations of this abstract Weyl algebra. One such representation is the standard Fock representation for the free quantized Maxwell field. But Ashtekar and Isham (1992) also show that the choice $A(\bullet) = A(\gamma)$ has a Poisson algebra that corresponds to a different Weyl algebra,

---

[6] The Poisson bracket of two classical fields is a natural generalization, to systems of an infinite number of degrees of freedom, of the Poisson bracket of a pair of functions on the finite-dimensional phase space of a system of particles, as defined in appendix C. The associative algebra generated by the Poisson bracket is called a Poisson algebra. On quantization, Poisson brackets of classical variables are formally replaced by appropriate commutators of corresponding operators. This leads to a muliplication law for the associated Weyl operators (cf. equation 7.9). Together with a definition of the * operation, this defines the Weyl algebra corresponding to the classical Poisson algebra. See appendix E for further details.

whose unitary representations do *not* include this Fock representation! As they say (1992, p. 396),

... although the two classical Poisson algebras are on the same footing, the corresponding Weyl algebras are not. If one chooses to quantize the system using $\hat{W}[\gamma,f]$ instead of $\hat{W}[g,f]$, we have to forego the possibility of using the standard Fock representation purely on kinematical grounds.

Thus choosing an algebra involving "loop" variables as the starting point for quantizing the free Maxwell field apparently excludes thinking of photons as the quanta of that field! The interpretative significance of this non-standard representation will be assessed in chapter 8, section 8.2.

But it would be wrong to conclude that *no* algebra involving "loop" variables admits a Fock representation. Ashtekar and Isham (1992) themselves also consider a different choice of "loop" and electric field variables whose Poisson algebra corresponds to a Weyl algebra that *does* admit the standard Fock representation. And in another paper, Ashtekar and Rovelli (1992) show how a closely related choice leads to a representation of quantum states as function(al)s of holonomic loops (hoops).[7] States in this loop representation may be arrived at either by taking the so-called loop transform of state functionals of a corresponding connection representation, or *ab initio* by canonical quantization of the "loop" variables and electric fields. Specifically, Ashtekar and Rovelli, (1992) choose as variables $h[\gamma] = \exp \oint_{C \in \gamma} A_j^{(-)}(x).dx^j$, where $A_j^{(-)}$ is the negative frequency part of the transverse connection A_j^T, and, for $E[f]$, the positive frequency part $E^{k(+)}$ of the transverse electric field E_T^k "smeared" by f_k.[8] These variables are complex, and so $h[\gamma]$ may be viewed as a holonomy closely related to $H_A(\gamma)$.

On quantization, classical Poisson brackets give rise to a corresponding multiplication rule for Weyl operators, and the Poisson algebra of $h[\gamma]$ and $E[f]$ yields an abstract Weyl algebra of operators $\hat{h}[\gamma]$, $\hat{E}[f]$ that may be represented on a Hilbert space of function(al)s of the form $\Psi[\gamma]$, whose inner product is determined by the condition that the classical transverse connection and electric field are real-valued functions. The resulting loop representation is equivalent, via a loop transform operation, to the Bargmann connection representation.[9]

---

[7] A wave-function $\Psi[A(\mathbf{x})]$ in the connection representation is clearly a function*al*, rather than a function, of the connection $A(\mathbf{x})$. Despite the common practice in the literature of referring to a wave-function $\Psi(\gamma)$ in the loop representation as a functional, it seems more properly called a function of the loop variable γ, since γ is not itself a function of the coordinate $\mathbf{x}$.

[8] Chapter 5, section 5.2 explains the terminology of positive and negative frequencies, and transverse components. While restriction to *transverse* components may appear to make the definitions of $h[\gamma]$ and $E[f]$ gauge dependent, Ashtekar and Rovelli (1992) note that this restriction is inessential. Dropping the restriction to the transverse parts of the positive and negative frequency fields in their definitions would not affect the final result.

[9] The Bargmann connection representation expresses wave-functions in the form $\Psi[\zeta_i(\mathbf{k})]$, where each $\zeta_i(\mathbf{k})$ ($i = 1, 2$) is a complex-valued function of the momentum variable $\mathbf{k}$. The $\zeta_i(\mathbf{k})$ appear

The transform defines loop-representation counterparts of the n-photon states of the Bargmann representation. These counterparts form the basis for the Fock space of a Fock representation for the free quantized Maxwell field that is unitarily equivalent to the standard Fock representation.

Ashtekar and Rovelli note that simply requiring operators $\hat{h}[\gamma]$, $\hat{E}[f]$ to satisfy this abstract Weyl algebra does not by itself define a loop representation: it is necessary also to specify an interpretation of these operators, and more than one such interpretation is available, leading to a multiplicity of loop representations. In particular, taking $\hat{h}[\gamma]$ to represent the holonomy of the real-valued connection A (corresponding to the classical function $H_A(\gamma)$) and $\hat{E}[f]$ to represent the real-valued electric field, one arrives at a different loop representation obtained by Gambini and Trias (1981, 1983) that can also be shown to be equivalent to the standard Fock representation. This last loop representation is constructed by Gambini and Pullin (1996) in a slightly different way by quantizing an algebra based on the gauge-invariant variables $H_A(\gamma)$ and $E^k(x).H_A(\gamma)$, where $A(x)$ is the full (real-valued) connection. They use a formal loop transform to demonstrate its equivalence to the standard Fock representation. This last approach has a natural generalization to non-Abelian gauge fields which will be considered in the next section.

It is interesting to reflect on the interpretation of particular states in a loop representation of the free quantized Maxwell field. Consider first the representation analyzed by Ashtekar and Rovelli. There are two states naturally associated with each hoop γ. One of these may be written as a Dirac ket $|\gamma\rangle$: it is an element of an (overcomplete, non-orthogonal) basis of coherent states[10] for the space, and represents a state in which the connection is concentrated on the hoop γ. This is because the quantity $\langle\gamma\,|\,\Psi\rangle$ gives the amplitude for observing a loop-like classical excitation of the connection along γ when the system is in state Ψ. The other state may be written as $\Psi_{\gamma_0}(\gamma)$. It is the characteristic function of the hoop γ, with value 1 for $\gamma = \gamma_0$ and 0 otherwise. Though not normalizable, it is a simultaneous (generalized) eigenstate of the electric field operators $\hat{E}[f]$, and so represents the simplest excitation of the electric field, in which the lines of electric flux are wholly determined by the flux around γ_0. Not surprisingly, these states are in a sense complementary: a state in which the holonomy is maximally definite around γ is a state in which the electric flux around γ is maximally indefinite, and vice versa.

---

as coefficients in a Fourier expansion of a positive-frequency complex connection $\mathbf{A}^{(+)}$ constructed from the real transverse fields $\mathbf{A}^T, \mathbf{E}^T$. The Bargmann form of the connection representation has the advantage that the loop transform connecting it to the loop representation is a well-defined mathematical operation, involving no divergences.

[10] A coherent state in a Fock representation of a quantum field is an eigenstate of an annihilation operator. The number of quanta in a coherent state is, in a sense, maximally indefinite. Coherent states of the quantized free Maxwell field in the standard Fock representation (in which the quanta are photons) provide the closest analog to states of a classical electromagnetic wave.

7.3 Loop representations of other free Yang–Mills fields

Gambini and Pullin (1996) have described a general method for creating loop representations of free Yang–Mills fields that applies to non-Abelian as well as Abelian fields. It follows Dirac's prescription for quantizing a constrained Hamiltonian system, but for the first step. The method begins instead with an algebra of classical variables that is non-canonical, in the sense that it is not simply read off the Poisson bracket relations of the usual canonical variables.

To construct a loop representation of the quantized theory, it is necessary to set up a suitable algebra of gauge-invariant variables. The $\vec{E}^a$ are gauge invariant, but the A_μ^a are not.[11] In a non-Abelian gauge theory, the holonomies $\mathbf{H}_A(\gamma)$ are not quite gauge invariant, since they undergo a common similarity transformation under a change of gauge (see appendix B, equation B.48), but their traces, Wilson loops, are gauge invariant, as required.

The holonomy of a curve in M may be defined either on a principal fiber bundle or on an associated vector bundle.[12] But Wilson loops are defined only for holonomies on a vector bundle, since the trace operation is well defined only on a matrix representation of an element of the structure group. So consider a vector bundle $<E, M, G, \pi_E, V, P>$ associated to the principal fiber bundle P on which a Yang–Mills gauge field is defined. The typical fiber V is a vector space, and the holonomies of the bundle are representations of elements of G on V—linear transformations (matrices) acting on vectors. One can define the holonomy of a curve C in Σ with base point m relative to a section σ of P as follows:

$$\mathbf{H}_\sigma(C) = \wp \exp\left(-\oint_C \mathbf{A}_j dx^j\right) \quad (j=1,2,3) \tag{7.11}$$

where $\mathbf{A}_j$ is a square matrix whose elements are the "components" of the connection on E that are determined (relative to σ) by the connection ω on P (see appendix B). The holonomy of a hoop γ is then

$$\mathbf{H}_\sigma(\gamma) = \wp \exp\left(-\oint_{C \in \gamma} \mathbf{A}_j dx^j\right) \quad (j=1,2,3) \tag{7.12}$$

and these transform by a common similarity transformation under the change of section $\sigma \to \sigma'$:

$$\mathbf{H}_{\sigma'}(\gamma) = [\rho(h)]^{-1} \circ \mathbf{H}_\sigma(\gamma) \circ \rho(h) \tag{7.13}$$

---

[11] This section follows the notation of section 5.7 for the components of non-Abelian fields.
[12] For further details, see appendix B.

7.3 LOOP REPRESENTATIONS OF OTHER FREE YANG–MILLS FIELDS

where $\rho(h)$ is an element of a representation of G in V; and also under gauge transformations—vertical automorphisms of P:

$$\mathbf{H}'_\sigma(\gamma) = [\rho(g)]^{-1} \circ \mathbf{H}_\sigma(\gamma) \circ \rho(g) \tag{7.14}$$

A holonomy in the associated vector bundle is just a square matrix with n^2 elements $[\mathbf{H}]_{ab}$ ($a, b = 1, \ldots, n$), and so its *trace* may be defined as follows:

$$\mathrm{Tr}(\mathbf{H}) \equiv \sum_{a=1}^{n} [\mathbf{H}]_{aa} \tag{7.15}$$

Even though these holonomies themselves depend both on gauge (through the dependence of ω on vertical automorphisms of P) and on section σ, equations 7.14 and 7.13 imply that their traces do not depend on section and *are* gauge invariant. These are the *Wilson loops*

$$W_A(\gamma) = \mathrm{Tr}[\mathbf{H}(\gamma)] = \mathrm{Tr}\left[\wp \exp\left(-\oint_{C \in \gamma} \mathbf{A}_j \mathrm{d}y^j\right)\right] \tag{7.16}$$

Moreover, unlike the holonomy of the hoop γ, its Wilson loop defined by 7.16 is not only gauge invariant, but also independent of the base point m of γ, as a consequence of equation B.49. Note that the Wilson loop of a hoop does depend on the dimension of the representation ρ of G.

Now consider the hoop group L_o with base point o described in chapter 3 (section 3.1.2). The base point m of hoop γ will typically not coincide with o. Nevertheless, the hoop γ may be represented by any hoop γ_o in L_o that includes an unparametrized curve C whose image differs from that of a non-self-intersecting curve in γ only by a "tree" formed by tracing and retracing a path connecting o to m. The holonomies of all such curves are related to each other and to the holonomy of γ by a similarity transformation, and so their Wilson loops are all equal to each other, and equal to the Wilson loop of γ. Moreover, all their Wilson loops are independent of the base point o of L_o. Now consider an arbitrary loop L—a one-dimensional region of space-time that is represented in a manifold (Σ in this case) by the oriented image of a continuous, and piecewise smooth, non-self-intersecting, closed curve C from a hoop γ whose base point m lies in the image of C. While the Wilson loop of C depends only on L, it is equal to the Wilson loop of any element of a hoop group that represents γ, and thereby also represents L.

In this way, the Wilson loops of elements of an arbitrary hoop group L_o determine the Wilson loop that pertains to any and every loop in Σ. And as mentioned in chapter 3 (section 3.1.2) there are so-called reconstruction theorems showing that, at least for a gauge field with one of a wide class of structure groups, all the gauge-invariant information contained within the holonomies is derivable from the Wilson loops of the gauge field. This makes Wilson loops an important focus for the interpretation of a gauge theory, as we shall see in chapter 8. It also gives them a privileged role in constructing a loop representation of a Yang–Mills field gauge theory directly by loop quantization.

We may therefore take as the starting point for loop quantization of a Yang–Mills field gauge-invariant classical variables including $\vec{E}^a$, and $W_A(\gamma)$ for some A^a. Further gauge-invariant information involving the electric field variables is contained in additional variables formed by breaking up the holonomy of a hoop by inserting electric field variables at intermediate points. For convenience, Gambini and Pullin label these (gauge-invariant) variables as follows:

$$T(\gamma) \equiv W_A(\gamma) \tag{7.17}$$

$$T_i(\gamma_x^x) \equiv \mathrm{Tr}(\mathbf{H}_A(\gamma_o^x)\mathrm{E}_i(x)\mathbf{H}_A(\gamma_x^o)) \tag{7.18}$$

$$\ldots\ldots$$

$$T_{ijk\ldots}(\gamma_{x_1}^{x_2},\ldots,\gamma_{x_n}^{x_1}) \equiv \mathrm{Tr}(\mathbf{H}_A(\gamma_o^{x_1})\mathrm{E}_i(x_1)\mathbf{H}_A(\gamma_{x_1}^{x_2})\ldots$$
$$\mathbf{H}_A(\gamma_{x_{n-1}}^{x_n})\mathrm{E}_{i_n}(x_n)\mathbf{H}_A(\gamma_{x_n}^o)) \tag{7.19}$$

Here γ_y^x is what I shall call a *hath* (not a path!) from y to x—an equivalence class of open curves from y to x whose images differ by at most the image of one or more trees, γ_x^x is a hoop that begins and ends at x after passing through the base point o, and $\mathbf{H}_A(\gamma_o^{x_1})$ (for example) is the obvious generalization of 7.12 to a hath beginning at the base point o and ending at x_1. The Poisson brackets of these variables may now be evaluated, and the corresponding operator algebra for their associated $\hat{T}$'s derived. The algebra of the operators corresponding to just 7.17 and 7.18 is closed: its exact form depends on the structure group of the theory. But the commutators of other $\hat{T}$'s generally yield "higher-order" $\hat{T}$'s, in which case no larger finite set of these commutation relations is closed.

For the simple Abelian U(1) case the Poisson algebra of non-canonical variables is specified by

$$\{T(\eta), T(\gamma)\} = 0 \tag{7.20}$$

$$\{T^j(\gamma_x^x), T(\eta)\} = -\mathrm{i}X^{jx}(\eta)T(\eta \circ \gamma) \tag{7.21}$$

$$\{T^j(\gamma_x^x), T^k(\eta_y^y)\} = -\mathrm{i}X^{jx}(\eta)T^k(\eta \circ \gamma)_y^y + \mathrm{i}X^{ky}(\eta)T^j(\eta \circ \gamma)_x^x \tag{7.22}$$

Operators defined by the quantized analogs of 7.17, 7.18 then act on loop-representation wave-functions as follows:

$$\hat{T}(\eta)\Psi(\gamma) = \Psi(\eta^{-1} \circ \gamma) \tag{7.23}$$

$$\hat{T}^j(\eta_x^x)\Psi(\gamma) = X^{jx}(\gamma)\Psi(\eta^{-1} \circ \gamma) \tag{7.24}$$

Here $X^{jx}(\gamma)$ is a distributional vector density corresponding to the tangent vector to the hoop γ (here represented by a curve with parameter s) at point x

$$X^{jx}(\gamma) = \oint_{C \in \gamma} ds \dot{C}^j(s) \delta^3(x, C(s)) \qquad (7.25)$$

where $C(s)$ is the point on the curve representing γ at parameter s, and $\dot{C}^j(s)$ is the tangent vector to the curve at $C(s)$.

This is one example of how the algebra of gauge-invariant operators for a general Yang–Mills theory may be represented on a space of wave-functions of hoops. Gambini and Pullin (1996) also give the example of an SU(2) gauge theory. In simple cases like these, loop-representation wave-functions may be restricted to those with single hoops as arguments, of the form $\Psi(\gamma)$. In other cases, such as SU(3), they must include those with more than one hoop as arguments, e.g. $\Psi(\gamma, \eta)$. The wave-functions also satisfy various identities that follow from the Mandelstam identities obeyed by the Wilson loops (for which, see chapter 3, section 3.1.2). For example, they are symmetric functions of their arguments, $\Psi(\gamma, \eta) = \Psi(\eta, \gamma)$, and they satisfy $\Psi(\gamma \circ \eta) = \Psi(\eta \circ \gamma)$. Other identities depend on the particular structure group of the theory: for an SU(2) group we have, for example, $\Psi(\gamma, \eta) = \Psi(\gamma \circ \eta) + \Psi(\gamma \circ \eta^{-1})$.

7.4 Interacting fields in loop representations

To be regarded as a serious candidate for a gauge-free formulation, a loop representation of a quantized Yang–Mills gauge field theory in the Standard Model must recapture the empirical success already achieved by the usual gauge-dependent Lagrangian formulations. To date, this empirical success has largely flowed from their use to treat interactions with fermion fields representing elementary particles, and specifically quarks and leptons (including electrons). In the conventional Lagrangian formulation, such interactions are represented by adding interaction terms to a Lagrangian density containing terms representing the free gauge and fermion fields. This approach was sketched in chapter 5, section 5.9: the QED Lagrangian (density), for example, included the interaction term $e\bar{\psi}\gamma^\mu\psi A_\mu$, coupling the fermion field ψ to the quantized electromagnetic potential A_μ at each space-time point x^μ. How can one represent fermion fields and their interactions with gauge fields in a loop representation?

The basic idea is to generalize the representation to include gauge-invariant quantities defined on *open* paths as well as closed loops, where each open path connects two points at which a fermion field is defined. To construct such a loop/path representation of the quantized theory *ab initio*, it would be necessary to set up a suitable algebra of gauge-invariant variables. To gain some insight into the difficulties this would involve, consider an alternative approach that begins with a connection representation for the interacting theory, and seeks to arrive at a loop/path representation via a transformation that generalizes the loop transform Ashtekar and Rovelli (1992) used to arrive at their loop representation of the free Maxwell field (see section 7.2). Wave-functions in the connection representation would be written as functionals $\Psi(A_\mu^a, \psi)$ of the gauge potential A_μ^a and the lepton field ψ. To arrive at a loop/path representation via a loop transform, one would like to express such a wave-function as a functional integral with respect to the configuration variables A_μ^a, ψ that "expands" it over a "basis" of gauge-invariant quantities. But, as Gambini and Pullin (1996) point out (page 157), there do not exist, in general, gauge-invariant quantities associated with a single open path that are functions only of the configuration variables. For example, the gauge-invariant quantity

$$W(\pi_x^y) = (\psi^a(x))^\dagger H(\pi_x^y)_a^b \psi_b(y) \tag{7.26}$$

associated with a single hath π_x^y from x to y (where a,b are indices of a representation of the structure group of the theory) is the *only* gauge invariant quantity in the case of quantum electrodynamics, the theory of a U(1) gauge field interacting with a fermionic matter field (for which the indices take on only a single value). But the first term on the right of this equation is not a configuration variable, but a canonical momentum variable corresponding to the configuration variable $\psi_a(x)$.

For an SU(2) gauge theory, the following expression does define a gauge-invariant quantity involving only configuration variables:

$$W(\pi_x^y) = \psi_a(x) \epsilon^{ab} H(\pi_x^y)_b^c \psi_c \tag{7.27}$$

where $\epsilon^{ab} = -\epsilon^{ba} = 1$. But even in this case, Gambini and Setaro (1995) found it necessary also to make use of quantities like 7.26. In doing so, they followed the approach of Fort and Gambini (1991), who decomposed the fermionic degrees of freedom in order to provide a loop/path representation of quantum electrodynamics capable of treating interactions with fermions in a lattice approximation. This involves the added complexity of writing wave-functions in the connection representation in the form $\Psi(A_\mu, \psi_u^\dagger, \psi_d)$, where u corresponds to the upper part of a Dirac spinor, and d to the lower part. The resulting path/loop representation is well suited for application in the lattice approximation, where the ends of the paths connect fermions on "staggered" lattice sites.

This application of a loop/path representation may prove a valuable alternative way of implementing the lattice approximation when performing calculations in quantum electrodynamics and quantum chromodynamics. But it does not show that there is a general technique for constructing a loop/path transform for passing from a connection representation to a loop/path representation of an interacting gauge theory. Even if such a technique were forthcoming, the transform itself may prove to exist only in a formal sense, while lacking a rigorous mathematical definition. The possibility of constructing a loop/path representation of the quantized interacting theory *ab initio* by setting up a suitable non-canonical algebra of gauge-invariant variables does not appear to have been explored, let alone successfully implemented. It remains to be seen whether a loop/path representation exists for a general Yang–Mills gauge field interacting with fermion fields. So perhaps it is premature to assess the conceptual significance of any such formulation. But I think the potential significance is sufficiently interesting to make it worthwhile to entertain the possibility of such a formulation in the final chapter's inquiry into the interpretation of quantized Yang–Mills gauge theories.

7.5 The θ-vacuum in a loop representation

The availability of loop representations of quantized Yang–Mills theories has interesting implications for the nature of the θ-vacuum discussed in chapter 6, section 6.6. Recall that when the theory is non-Abelian, "large" gauge transformations with non-zero winding number connect potential states with different Chern–Simons numbers, including different candidates for the lowest-energy, or vacuum, state of the field. Requiring that the theory be symmetric under such "large" gauge transformations implies that the actual vacuum state is a superposition of all these candidate states of the form 6.47:

$$|\theta\rangle = \sum_{n=-\infty}^{+\infty} \exp\{-in\theta\} |n\rangle \tag{7.28}$$

where θ is an otherwise undetermined dimensionless parameter—a fundamental constant of nature. Associated with the θ-vacuum is an additional term that enters the effective Lagrangian density for quantum chromodynamics, as mentioned in chapter 6, section 6.3—unless the value of θ is zero, in which case this term itself becomes zero. It turns out that certain empirical consequences of quantum chromodynamics are sensitive to the presence of this extra term: if it were present, then strong interactions would violate two distinct discrete symmetries, namely parity and charge conjugation symmetry. Experimental tests have shown that $|\theta| \leq 10^{-10}$, making one suspect that in fact $\theta = 0$. This fact—that of all the possible real number values it could take on, θ appears to be zero—is known as the *strong CP problem*. Various solutions have been offered, several of which appeal to some new physical mechanism that intervenes to force θ to equal 0. But from the perspective of a loop representation, there is no need to introduce θ as a parameter in the first place. I quote (Fort and Gambini 2000, p. 348):

It is interesting to speculate what would happen if from the beginning holonomies were used to describe the physical interactions instead of vector potentials. Probably we would not be discussing the strong CP problem. This would simply be considered as an artifact of an overdescription of nature, by means of gauge potentials, which is still necessary in order to compute quantities by using the powerful perturbative techniques. From this perspective, the strong CP problem is just a matter of how we describe nature rather than being a feature of nature itself.

As Fort and Gambini (2000) explain, when a theory is formulated in a loop/path representation, all states and variables are automatically invariant under both "small" and "large" gauge transformations, so there is no possibility of introducing a parameter θ as in equation 6.47 to describe a hypothetical superposition of states that are not so invariant. While the conventional perspective makes one wonder why θ should equal zero, from the loop perspective there is no need to introduce any such parameter in the first place. Once formulated, the loop representation will be equivalent to the usual connection representation with $\theta = 0$. One can introduce an arbitrary parameter θ into a loop representation of a more complex theory, as Fort and Gambini (2000) show. But from the holonomy perspective there would have been no empirical reason to formulate such a more complex theory, and the fact that even more precise experiments do not require it would be considered a conclusive reason to prefer the simpler theory—the one that never introduced an empirically superfluous θ parameter. Here we have another instance of the general epistemological situation discussed extensively in chapter 4, and especially in section 4.4.1.

7.6 Conclusion

The discussion in chapters 5 and 7 has made it abundantly clear that, for each classical Yang–Mills theory, there is not just one corresponding quantized theory, but many. Still further proliferation results from additional technical choices (e.g. of factor ordering), the need for which has not even been considered. Are the various candidates for *the* quantized theory corresponding to a classical theory of a given Yang–Mills gauge field equivalent, and if so in what sense?

We have already noted some equivalences and inequivalences. Ashtekar and Isham showed that two loop representations of the free quantized Maxwell field are inequivalent, in so far as one can be represented on a Fock space

7.6 CONCLUSION 199

on which the other cannot. The empirical success of the description of quantized electromagnetism in terms of photons may appear to rule out the latter representation from further consideration. Surely, no candidate should be granted the title of quantization of a classical gauge field if it is not even empirically equivalent to another empirically successful quantized theory of that field. But its dismissal on these grounds may be premature, as the next chapter will argue.

Path-integral quantizations and connection representations based on canonical quantization appear at least to be empirically equivalent. The existence of a loop transform relating a connection to a loop representation of a canonically quantized free Yang–Mills field shows that these are equivalent in a stronger sense than that—arguably, they are simply alternative formulations of the same theory. There is an obvious parallel with the alternative wave-function representations of a quantum particle theory (e.g. position and momentum representations) which are related by a unitary transformation, as explained in appendix D. But only in certain cases has the loop transform been show to exist as a well-defined mathematical object, so this conclusion may prove premature. Nevertheless, I assume in the next chapter that this technical problem can be overcome, so that any loop representation of a Yang–Mills gauge field that admits a Fock representation may be regarded as simply an alternative formulation of a corresponding connection representation (and vice versa).

8

Interpreting quantized Yang–Mills gauge theories

What should we believe the world is like, given the empirical success of the quantized Yang–Mills theories of the Standard Model? This is the question to pose to their would-be interpreters. Despite the theories' contemporary prominence in fundamental physics, the question has rarely been squarely faced. After considering, but rejecting, one interesting answer offered by Auyang (1995), I go on to see how much light is shed on this question by the existence of loop representations. One can view this as an initial survey, from one point of view, of the ontology and "ideology" (in Quine's sense: see e.g. Quine (1966)) of quantized Yang–Mills gauge theories. The motivation is to explore the prospects for extending the non-localized holonomy properties interpretation of classical Yang–Mills gauge theories to their quantum counterparts.

8.1 Auyang's event ontology

In her 1995 book, Auyang proposes an interpretation of the quantized Yang–Mills gauge theories that figure in the Standard Model in terms of what she calls events. Each such event appears as a localized occurrence, in so far as it is indexed by a single space-time point. But, for her, these space-time indices are not labels for points of an independently existing structure. Rather, the space-time structure is itself abstracted from a prior structure of events, each represented by bundle points in a fiber bundle formulation. The space-time structure is characterized by the symmetries of the bundles' base manifold, which Auyang takes to be those of the Poincaré group—spatial and temporal translations, as well as the uniform (special relativistic) velocity boosts of the Lorentz group.

The events themselves are thought of as local interactions between a gauge potential, represented by the connection on a principal fiber bundle, and one or more matter fields, each represented on an associated vector bundle. Auyang writes (a point of) a single free matter field as $\psi(x)$, but she uses this to symbolize the fiber above base point x rather than a particular element of that fiber. In

8.1 AUYANG'S EVENT ONTOLOGY

a fixed gauge, distinct elements $\theta(x), \theta'(x) \in \psi(x)$ then correspond to different possible phases of the matter field at x, one of which is actual. She takes a choice of gauge to correspond to a choice of section for the principal bundle, yielding a representation of its connection by a gauge potential with coordinate representation $A_\mu(x)$ on the base manifold. In a fixed gauge, each matter field $\psi(x)$ is represented by a particular section $\theta(x)$ of an associated vector bundle. She understands a gauge transformation as a change from one coordinated set $A_\mu(x)$, $\theta(x)$ to a gauge-equivalent set $A'_\mu(x), \theta'(x)$ induced by a change of principle bundle section, rather than as a vertical automorphism of the principal fiber bundle that transforms its connection. The structure group of the bundles gives the structure of possible events located at each point x, just one of which is actual.[1]

But it is unclear what Auyang takes an actual event to consist in. Here is what she says (1995, p. 129). Italics are in the original, but I have added "hats" to all operator expressions.

Consider an interacting field system with N free fields represented by local field operators $\hat{\psi}_i(x)$, $i = 1, \ldots, N$, where x is the four-dimensional spatiotemporal parameter. The field system is characterized by the Lagrangian in which all interaction terms are of the form $\hat{\psi}_i(x)\hat{\psi}_j(x)\hat{\psi}_k(x)$; that is, all interactions among the free fields occur at a point. The interacting system itself must be considered as an integral whole, in which the free fields are approximations, for the charges of the free fields are also the sources of interaction. *All local fields with the same parameter x and their products constitute an event. An event is an entity in an interacting field system.* An event is extensionless in all four dimensions; hence it is spatially and temporally indivisible. However it is analyzable into free fields and their interaction. …

An event is a dynamical quantity; it is the transformation of the state of the field system at a certain point. For example, the event $\hat{\psi}_i(x)$ may represent the excitation of certain modes of type i, $\hat{\psi}_i(x)\hat{\psi}_j(x)\hat{\psi}_k(x)$ may represent the excitation of certain modes of type i and deexcitation of certain modes of types j and k. Our events are concrete; they are distinct from the events in general relativity texts, which are merely points of a bare manifold. Our technical "event" differs from the ordinary usage of "event," which means a happening to enduring things. However the two are related. The concept of enduring things can be constructed from our events. Conversely, if we regard things as the basic concept, then our events become incidents.

Further attempts at clarification follow:

The properties of an event, which is designated by a single identity or absolute position, are explicitly analyzed into matter and interaction fields and their coupling.

---

[1] As chapter 1 and appendix B make clear (see figure 1.6), it would be better to regard the matter field as represented by a *single* section of an associated vector bundle. Depending on what section is chosen for the principal bundle, this section may be taken to represent a field $\psi(x)$, with generalized phase $\theta(x)$, or a field $\psi'(x)$, with generalized phase $\theta'(x)$. Either way, changing the principal fiber bundle section changes ψ, θ, and A_μ—the representation of the gauge potential that enters into the coordinate form $D_\mu = \partial_\mu + A_\mu$ of the covariant derivative on the vector bundle.

Since dynamical coupling becomes a characteristic of an event, the concept of causal relation is built into the concept of individual events. (*ibid.* pp. 183–4)

Phases and potentials are both what I call *relational properties* or *gear qualities*, in the sense that an isolated gear is incomplete; gears are designed to be coupled into a system. The two relational properties complement each other to form an interacting system. ... *Both relational properties and their coupling belong to a single event designated by the same x. Thus the concept of events is enriched.* (ibid. p. 186)

and

The potential of the interaction field is part of the structure of the event, not something added on to ready-made events. ... *the general concept of interactions is inherent in the general concept of objects.* This is the result of field theories with local symmetries. The explicit spatiotemporal and causal relations among events are not external but internal and well-founded. Events have not lost their identities with the introduction of explicit relations. Rather, they have individually acquired richer structures accounting for their being members of the community that is the interacting field system. And it is within the system that individual events become fully intelligible. (ibid. p. 189)

Clearly each of Auyang's "events" is supposed to be a localized interaction at a single space-time point. But while it is easy to say that an "event" is a local interaction between quantum fields, it is not at all clear what in a quantized Yang–Mills gauge theory represents such a local interaction, or how it is supposed to represent it. A matter field operator such as the Dirac $\hat{\psi}(x)$ has no localized eigenstates with support confined to x, so it is unclear in what sense it could represent the excitation of any localized field mode. Moreover, such an operator is not self-adjoint, so that even if it did have a localized eigenstate, the corresponding eigenvalue could not represent a real localized field value: if one tried to think of $\hat{\psi}(x)$ as representing the excitation of a particle, then it would equally have to represent the deexcitation of an antiparticle. More fundamentally, as Teller (1995) stresses, the state of a system of interacting fields is specified not by any field operators, but by the state vector on which these act (or perhaps by the operators' expectation values in that state), just as the state of a system in non-relativistic quantum mechanics is specified by a state vector or wave-function, not by operators such as $\hat{x}, \hat{p}_x$ that represent dynamical variables. Similarly, it is at best a metaphor to say that an interaction term in a Lagrangian (such as the term $-e\bar{\psi}\gamma^\mu\psi A_\mu$ in 5.54) represents a local interaction in an interacting field system.

When Auyang says that an event is the *transformation* of the state of the field system at a point she in effect acknowledges that field operators and their products do not represent the state of the field system itself at that point. But since it is not clear how or whether the state vector or anything else in the theory represents the system's state at a space-time point, it is not clear how an event (characterized by local operators and their products) can be the transformation of anything acknowledged by the theory into anything else within its domain.

One can make some sense of such talk *globally*. For example, in a free field theory with a Fock representation, the total number operator $\hat{N}$ represents a *mathematical* transformation of the state of the field into another state, which differs from it unless the original state was interpretable as one containing a determinate number of (non-localized!) quanta. But this was an example of a purely mathematical operation, with no relation to any physical interaction involving the field. Confusion between mathematical and physical operations may be fostered by reference to operators such as $\hat{a}_k^\dagger, \hat{a}_k$ as creation and annihilation operators—a confusion noted by Feynman (1965) in his Nobel lecture with the remark "How do you create an electron? It disagrees with conservation of charge."

Despite her interesting and suggestive remarks, Auyang has not presented a coherent ontology for a Yang–Mills gauge theory that tells us what we should believe the world is like, given the empirical success of the quantized Yang–Mills theories of the Standard Model. The search for an interpretation must be pursued elsewhere.

8.2 Problems of interpreting a quantum field theory

The empirically successful Yang–Mills gauge theories of contemporary physics's Standard Model are quantum field theories. To say what the world is like if these theories are true it is therefore necessary to engage in the project of interpreting a quantum field theory—a project that Stein (1970) once called the contemporary locus of metaphysical research. In recent years philosophers have begun to pursue this project in earnest[2]. But no consensus has yet emerged, even on how to interpret the theory of a free, quantized, real scalar field. Key ontological questions provoke dispute and ongoing research: "Can a quantum field theory be considered to describe particles, and if so in what sense?," "Is quantum field theory really a theory of fields, and if it is, then how are these related to the fields described by a corresponding classical theory?" These are over and above the familiar interpretative issues faced by any quantum theory, including the measurement problem and the nature of quantum mechanical non-locality.[3]

The additional interpretative problems raised by quantum field theories are not only ontological. There is an important difference between the mathematical structures involved in quantizing a field and quantizing a particle theory.

---

[2] See, for example, Redhead (1983, 1988), Brown and Harré (1988), Teller (1995), Auyang (1995), Huggett (2000), Kuhlmann, Lyre, and Wayne (2002), Ruetsche (2002), Clifton (2004).

[3] See, for example, van Fraassen (1991), Albert (1992), Bub (1997), Maudlin (1994), Healey (1998), Ghirardi (2004).

8 INTERPRETING QUANTIZED YANG—MILLS GAUGE THEORIES

A classical particle theory describes a system containing only finitely many kinematically independent variables—so-called degrees of freedom. But a field theory describes a system with an infinite number of degrees of freedom. When applied to a classical theory, the canonical quantization procedure first replaces classical Poisson bracket relations among the canonical variables by commutation (or anti-commutation) relations among corresponding abstract operators, and then seeks to represent the resulting algebra by operators on a Hilbert space of states. As appendix D explains, the Stone—von Neumann theorem shows that all representations are essentially equivalent in the case of a particle theory in ordinary space. But the theorem fails to generalize to systems with an infinite number of degrees of freedom. Hence a basic abstract algebra of operators pertaining to a quantum field system has many inequivalent representations on a Hilbert space of states. An interpretation of a quantum field theory needs to account for the significance of these inequivalent representations, and the relations between them.

At the root of all these interpretative problems is the existence in any quantum theory of multiple representations of the space of possible states of a system, and of the operators on this state space representing dynamical variables pertaining to that system. This multiplicity of representations does not occur in a classical theory.

In a quantum theory, states and dynamical variables (here often called "observables") may be represented in various ways, as we saw in chapter 7, section 7.2. For the quantum mechanics of particles, a standard representation of the particles' state is by means of a vector in a Hilbert space, on which the observables are represented by self-adjoint operators. The canonical commutation relations 7.1, or rather their Weyl form 7.6, constrain these Hilbert space representations, but leave open an infinite set of alternatives. The Stone—von Neumann theorem shows that all of these are formally equivalent, in the sense that any pair of "well-behaved" representations[4] are related by a unitary mapping. This formal equivalence makes the task of interpreting the quantum mechanics of particles easier than that of interpreting a quantum field theory, to which the Stone—von Neumann theorem does not apply. In so far as the empirical content of a quantum theory is exhausted by its probabilistic predictions concerning the outcomes of measurements on a system, equivalence of representations entails empirical equivalence of theory formulations based on different representations: for the predictions flow from expectation values of operators, and unitary equivalence guarantees that these are the same in all representations.

But while this unitary equivalence of representations simplifies the problem of interpreting a quantum particle theory, it certainly does not solve it.

---

[4] A faithful representation of 7.6 is well behaved if it is strongly continuous, irreducible, and unitary: see appendix D.

An interpretation of quantum mechanics needs to say what it is about the measurement process that gives rise to outcomes, and how we are to understand the probabilities the theory prescribes for them. Not only is there no consensus on how to interpret such a theory, but also the interpretative significance of the equivalence of representations is itself disputed. Appendix F provides a rapid sketch of the main features of several prominent approaches to the interpretation of quantum mechanics as these apply to the quantum mechanics of non-relativistic particles: readers seeking a deeper understanding are urged to consult the sources referred to there or in footnote 3.

Some approaches, including Bohmian mechanics and certain modal interpretations, privilege a particular representation (typically a representation that represents states by wave-functions that are functions of position) when it comes to saying what the world is like according to the theory. Other interpretations, including the Copenhagen interpretation, Everett-inspired interpretations that appeal to environmental decoherence, and other modal interpretations, take a more democratic attitude toward alternative representations. They regard one representation as offering our best description of a system in one situation, while a different representation is preferable in another situation: which is preferable depends on the experimental arrangement, or (more specifically) the character of the interactions between the system and its measuring apparatus and/or its environment.

This chapter focuses on interpreting quantized Yang–Mills gauge theories, and more particularly on the interpretative import of loop representations of those theories. Two related questions are of particular interest in the light of the concerns of the first part of the book. What do loop representations imply about the existence of localized gauge properties and/or non-localized holonomy properties? What are the implications of loop representations for the locality or separability of processes described by quantized Yang–Mills gauge theories? At this stage in our attempts to interpret a quantum field theory the best one can do in trying to answer these questions is to explore the role of loop representations within various interpretative approaches.

8.2.1 Particle interpretations

A quantized Yang–Mills gauge theory is a quantum field theory. Interpretative studies of quantum field theory have concentrated on theories without gauge symmetries. Indeed, in his benchmark text, Teller (1995) does not deal with gauge theories at all. By contrast, many interpretative discussions have explored the prospects of either a particle or a field ontology for a quantum field theory, and examined the conceptual shift required to implement either or both of these ontologies. Such issues clearly arise in the case of gauge theories of elementary "particles". Photons are commonly thought of as the massless quanta of the quantized electromagnetic field, while massive vector bosons mediate the weak interaction field, and the strong interaction is carried by gluons—the quanta of the nuclear force field. A minimal requirement for any kind of particle ontology for a quantum field is the existence of a Fock representation of its ETCRs, since this guarantees that the associated Fock space decomposes into a direct sum of Hilbert spaces, each associated with a definite number of quanta. This is surely one reason why Fock representations of a quantum field theory have figured prominently, both in textbook presentations of the theory and in attempts to say what the world would be like if such a theory were true.[5]

The Hilbert space of a Fock representation for a boson field is the infinite direct sum of n-dimensional Hilbert spaces, each spanned by symmetrized vectors. The free real Klein–Gordon field provides a simple example. The total energy of a free real Klein–Gordon field is represented by a Hamiltonian operator $\hat{H}$ acting on a Fock space of states. The total field energy may be decomposed into a discrete amount $E(\mathbf{k}) = \hbar\omega_\mathbf{k}$ of energy associated with each so-called number operator $\hat{N}(\mathbf{k}) = \hat{a}^\dagger(\mathbf{k})\hat{a}(\mathbf{k})$:[6]

$$\hat{H} = \int \hbar\omega_\mathbf{k} \hat{a}^\dagger(\mathbf{k})\hat{a}(\mathbf{k}) \mathrm{d}^3 k \qquad (8.1)$$

and the total field momentum may be similarly decomposed:

$$\hat{\mathbf{p}} = \int \hbar\mathbf{k} \hat{a}^\dagger(\mathbf{k})\hat{a}(\mathbf{k}) \mathrm{d}^3 k \qquad (8.2)$$

These decompositions suggest that the free field contains various numbers of quanta, each of momentum $\hbar\mathbf{k}$ and energy $\hbar\omega_\mathbf{k}$, and lends some support to a particle interpretation based on the Fock space representation, though this must be qualified in a number of ways. First, eigenstates of the total number operator $\hat{N} = \int \hat{N}(\mathbf{k}) \mathrm{d}^3 k$ superpose to give states with no determinate number of quanta present. Second, there is no particle position operator in the representation whose orthogonal, Lorentz-covariant eigenstates could be thought to represent states in which one or more quanta have precise positions. Third, nothing in the representation can represent interchanging the states of quanta with the same momentum. A state can be thought to represent only the total number of quanta of a particular momentum, not which quantum is which. Teller (1995) expresses this fact by saying that quanta, unlike particles, can be aggregated but not counted.

The states and observables of other free fields also have Fock representations, including the free Maxwell field and free Yang–Mills fields considered in

---

[5] Fock representations are more fully described in appendix E.
[6] This assumes that the terms entering into the definition of the Hamiltonian operator have been normal-ordered, with annihilation operators placed to the right of creation operators, thus effectively "subtracting" the infinite zero-point energy associated with the $\delta^3(0)$ term in equation E.21.

chapter 5, whose canonical quantization began with ETCRs like 5.33–5.35. It is presumably within such representations that one can hope to find a place for talk of photons, gluons, and other gauge bosons such as the massive intermediate vector bosons $W^{\pm}$, Z^0. But, as Ashtekar and Isham (1992) point out, the relation between Fock representations and loop representations of free quantized Yang–Mills theories is problematic. As they put it (p.398) "... it would appear that there are kinematic obstructions to the use of Fock representations in the non-abelian context." They take this conclusion to follow from their preceding analysis, which they summarize as follows (p. 397).[7]

The text-book treatments of the quantum Maxwell field use the conventional Weyl algebra $\mathcal{A}$. However, there is no a priori reason for preferring it to the Weyl algebra of loops and one-forms, or of loops and surfaces. On the contrary, in non-abelian theories—including general relativity—the conventional algebra is less natural because the smeared-out connection operators $\hat{A}(g)$ are not gauge-invariant. On the other hand, the holonomies *are* gauge-invariant. Hence it is the algebras based on closed loops that seem to admit the most useful extension to the non-abelian context.

So there is an apparent tension between two approaches to the interpretation of quantized (non-Abelian) Yang–Mills theories, one based on Fock representations, the other based on loop representations. By insisting on the priority of loop representations one may hope to reach an understanding of these theories as describing a world without gauge: but this would also be a world without gauge bosons! But there are reasons to believe that closer examination will relieve this tension.

Ashtekar and Isham (1992) show that in the quantum theory of the source-free Maxwell field there is an additional ambiguity over and above that flowing from the existence of inequivalent representations of the fundamental ETCRs (or more precisely, the Weyl algebra of operators). Moreover, they construct algebras which in the classical theory are on the "same footing" as the one normally used, but which in the quantum theory cannot be represented by operators on the standard Fock space.

However, Ashtekar and Isham do not claim that *no* algebra pertaining to a quantized Yang–Mills gauge theory which is based on closed "loops" admits *any* Fock representations. Indeed they show that in the case of the source-free Maxwell field one such algebra may be represented on the *standard* Fock space whose basis states have the usual interpretation in terms of photon occupation numbers. This is the algebra based on negative frequency connections and positive frequency electric fields whose gauge-invariant operators Ashtekar and Rovelli showed to be related to operators corresponding to those fields via

---

[7] See appendix E for background on Weyl algebras, their representations, and the Gelfand–Naimark–Segal theorem.

a well-defined loop transform. It is even consistent with their claims that *every* algebra pertaining to a quantized Yang–Mills gauge theory which is based on closed "loops" admits a Fock representation (though this is not always unitarily equivalent to the standard Fock representation).

Now Kay and Wald (1991) show how the Gelfand–Naimark–Segal theorem may be used to arrive at a Fock representation of a Weyl algebra $\mathcal{A}$ starting from what they call a *quasi-free* state s_μ on $\mathcal{A}$. Can their result be used to arrive at a Fock representation even for those algebras based on closed "loops" that Ashtekar and Isham show not to be representable on the standard Fock space? If it can, then we would have a situation reminiscent of that occurring in the much-discussed case of the unitarily inequivalent Fock representations that give rise to Minkowski and Rindler quanta (see e.g. Clifton and Halvorson 2001). But it appears that the Proposition 3.1 of Kay and Wald (1991) is not applicable to the Weyl algebras that are relevant in the present case. Specifically, to use the result of that proposition to generate a Fock space it is necessary to assume that the symplectic form figuring in the Weyl relations is defined on a vector space S of classical real C^∞ solutions to a field equation on a spacelike hypersurface. But in the case of Ashtekar and Isham's algebras of "loops" and one-forms, or of "loops" and surfaces, this is not so: their "loop" operators $\hat{A}[\alpha] = \hat{A}(g[\alpha])$ are "smeared" not by real C^∞ functions of compact support, but rather by *distributions*: the (transverse) vector $g[\alpha]$ satisfies

$$g^a[\alpha](\mathbf{x}) = \oint_\alpha dt \dot{a}^a(t) \delta^3(\mathbf{x}, a(t)) \tag{8.3}$$

where t is any parameter along a curve from the hoop α and $\dot{a}^a(t)$ is the tangent to the curve. If this is right, then it remains an open question whether all the algebras discussed by Ashtekar and Isham (1992) have Fock representations.

In evaluating the significance of Ashtekar and Isham's conclusion it is important to distinguish their motivations for exploring the need for non-Fock representations from those lying behind the present interpretative project. Ashtekar and Isham are mainly interested in the lessons to be learned from source-free Maxwell theory for those interested in developing theories of quantum gravity based on "loop" variables, especially those related to Ashtekar's reformulation of general relativity in terms of a self-dual connection. This explains their remarks on diffeomorphism invariance and their lack of interest in extending their analysis to interactions with charged matter fields, as well as their repeated references to general relativity. But general relativity and quantum gravity are tangential to the project of interpreting the Yang–Mills theories that are at the core of the Standard Model, while interactions are highly relevant.

They do offer one reason for thinking that the existence of loop representations of Yang–Mills theories that are not representable in the standard Fock space is not merely a mathematical curiosity; namely, that the holonomies are gauge invariant, unlike the connection operators. Of course, if gauge symmetry is a purely formal symmetry, then a formulation of a theory in which no gauge-dependent magnitudes appear *ipso facto* offers a more intrinsic representation of its subject matter. But what Ashtekar and Isham here refer to as holonomies are operators, not magnitudes that take on measurable values. Their relation to intrinsic features of the theory's domain are therefore already quite indirect. We have yet to see any reason why gauge-invariant holonomy operators should offer a more faithful or intrinsic depiction of that domain than gauge-dependent connection operators. Attempts to secure a particle interpretation of a quantized non-Abelian Yang–Mills theory seem unlikely to reveal such reasons, even though they are not blocked by the unavailability of Fock representations of "loop" algebras. Perhaps a Bohmian interpretation will show why it is holonomy operators that faithfully represent the world of the Standard Model.

8.2.2 Bohmian interpretations

Consider now a Bohmian approach to the interpretation of the quantized Maxwell field. In his seminal paper Bohm (1952) already presented an interpretation in terms of a representation that privileged the transverse vector potential in Coulomb gauge:

$$\mathbf{A}^T(\mathbf{x}) = \frac{1}{(2\pi)^{3/2}} \sum_{l=1}^{2} \int d^3k e^{i\mathbf{k}\cdot\mathbf{x}} \boldsymbol{\varepsilon}^l(\mathbf{k}) q_l(\mathbf{k}) \qquad (8.4)$$

where $\boldsymbol{\varepsilon}^l(\mathbf{k}), l = 1, 2$ are two polarization vectors orthogonal to one another and to $\mathbf{k}$, and q_1, q_2 are two complex-valued fields in momentum space. States in the representation are functionals of the form $\Psi(q_1, q_2)$, and the operator $\hat{\mathbf{A}}^T(\mathbf{x})$ is diagonal in the $|q_1, q_2 >$ basis. Consequently the magnetic field $\mathbf{B} = \nabla \times \mathbf{A}^T$ of the quantized free Maxwell field is always well defined, while measurements of the non-diagonal electric field operator $\hat{\mathbf{E}}^T(\mathbf{x})$ do not reflect the actual values of the electric field (defined as $-\partial_t \mathbf{A}^T(\mathbf{x}, t)$). The contrasting roles of the magnetic and electric fields here is analogous to the contrasting status of position and momentum in Bohmian particle mechanics.

Accepting this interpretation commits one to localized gauge potential properties represented by the theory of the free quantized Maxwell field. But these properties are never locally manifested, in the Aharonov–Bohm effect or elsewhere. Consequently their epistemological status is just as shaky as the analogous localized gauge potential properties of one interpretation of classical Maxwell theory studied in chapters 2 and 4. Moreover, the evolution of these properties singles out a preferred Lorentz frame, which, however, remains empirically inaccessible, in just the same way as the evolution of particle positions in Bohmian mechanics.

A loop representation of the free Maxwell field corresponding to Bohm's connection representation would diagonalize holonomy operators of the form

$$\hat{H}_{\mathbf{A}}(\gamma) = \exp\left(i \int_{C \in \gamma} \hat{\mathbf{A}}^T(\mathbf{x}).d\mathbf{x}\right) \quad (8.5)$$

where γ is a hoop in a spacelike hyperplane of simultaneity in the Lorentz frame picked out by the Coulomb gauge for $\mathbf{A}^T$. These holonomy operators are invariant under gauge transformations $\mathbf{A}^T \to \mathbf{A}^T - \nabla \Lambda$, so this loop representation would support a Bohmian interpretation in which loop wavefunctions $\Psi(\gamma)$ define a guidance equation for the evolution of gauge-invariant holonomies $H_{\mathbf{A}}(\gamma)$ representing non-localized holonomy properties. Accepting this interpretation would commit one to an ontology of non-separable holonomy properties analogous to those of the holonomy interpretation of classical electromagnetism defended in chapter 4. The difference would be that these would pertain not to arbitrary space-time loops, but only to loops on spacelike slices represented by the preferred hyperplanes of the interpretation. Though gauge invariant, the interpretation would still violate fundamental Lorentz invariance. But the metaphysical implications would remain analogous to those discussed in chapter 4, section 4.5.

Bohm's interpretation of the quantized free Maxwell field has been extended by Struyve and Westman (2006) to a novel Bohmian interpretation of quantum electrodynamics. After noting problems faced by previous Bohmian interpretations of fermionic field theories, they show how these may be avoided by treating fermion fields in much the same way that Bell (1987) treated spin in his Bohmian model for the non-relativistic spin 1/2 particle. The idea is to assign *no* ontological significance to fermionic fields, regarding them as simply adding further degrees of freedom to the quantized Maxwell field which modify its dynamics but are then integrated over before attempting to interpret the result. The resulting interpretation is a radically "minimalist" extension of Bohm's interpretation based on the Coulomb transverse connection according to which quantum electrodynamics describes *only* the dynamics of this transverse connection, while nothing in reality corresponds to the electron field—neither electrons nor anything else.

The loop-representation version of this extension would interpret quantum electrodynamics as a theory describing *only* how non-separable holonomy properties associated with electromagnetism evolve—not how electrons interact electromagnetically. If acceptable, this promises a completely local, but non-separable, account of phenomena like the Aharonov–Bohm effect. It would not be an account of how the electromagnetic field in a region influences the behavior of electrons that never enter that region. Instead, it would be an account of how the evolution of magnetic holonomies in a region can fool us into believing we observe electrons passing through that region and then producing an interference pattern when they reach a screen! Perhaps this takes minimalism too far, though the idea that the *esse* of "ordinary" fermionic matter is its *percipi* might appeal to a contemporary follower of Bishop Berkeley.

One might try to generalize this Bohmian interpretation of a loop representation of the source-free quantized Maxwell field to yield an interpretation of other pure Yang–Mills gauge theories. This would privilege the loop representation by analogy to the way in which Bohmian mechanics privileges the position representation of ordinary non-relativistic quantum mechanics. To any operator that is diagonal in this representation, including those representing Wilson loops (traces of holonomies), there will be a corresponding magnitude which always takes on a definite value. The wavefunction in this representation is defined on each spacelike hypersurface of a foliation of space-time, and its evolution between hypersurfaces obeys a Schrödinger equation. The value of a Wilson loop magnitude on a hypersurface will be an assignment of the trace of a holonomy to every loop in the hypersurface.

The basic ideology of the theory would include non-localized properties represented by Wilson loops and any other gauge-independent magnitudes represented by operators that commute with them, defined on all loops within each spacelike hypersurface of the preferred foliation. It would *not* include local generalized electric field properties represented by a determinate vector at each space-time point; nor would gluons or other field quanta corresponding to a Fock representation of the relevant Weyl algebra figure in the basic ideology or ontology. Just as measurements of magnitudes other than position in non-relativistic Bohmian mechanics come down in the end to measurements of position, so also in the proposed Bohmian strategy for interpreting a source-free quantized Yang–Mills theory, measurements of generalized electric field magnitudes or Fock space magnitudes like photon number would come down in the end to measurements of gauge-invariant holonomy magnitudes.

Alternatively, following up a concluding suggestion by Struyve and Westman (2006), one might extend the minimalism by denying *any* ontological significance to Yang–Mills theories other than that of the source-free Maxwell field. No such additional ontological commitment is required to explain our experience, since all our experimental access to phenomena supposedly involving weak and strong interactions is ultimately mediated by electromagnetic processes—and the minimalist Bohmian interpretation provides all the ideology we need to account for them by way of the magnetic holonomy properties. Adopting such extreme minimalism would have the paradoxical effect of reducing the quest for an interpretation of non-Abelian Yang–Mills gauge theories to a wild goose chase.

8.2.3 Copenhagen interpretations

Consider instead a complementarity strategy for interpreting a quantized Yang–Mills theory. This would view alternative representations of a basic Weyl algebra as complementary to one another, offering mutually exclusive but jointly necessary perspectives on the world the theory models. On this approach, loop representations add an essential additional perspective. There are precedents for this strategy for interpreting a quantum field theory.

Teller (1995) speaks of the field and quantal aspects of quantum field theory, and takes each aspect to involve propensities for the manifestation in mutually exclusive experimental arrangements of properties characteristic of fields and quanta respectively. He even uses the language of complementarity in this context.

> If one is willing to identify quantal descriptions with descriptions having exact values in eigenstates of the number operator and wave descriptions with descriptions having exact values in eigenstates of at least some operators that do not commute with the number operator, then we have a precise statement of the so-called complementarity or duality of "wave" and "quantal" descriptions: These descriptions are complementary or dual exactly insofar as the operators in question do not commute, in precisely the way that position and momentum are complementary in conventional quantum mechanics. (p. 113)

He discusses quantum field-theoretic analyses of the Unruh effect, in which an accelerated observer would detect field quanta even in the Minkowski vacuum state. Such quanta are known as Rindler quanta. Teller regards this as an example involving wave descriptions, and groups it with field phenomena such as vacuum fluctuations because "the considerations that resolve the dispute about [Rindler quanta] are exactly the same as those which resolve the felt conflict about vacuum fluctuations." (p. 110)

But phenomena like the Unruh effect show that there is more than one quantal aspect or quantal description in a case involving unitarily inequivalent Fock representations of the same basic Weyl algebra, none of which is naturally regarded as providing a wave description or illustrating a field phenomenon. Moreover, while each of these inequivalent Fock representations has a (total) number operator, these operators neither commute nor fail to commute, since they are not defined in a common Hilbert space. So for Teller (1995) a quantal description in the Minkowski Fock space does not count as complementary to a quantal description in Rindler Fock space, despite the fact that there are mutually exclusive experimental arrangements in which one or other description is appropriate.

While critiquing the interpretation of Rindler quanta offered by Teller (1995), Clifton and Halvorson (2001) nevertheless defend the idea that unitarily inequivalent Fock representations here offer complementary perspectives on the quantum field. Each representation is appropriate for an experimental arrangement involving apparatus in a different state of motion—uniform for the Minkowski representation, uniformly accelerated for the Rindler representation. The perspectives are complementary in a physical sense, since the experimental arrangements are mutually exclusive. But they are not complementary in the formal sense that all non-trivial operators diagonal in one representation fail to commute with any non-trivial operator diagonal in the other. For the two representations are disjoint (no state defined on the Fock space of one is also defined on the Fock space of the other), and certain operators that are diagonal in one representation (such as its total number operator) are not even unitarily equivalent to any operator defined in the other.

An important residue of Teller's formal notion of complementarity remains. As Clifton and Halvorson (2001) show, one can construct a Rindler number operator $\hat{N}_R(f)$ for quanta *of wave function f* within a Fock representation for Minkowski quanta (and vice versa), and such a number operator in one representation fails to commute with any number operator in the other representation. Indeed, the same is true for number operators for any finite-dimensional subspace of wave-functions. When it comes to *total* number operators, the formal sense in which Minkowski and Rindler Fock representations are complementary is more subtle, since the total number operator of one Fock representation is not definable on the Fock space of the other representation. But Clifton and Halvorson (2001) offer two reasons why this indefinability does not refute a complementarity approach to Rindler vs. Minkowski representations.

First, they prove that either representation predicts dispersion in the total number operator of the other, in the sense that every extension of an abstract vacuum state to a state on all the observables of one representation produces dispersion in at least some bounded function of the total number operator of the other representation.

Second, they show that there is a close analogy between the present situation and that involving the complementary magnitudes position and momentum for a one-dimensional non-relativistic quantum system. While neither $\hat{x}$ nor $\hat{p}$ is diagonal in the usual representations (Schrödinger position, and momentum, respectively), one can define representations of the Weyl form of the CCRs in which either one or the other *is* diagonal. But $\hat{p}$ is not definable in a representation in which $\hat{x}$ is diagonal, and $\hat{x}$ is not definable in a representation in which $\hat{p}$ is definable.[8] This is analogous to the case of Rindler and Minkowski quanta, since $\hat{N}_M$ but not $\hat{N}_R$ is definable in the Minkowski Fock representation, while $\hat{N}_R$ but not $\hat{N}_M$ is definable in the Rindler Fock representation. The analogy provides a second reason for

---

[8] These representations are not unitarily equivalent. There is no conflict with the Stone–von Neumann theorem, since neither is strongly continuous.

considering descriptions in terms of Rindler quanta to be complementary to descriptions in terms of Minkowski quanta.

How can these considerations be applied to the interpretation of a quantized Yang–Mills gauge theory? One result of the analysis of Clifton and Halvorson (2001) is directly applicable, on our working assumption that the free theory has a Fock representation defined as a GNS representation of an abstract vacuum state of its Weyl algebra. The result is that while the total number operator of the Fock representation is not representable as an operator in any disjoint representation, one can still give a sense to the claim that (with probability one) there are infinitely many field quanta in any state of the disjoint representation, and indeed this claim is *true* in that sense.

There are then two cases to consider. Either a loop representation of a quantized Yang–Mills gauge theory is unitarily equivalent to the standard Fock representation whose quanta are photons, gluons, or vector bosons carrying the weak interaction, or it is disjoint from that representation.

In the former case, the loop representation may be taken to provide a description complementary to that offered by the standard Fock representation, much as the momentum representation of ordinary non-relativistic quantum mechanics provides a complementary description to that provided by the Schrödinger position representation. Then Teller's formal analysis applies, with the result that the loop representation reveals a further set of potentialities for the exhibition of non-quantal behavior characterized by well-defined gauge-invariant non-localized holonomy-type magnitudes as in the Bohmian approach, but only in an appropriate experimental arrangement.

In the latter case, Clifton and Halvorson's analysis reveals the surprising result that there are states of the free Yang–Mills field in which a global experimental arrangement suitable for detecting photons (gluons, intermediate vector bosons) would register more than any finite number of such quanta; while an experimental arrangement suitable for detecting a *different* kind of quanta, associated with a disjoint Fock representation unitarily equivalent to the loop representation, would register some finite number of "loop" quanta (perhaps zero).

It remains an interesting project for one pursuing the complementarity approach to the interpretation of a quantized Yang–Mills gauge theory to explore the ramifications of both cases to see whether we have good reason to exclude either case. But a more pressing task is to investigate the kind of detector suitable for manifesting the potential for looplike and/or non-standard quantal behavior of the quantized field. The general character of any resulting interpretation will then depend on the spirit in which that task is undertaken.

For Bohr, the founding father of the Copenhagen interpretation, it was essential to describe the experimental arrangement involved in any manifestation of a quantum phenomenon classically—i.e. in ordinary language, suitably enriched by the concepts of classical physics. Following his lead, any detector suitable for manifesting the potential for looplike and/or non-standard quantal behavior of the quantized field must itself be described classically. That would rule out any attempt to model such a detector in this context as itself just a part of the world represented by the theory being interpreted. It would also curb the metaphysical ambitions of anyone wishing to extract an account of what that world is like from a Copenhagen interpretation of quantized Yang–Mills theories. This should come as no surprise to those familiar with the Copenhagen tradition of dismissing talk of a reality behind and independent of our observations as empty speculation, if not literally meaningless.

The alternative would be to propose and investigate the features of quantum models of a detector and its operation. This is an approach that has appealed to some interested in understanding the analogous Unruh effect. As a way of implementing a Copenhagen interpretation of quantized Yang–Mills theories, it poses significant challenges. At a fundamental level, a detector would need to be built out of or realized by quantized fields. The natural candidates are fermionic matter fields. But it is unclear how one would go about modeling a localized detector realized by quantized fermionic matter fields, or how one would certify its status, absent a prior interpretation of the theory of the quantized matter field.

It was Everett's idea of modeling the operation of a measurement device within the theory that motivated him to suggest what many now take to offer an attractive alternative to the Copenhagen interpretation of quantum mechanics. Let us see what role loop representations of a quantized Yang–Mills theory might play in an Everettian interpretation.

8.2.4 Everettian interpretations

On an Everettian interpretation, the universal quantum state never "collapses," and may be represented equally well in any representation. But in understanding how such a state relates to the world we experience, certain representations play an important role. These privilege observables that are diagonal in a decoherence basis. Appendix F explains this notion for quantum particle theories. In that context, the interactions among subsystems of the universal system are such that the state of a complex subsystem S is almost always representable as a mixture $\hat{W}_S$ of pure states that is approximately diagonal in a basis of states $\{\psi_i\}$ where each ψ_i spans a one-dimensional subspace projected onto by a corresponding projection operator $\hat{P}_i$:

$$\hat{W}_S = \sum_i w_i \hat{P}_i \qquad (8.6)$$

Decoherence is almost never exact, and many different bases meet this condition to the same degree of approximation, so decoherence does not single out a unique preferred basis.

In the context of quantum field theory, it is the actual interactions among quantized fields that mark out certain bases for a Hilbert space representation of the universal state as preferred. But these interactions are not so readily modeled as occurring among a large number of localizable subsystems of the universal quantum system, for the only natural way to decompose a system of interacting quantum fields would be to treat each of these globally defined fields in its entirety as a subsystem. This is one reason why some contemporary proponents of Everettian interpretations have favored an alternative approach which focuses on what are called decoherent histories.[9]

One way to introduce decoherent histories is to start with the idea of some preferred basis for a subsystem in particle quantum mechanics (approximately) defined by environmental decoherence, and to progressively generalize. As Wallace (2002) explains, one can define the basis $\{\psi_i\}$ in terms of the projections $\{\hat{P}_i\}$—a family satisfying three conditions

1. $\quad\sum_i \hat{P}_i = \hat{1}$
2. $\quad\hat{P}_i.\hat{P}_j = 0 \text{ if } i \neq j$
3. Each $\hat{P}_i$ projects onto a one-dimensional subspace.

By dropping the third condition, one can associate an "approximate decoherence basis" with a "coarse-grained" family of projection operators. There will generally be many different ways of selecting one-dimensional projections to collectively span the range of each of the multi-dimensional ranges of projections in this family: each way corresponds to a different "fine-grained" decoherence basis.

One can now generalize further, to define a *decoherent set of histories*, where a particular decoherent history in the set corresponds to a sequence of projections $\{\hat{P}_{t_1}, \hat{P}_{t_2}, \ldots, \hat{P}_{t_n}\}$, where each $\hat{P}_{t_i}$ is itself an element of a family of projection operators satisfying the first two conditions, and $t_1, t_2, \ldots, t_n$ is a sequence of increasing times. If one now regards each $\hat{P}_{t_i}$ as corresponding to an event that may or may not occur at t_i, then it is at least mathematically consistent to assign a probability to each possible sequence of such events in this decoherent set of histories. The final generalization is to drop the restriction to projections only onto the Hilbert space of a subsystem of a universal particle system, and consider decoherent histories as histories of events that occur in a system of interacting quantum fields, so that the $\hat{P}_{t_i}$ project onto the universal Hilbert space on which the states of this entire system are represented.

The formal definition of a decoherent set of histories may be met by "toy" model quantum systems: in such a case it is common to refer merely

---

[9] See in particular Wallace (2002) and the papers by Saunders to which he refers there. I take the development and defense of an Everettian interpretation by Wallace and collaborators here and elsewhere (see the references) to represent the current state of the art.

to consistent histories. It is in the application to a fully fledged system of interacting quantum fields, thought of as offering a fundamental model of our universe, that the term 'decoherent histories' becomes appropriate. For the assumption behind the application is that the interactions among these fields give rise to a sufficiently rich decoherent history set to account for all the events we experience in the particular decoherent history in which we find ourselves. For it is now the decoherent histories that define the (coarse-grained) Everett branches, including the branch that constitutes the world of our experience.

The role of loop representations of quantized Yang–Mills fields in such an Everettian interpretation will depend on the relation between two families of operators. On the one hand are the Wilson loop and generalized electric field operators that define the non-canonical algebra leading to the loop representation. On the other hand are the projection operators that define the quasiclassical domain constituted by the decoherent set of histories that includes the history (or histories) in which we find ourselves. A *quasiclassical domain* consists of a decoherent set of histories, characterized largely by the same types of variables at different times, and whose probabilities are peaked about deterministic evolution equations for the variables characterizing the histories. Quantum mechanically, such a variable is represented by a self-adjoint Hilbert space operator in the Heisenberg picture, and this operator defines the family of projections in the decoherent histories at that time.[10]

Unfortunately, even advocates of the decoherent histories approach admit that considerable uncertainty remains about the identity of the latter family. Gell-Mann and Hartle (1993) argued that the variables typically characterizing the quasiclassical domain of a large complex system are the integrals over small volumes of locally conserved densities—hydrodynamic variables. These would include electric charge and other magnitudes associated with conserved Noether currents. Since the Gauss law connects the charge inside a volume to the electric field on its boundary, electric fields may turn out to be key variables in a quasiclassical domain, with possible extensions to the generalized Noether charges and associated generalized electric field magnitudes of non-Abelian Yang–Mills theories. But note that while such generalized electric field magnitudes are gauge invariant, they are represented by operators that do not commute with Wilson loops or other gauge-invariant operators involving

---

[10] The probabilistic predictions of Hilbert space quantum mechanics may be represented in either of two equivalent ways. In the *Schrödinger picture*, the state evolves unitarily in accordance with the Schrödinger equation: $\psi_0 \to \psi_t = \hat{U}_t \psi_0$, and the probability that observable A lies in Δ at time t is $\|(\psi_t, \hat{P}^A(\Delta)\psi_t)\|^2$, where $\hat{P}^A(\Delta)$ is a projection operator representing an observable with value 1 if $A \in \Delta$, and 0 if $A \notin \Delta$. In the *Heisenberg picture*, it is the observables that evolve, not the states; so the probability that observable A_t lies in Δ in state ψ is $\|(\psi, \hat{P}_t^A(\Delta)\psi)\|^2$, where $\hat{P}_t^A(\Delta) = \hat{U}_t^{-1}\hat{P}^A(\Delta)\hat{U}_t$ and $\hat{P}_t^A(\Delta)$ represents an observable with value 1 if $A_t \in \Delta$, and 0 if $A_t \notin \Delta$. These probabilities are clearly equal. The projections that figure in decoherent histories are Heisenberg picture operators like $\hat{P}_t^A(\Delta)$—an operator that is uniquely defined in terms of the spectral decomposition of the self-adjoint operator $\hat{A}_t$.

holonomies. So holonomy properties seem poor candidates for properties that emerge as basic features of a quasiclassical domain in the way that electric charge and other hydrodynamic variables supposedly emerge.

There are several characteristics of this kind of Everettian interpretation that temper this negative conclusion. First, it rests on little more than speculation as to what variables will end up as basic to the demarcation of a quasiclassical domain. Second, it ignores the role played by *non*-fundamental properties—those that supervene on the values of these basic variables—in constituting the world we experience. We know, for example, that classical physics gives a remarkably effective description of how particles interact under the influence of classical electromagnetism, and chapter 2 showed how this description can even be improved upon by describing these particles quantum mechanically. Presumably the Everettian interpretation must take both the particle ontology and the classical ideology of well-defined trajectories to supervene on the basic variables that demarcate a quasiclassical domain. There is no reason why holonomy properties should not share their status as relatively well-defined supervening structures. Third, whatever status non-separable holonomy properties end up with here, there is nothing in an Everettian interpretation that suggests that localized gauge potential properties will appear as either basic or supervening properties in such an interpretation. They seem gone for good.

8.2.5 Modal interpretations

Modal interpretations were first developed for the non-relativistic theory of quantum particles (see appendix F). Their extension to relativistic quantum field theory is a project in its infancy; moreover, the infant is showing signs of failure to thrive.

Consider the class of modal interpretations that rely on the biorthogonal decomposition theorem to pick out determinate properties on a subsystem S of a quantum system. The rules for assigning determinate dynamical properties to S assume that the quantum state of S may be represented by a density operator $\hat{W}_S$ on the Hilbert space H_S on which the self-adjoint operators representing S's observables act.[11] Since a system of quantum fields has no subsystems with states represented on Hilbert spaces that are factors of the Hilbert space of the system they compose, some modification of these rules is necessary to arrive at a modal interpretation of quantum field theory.

---

[11] The following rules are typical, though there are significant variations among different versions of this variety of modal interpretation. If dynamical property P_i is represented by projection operator $\hat{P}_i$, then S has P_i in state $\hat{W}_S$ with probability w_i just in case $\hat{W}_S = \sum_i w_i \hat{P}_i$; if A is an observable on S represented by self-adjoint operator $\hat{A}$, then it has precise (eigen)value a_j in state $\hat{W}_S$ with probability w_j just in case the projection $\hat{P}_j$ onto the eigenvector corresponding to a_j is one dimensional.

8.2 PROBLEMS OF INTERPRETING A QUANTUM FIELD THEORY 219

Pioneers have explored the prospects of dividing up a quantum field system spatiotemporally, and regarding each open region of space-time together with a local algebra of observables as a subsystem, in the spirit of algebraic quantum field theory.[12] But this raises a significant technical problem, since these local algebras of observables typically cannot admit a density operator—the essential ingredient in the modal rules. Dieks (2000) tries to finesse this problem by appealing to a technical property of local algebras that permits the interpolation of a local algebra that *does* admit a density operator between algebras associated with an open region and an infinitesimally larger open region that contains it.[13] But Clifton (2000) argues that the resulting property ascription rules are ill-defined, and proposes a rather different set of rules that reduce to those of the modal interpretation in circumstances where these are applicable.[14]

While acknowledging Clifton's achievement in formulating a well-defined modal interpretation of a quantum field theory captured by the axioms of local algebraic quantum field theory, Earman and Ruetsche (2005) raise a serious difficulty for this interpretation. The difficulty is that, for at least some physically interesting models of these axioms, the rules of this modal interpretation assign determinate values to *no* non-trivial observables on systems associated with physically interesting space-time regions.

To date, the prospects of extending a modal interpretation along the lines of Bub (1997) or Berkowitz and Hemmo (2006) remain unexplored. In the absence of any well-defined modal interpretation for quantum field theory not open to very serious objections, it is premature to inquire into the significance of loop representations of quantized Yang–Mills theories for such an interpretation. Whatever status non-separable holonomy properties might end up with in some future modal interpretation, there is currently no sign that localized gauge potential properties will appear as either basic or supervening properties in such an interpretation.

---

[12] See, in particular, Dieks (2000), Clifton (2000), Earman and Ruetsche (2005).

[13] The split property concerns diamond-shaped spacetime regions, each defined by the interior of the intersection of the forward and backward light cones of two timelike separated points. It is that, for any two such diamond-shaped regions $\mathcal{R}_r, \mathcal{R}_{r+\epsilon}$, with radii r and $r + \epsilon$ respectively, there is a type I "interpolating" factor $\mathcal{N}_{r+\epsilon}$—a local algebra that includes that of $\mathcal{R}_r$ and is included by that of $\mathcal{R}_{r+\epsilon}$. Since $\mathcal{N}_{r+\epsilon}$ is of type I, a Hilbert space on which it acts *does* admit a density operator $\rho_{r+\epsilon}$ consistent with the state of the quantum field; $\rho_{r+\epsilon}$ may then be inserted into the modal rules to specify a set of determinate dynamical properties associated with the region $\mathcal{R}_{r+\epsilon}$.

[14] These circumstances obtain only if the local algebra is of type I—the exception rather than the rule in models of algebraic quantum field theory.

9

Conclusions

The introduction to this book began by asking what kind of world our gauge theories reveal to us. That has turned out to be a remarkably difficult question to answer with any confidence or finality, and it is worthwhile reflecting on the reasons for and consequences of that difficulty. But a reader who has made it this far is first owed at least a tentative answer, whether or not (s)he has skipped some or all of the intervening chapters!

We have reasons to believe there are physical processes involving properties that are neither localized at or near a point nor determined by properties localized at or near a point, and that some of our theories of fundamental "forces" succeed in capturing significant features of these processes. These reasons are defeasible, both because we do not fully understand our best current theories of these "forces" (the quantized Yang–Mills gauge theories of the Standard Model), and because these represent just the latest (though also most successful!) stage in the evolution of physical theorizing on these matters. We do not currently have reasons to believe that gravitational processes involve such non-localized properties.[1]

While the Aharonov–Bohm effect and related effects provide vivid examples of physical processes that seem best accounted for in terms of non-localized holonomy properties, this does not establish the existence of such properties. Accounts of such phenomena that introduce classical electromagnetism or other gauge fields into models of the quantum mechanics of charged particles hold out the promise of a description in terms of non-separable holonomy properties involving no mysterious action at a distance. But without an agreed interpretation of the theory of quantum particle mechanics there can be no consensus on what that description is, or whether it really avoids action at a distance. Moreover, even if we had such a description, it would not establish the existence of the holonomy properties to which it appeals. For the description would be in terms of theories we believe to be empirically inadequate, though

---

[1] The loop quantum gravity program may provide such reasons. See Smolin (2001), Rovelli (2004). But, like other approaches to quantum gravity, this program has not yet produced an empirically supported theory.

still able to deliver remarkably accurate predictions of what will happen in a wide variety of circumstances.

Known empirical inadequacies of these theories are removed by the quantized gauge theories of the Standard Model, which are currently our most fundamental, empirically confirmed, theories of this domain. But we are even further from an agreed understanding of these theories than we are from an agreed interpretation of the theory of quantum particle mechanics. As quantum field theories, they inherit the measurement problem from quantum particle theories. But while that is a problem for quantum particle theories in so far as they leave it quite unclear what properties the particles have and when they have them, the problem becomes much worse in a quantum field theory. For a quantum field theory removes even the basic particle ontology, while leaving it quite unclear what is to replace it. There is no agreement as to what object or objects a quantum field theory purports to describe, let alone what their basic properties would be. It is not surprising, then, that we lack an agreed account of phenomena like the Aharonov–Bohm effect formulated entirely within the concepts of the quantized gauge theories of the Standard Model.

Still, one may hope to learn something by studying the implications of the rival approaches to interpreting these theories, a task begun in the previous chapter. I take one lesson to be this. Nothing in our current understanding of quantized gauge theories suggests that an account in their terms of phenomena like the Aharonov–Bohm effect will appeal to gauge potential properties localized at or near points. Lacking the need to refer to or represent such properties, an intrinsic formulation of a gauge theory need not involve terms that might be thought to permit such reference or representation. It may be mathematically convenient and/or heuristically useful to continue to formulate our gauge theories using such terms, just as it remains useful to introduce terms for the electric, magnetic, and electromagnetic potentials into the theory of classical electromagnetism. But if we do continue to employ gauge-dependent terminology, we must constantly bear in mind that we do so purely for convenience, and not be fooled into thinking that the consequent gauge symmetry of our theories has any empirical content.

Should we believe that non-separable processes involving non-localized holonomy properties are responsible for phenomena like the Aharonov–Bohm effect? This belief may be encouraged by the predictive successes consequent upon introducing classical electromagnetism into the quantum mechanics of particles. But our limited understanding of the quantized gauge theories of the Standard Model (including quantum electrodynamics) that improve on that success does not unequivocally link these latter theories to any non-localized holonomy properties. Still, in so far as this understanding excludes an interpretation in terms of interactions among particles or localized field structures, it does at least suggest that we look for an account of Aharonov–Bohm-type phenomena in terms of some kind of non-separable process.

9 CONCLUSIONS

Even though gauge theories are at the heart of contemporary physics, perhaps they are *not* our surest guide to the basic structure of the world. Someone may wonder whether there is some basic aspect of the world (consciousness?) they do not cover, but a worry about incompleteness is not the only reason for skepticism. I can think of four grounds for mistrusting the ontological pretensions of contemporary gauge theories.

Like van Fraassen (1991), one may point to the diversity of interpretations of which a physical theory is capable, with the consequent plurality of alternative ontologies. Indeed the previous chapter canvassed several alternative approaches to understanding the quantized gauge theories of fundamental "forces" that form the basis of the Standard Model. But that brief sketch revealed not a thousand blooming flowers but a few struggling shoots. We do not face a conflicting set of equally forceful ontological claims here, but a few tentative ontological suggestions. The tentative character of these suggestions was duly noted in the qualified answer with which this chapter began.

Rather than pointing to the diversity of alternative interpretations, one may query the connection between the task of interpretation and the enterprise of ontology. Teller (1995), describes his task in offering an interpretation of quantum field theory as follows:

I take an interpretation to be a relevant similarity relation hypothesized to hold between a model and the aspects of actual things that the model is intended to characterize. By extension, an interpretation of a theory is a programmatic sketch for filling out interpretations of the theory's constitutive models. (p. 5)

If that is its task, an interpretation must *assume* an ontology of "actual things" in order to be (hypothesized to constitute) the relevant similarity relation. Having taken its ontology for granted, the interpretation need not itself be expected to deliver any further ontological insight. Such a view of the interpretative task may correctly characterize what is involved in understanding a non-fundamental theory, for then an independent characterization of the "actual things" the theory is intended to model will already be to hand—either in pre-theoretical discourse ("pendulum", "continent," "climate," "animal population") or through a more fundamental theory ("neutral quark-gluon bound state of mass m," "bound state of six carbon atoms and six hydrogen atoms," 'collection of neutrons collapsed into a degenerate quantum state under mutual gravitational attraction").

But an interpretation of a fundamental theory like quantum field theory cannot take for granted an ontology of particles, measuring devices, etc. since these must themselves either be composed from or otherwise supervene upon the basic ontology postulated by that quantum field theory itself. An important part of interpreting the theory is to say how such "actual things" do, or even can, arise given the basic ontology of the theory. Thus an interpretation of relativity must explain how there can be rods and clocks to measure space-time intervals, and an interpretation of quantum theory (notoriously!) must explain how there can be measuring instruments capable of revealing the outcomes of observations on quantum systems whose statistics the theory predicts. Teller (1995) frankly admits that nothing he says in his book will help in any way with the quantum measurement problem, faced by all quantum theories including quantum field theories. He is in good company here. Given the intractability of that problem, it is a perfectly reasonable interpretative strategy to look for progress elsewhere. But it would be unreasonable to dismiss the measurement problem as not an interpretative problem at all.

A third ground for ontological skepticism stems from an attitude toward quantum field theories associated with what has come to be known as the effective field theory program. While this book has focused on those quantum gauge field theories that are central to the Standard Model, quantum field theory is a tool that is applied in many other areas of contemporary physics. It is widely used in statistical physics and the physics of condensed matter, and also in domains of elementary particle and nuclear physics that are not regarded as fundamental since they do not explicitly involve the basic strong and electroweak interactions among their hypothesized quark and lepton constituents. Such applications are often justified as resulting from a controlled approximation based on the energy scale of the processes concerned. Very roughly, the approximation involves neglecting terms in the overall Lagrangian density for the more fundamental quantum fields involved that are (believed to be) unimportant at the energy scales under investigation in favor of the remaining terms of a suitably modified Lagrangian. This leads to a revised quantum field theory whose consequences are then developed independently of the original theory. The revised theory is known as an effective quantum field theory, since its effects are the ones that matter at the energy scale being studied.

The quantum field theories of the Standard Model have the important property of being renormalizable.[2] Most significant predictions of a quantum field theory may be developed by a calculational technique (perturbation theory) involving the addition of what are supposed to be progressively smaller terms to a series, leading to successively more precise approximations. But only if a theory is renormalizable can this procedure be used to yield successive terms that really are smaller and smaller. Until recently, renormalizability was therefore taken as a *sine qua non* of a well-defined quantum field theory. Nevertheless, many effective theories are *not* renormalizable. How then can they be *effective*?

---

[2] Indeed, it was the discovery that, because of their gauge symmetry, non-Abelian Yang–Mills theories are renormalizable even after the Higgs mechanism permits their gauge bosons to be massive, that led to the development of the Standard Model. My discussion of effective theories is indebted to Hartmann (2001). For an introduction to early approaches to renormalization, see Teller (1995). For a gentle introduction to more modern views, see (for example) http://math.ucr.edu/home/baez/renormalization.html.

Renormalizing a theory means re-scaling the fundamental parameters appearing in the theory (representing magnitudes including masses and charges) in a systematic way to ensure that all terms are well defined, in the perturbation series whose sum represents the predicted value of some empirical magnitude. A traditional way to do this involves first introducing an arbitrary parameter Λ_0 representing a large, fixed energy. All integrals that figure in terms of the perturbation expansion are then evaluated with this parameter as upper limit of integration (the "cut-off"), instead of infinity. They are all then well defined: none of the integrals diverge. Since Λ_0 was arbitrarily chosen, one may ask what the consequences would have been had it been chosen somewhat smaller—say, $\Lambda < \Lambda_0$. In the case of a renormalizable theory, one can show that choosing a smaller cut-off will lead to the same predictions via perturbation theory provided that the fundamental parameters are re-scaled in a way that depends only on the ratio Λ/Λ_0. The renormalized theory is then defined by taking the limit $\Lambda_0 \to \infty$. This renders its predictions finite and well defined to all orders in perturbation theory, but only if the fundamental parameters are re-scaled by an infinite amount—a consequence that even some of the founders of quantum field theory considered unsatisfactory.

But suppose renormalization is reconceptualized by interpreting the cut-off parameter not as a formal device, but rather as part of the theory itself, in the sense that it defines its upper limit of applicability, above which new phenomena may be expected which the theory is not adequate to model. The idea is that an effective theory is only supposed to yield reasonable predictions below a certain maximum energy known as the cut-off. Such a theory need not be renormalizable, provided that the effect of the terms that *prevent* it from being renormalizable becomes progressively less as one considers lower and lower energies below the cut-off. At sufficiently low energies, the non-renormalizable theory may still yield very accurate predictions. In some circumstances, the problematic terms may simply be dropped, while their presence is still registered by constants multiplying remaining terms in the theory's Lagrangian density. After dropping these terms, the effective theory may then become renormalizable, and so yield well-defined predictions to all orders of perturbation theory.

Suppose then that *every* quantum field theory comes equipped with an energy cut-off that has some physical significance. In quantum electrodynamics and the quantum field theories of the Standard Model, the high-energy cut-off would naturally be associated with some fundamental, very small length scale, perhaps the result of quantum fluctuations in gravity. Whatever this scale is, it lies far beyond the reach of present-day experiments. This circumstance explains the renormalizability of quantum electrodynamics and other quantum field theories of particle interactions. Whatever the Lagrangian of quantum electrodynamics may be at the fundamental scale, as long as its couplings are sufficiently weak, it must be described at the energies of our experiments by a renormalizable effective Lagrangian.

If one views even the gauge theories of the Standard Model as effective theories, their fundamental status appears as merely provisional. Even the fact that they are renormalizable does not single them out as special, if this is simply a consequence of the fact that they arise as low-energy limits of some more fundamental, as yet unknown, quantum field theory. Georgi (1993) expresses an extreme form of this view as follows:

> The philosophical question underlying old-fashioned renormalizability is this: How does the process end? It is possible, I suppose, that at some very large energy scale, all nonrenormalizable interactions disappear, and the theory is simply renormalizable in the old sense. This seems unlikely, given the difficulties with gravity. It is possible that the rules change dramatically, as in string theory. It may even be possible that there is no end, simply more and more scales as one goes to higher and higher energy. Who knows? Who cares? In addition to being a great convenience, effective field theory allows us to ask all the really scientific questions that we want to ask without committing ourselves to a picture of what happens at arbitrarily high energy. (p. 215)

The view has given rise to the metaphor of a tower of effective quantum field theories, ascending to higher and higher energy scales until it disappears into the clouds beyond which our instruments cannot see. Why take seriously any ontological conclusions based merely on the highest part of the tower we happen to have climbed, especially because the features of an effective theory at one level are so independent of what happens at higher energies as to give us almost no clue as to what to expect if and when we get there?

There is one negative conclusion that should certainly not be taken seriously—that there are no other fields or "particles" in addition to those described by the quantized gauge theories of the Standard Model. Such a conclusion is not warranted, since there is no reason to expect the effects of any such novel structures to be manifested at the energy scales where these theories have been successfully tested. But I see nothing specific to the effective field theory program that would defeat an inference to the tentative conclusions drawn on the basis of the investigation undertaken in this book. Indeed, it is striking that even when quantum electrodynamics is viewed as an effective field theory, renormalizable only after neglecting high-energy terms representing the massive vector gauge bosons associated with the weak interaction, a quantum field theory above it in the tower (the Weinberg–Salam unified electroweak theory of the Standard Model) is itself a renormalizable gauge theory. This reinforces the conclusion that the evidence for contemporary gauge theories lends credence to the belief that these describe non-separable processes, while nothing in the world corresponds to or is represented by a locally defined gauge potential.

226   9 CONCLUSIONS

Proponents of the effective field theory program recommend a policy of epistemic caution in the light of their view of the development of quantum field theories in physics. From a wider perspective, quantum field theory itself represents only a brief recent episode in the history of physical theorizing on these matters. This wider perspective reveals a fourth and final ground for mistrusting the ontological pretensions of contemporary gauge theories.

I began by introducing electromagnetism as a paradigm example of a gauge theory. The evolution of physical theorizing about electromagnetism has taken many twists and turns since the days of Faraday and Maxwell. After Maxwell, physicists no longer thought of lines of force as real structures; after Einstein, they no longer took electromagnetic radiation to propagate energy continuously through a material medium since it could convey energy in discrete quanta through empty space. These are not isolated examples of how apparent ontological commitments have changed with changing theories. They recall Kuhn's famous remarks on changing views of motion under gravity Kuhn (1970):

Newton's mechanics improves on Aristotle's and ... Einstein's improves on Newton's as instruments for puzzle-solving. But I can see in their succession no coherent direction of ontological development. On the contrary, in some important respects, though by no means in all, Einstein's general theory of relativity is closer to Aristotle's than either of them is to Newton's. (pp. 206–7)

One could interpret the following remarks by Ashtekar and Rovelli (1992) as supporting an analogous view of the development of theories of electromagnetism:

In a sense, however, the tradition of using loops as basic objects goes back substantially further—in fact, all the way to Faraday! For, gauge theories can be said to have originated in Maxwell's work which formalized Faraday's intuitive picture of electromagnetism as a theory of "lines of force" trapped in space. In absence of sources, each line of force is a closed loop. It turns out, quite remarkably, that this picture of a classical field has direct analogues in the loop formulation of the quantum theory. (pp. 1148–9)

Notoriously, Kuhn (1970) took his historical perspective to warrant skepticism concerning any ontological conclusions based on the empirical success of fundamental physical theories:

A scientific theory is usually felt to be better than its predecessors not only in the sense that it is a better instrument for discovering and solving puzzles but also because it is somehow a better representation of what nature is really like. One often hears that successive theories grow ever closer to, or approximate more and more closely to, the truth. Apparently generalizations like that refer not to the puzzle-solutions and the concrete predictions derived from a theory but rather to its ontology, to the match, that is, between the entities with which the theory populates nature and what is "really there." (p. 206)

In this passage, Kuhn sets up this "conventional wisdom" only to reject it, for philosophical as well as historical reasons. As a philosopher, he finds the notions of truth and reality so deeply puzzling that they are best avoided in our attempts to understand science and its evolution. The passage continues

> Perhaps there is some other way of salvaging the notion of "truth" for application to whole theories, but this one will not do. There is, I think, no theory-independent way to reconstruct phrases like "really there"; the notion of a match between the ontology of a theory and its "real" counterpart in nature now seems to me illusive in principle. (*ibid*. p. 206)

Later, in his Rothschild Lecture at Harvard in 1992, he remarked,

> I am not suggesting, let me emphasize, that there is a reality which science fails to get at. My point is rather that no sense can be made of the notion of reality as it has ordinarily functioned in the philosophy of science.

In the present context it seems best to lay aside any concerns a philosopher may have about how to understand a correspondence theory of truth, for these concerns arise whatever the particular domain about which empirical claims are made.[3] What makes Kuhn's skepticism distinctive and presently relevant is its connection to the historical development of fundamental physical theories.

Shorn of its semantic clothing, what remains of this skepticism is an induction, from the transience of fundamental ontologies in the history of physical theories, to the conclusion that we are not warranted in accepting ontological claims based on even our best-supported contemporary theories, including the gauge theories of the Standard Model. But this general, indirect, inductive argument must be balanced against the experimental evidence we have to back up specific ontological claims of those gauge theories themselves. Even Kuhn would surely be reluctant to argue that it was wrong for the Nobel committee to award the 1984 Nobel prize for Physics to Carlo Rubbia and Simon van der Meer for their discovery (only!) the previous year of the W and Z gauge bosons predicted by the unified electroweak theory of the Standard Model!

Consider instead the claim that there are physical processes involving properties that are neither localized at or near a point nor determined by properties localized at or near a point, and that some of our theories of fundamental

---

[3] In "The Revolution That Didn't Happen," *The New York Review of Books*, October 8, 1998, Weinberg put the point more bluntly: "Certainly philosophers can do us a great service in their attempts to clarify what we mean by truth and reality. But for Kuhn to say that as a philosopher he has trouble understanding what is meant by truth or reality proves nothing beyond the fact that he has trouble understanding what is meant by truth or reality."

"forces" succeed in capturing significant features of these processes. No experiment at CERN (or elsewhere) is going to provide such convincing evidence for this claim as to be worthy of the award of another Nobel prize. Moreover, reflection on the history of physical theorizing on gauge theories since Faraday and Maxwell *should* give one pause before committing oneself to full belief in this claim. But while it is a very general and very abstract ontological claim that can be related only distantly to observation, the evidence for contemporary Yang–Mills gauge theories does provide some reason to believe this claim. Or so I have argued.

APPENDIX A

Electromagnetism and its generalizations

Prior to their theoretical unification by Maxwell and then Einstein, classical physics treated electricity and magnetism as separate phenomena. A point charge q_1 produced an electric force $\mathbf{F}_{12}$ on another point charge q_2 in accordance with Coulomb's law

$$\mathbf{F}_{12} = \frac{q_1 q_2}{r_{12}^2} \hat{r}_{12} \tag{A.1}$$

where $\hat{r}_{12}$ is a vector of unit length in the direction from q_1 to q_2. The net force $\mathbf{F}$ on a charge q at $\mathbf{r}$ produced by other point charges q_i ($i = 1, \ldots, n$) located at positions $\mathbf{r}_i$ could then be thought to arise from the sum of their electric fields $\mathbf{E}(\mathbf{r}) = \sum_i \mathbf{E}_i(\mathbf{r})$, with $\mathbf{E}_i(\mathbf{r}) = \frac{q_i}{|\mathbf{r}-\mathbf{r}_i|^3}(\mathbf{r}-\mathbf{r}_i)$, according to $\mathbf{F} = q\mathbf{E}(\mathbf{r})$. Alternatively, and more conveniently, each charge q_i could be thought to give rise to an electric potential $\varphi_i(\mathbf{r})$ with $\mathbf{E}_i(\mathbf{r}) = -\nabla \varphi_i(\mathbf{r})$, so the total electric potential $\varphi(\mathbf{r}) = \sum_i \varphi_i(\mathbf{r})$ and $\mathbf{E}(\mathbf{r}) = -\nabla \varphi(\mathbf{r})$. One such electric potential is given by $\varphi_i(\mathbf{r}) = \frac{q_i}{|\mathbf{r}-\mathbf{r}_i|}$, but so also is every other potential of the form $\varphi'_i(\mathbf{r}) = \varphi_i(\mathbf{r}) + \varphi_0$, for arbitrary constant φ_0.

Just as the introduction of an electric scalar potential $\varphi(\mathbf{r})$ simplifies electrostatics, so also the introduction of a magnetic vector potential $\mathbf{A}(\mathbf{r})$ proves useful in treating stationary magnetic phenomena. In this case, the magnetic field $\mathbf{B}(\mathbf{r})$ at a point may be thought to arise as the curl of $\mathbf{A}(\mathbf{r})$: $\mathbf{B} = \nabla \times \mathbf{A}$, where $\mathbf{A}(\mathbf{r})$ itself is the vector sum $\mathbf{A}(\mathbf{r}) = \sum_i \mathbf{A}_i(\mathbf{r})$ of the potential due to each of a variety of constant electric currents in the neighborhood. Since $\nabla \times \nabla \Lambda = 0$, the transformation $\mathbf{A}(\mathbf{r}) \to \mathbf{A}(\mathbf{r}) - \nabla \Lambda(\mathbf{r})$ leaves $\mathbf{B}(\mathbf{r})$ unchanged, for arbitrary suitably differentiable $\Lambda(\mathbf{r})$. Note that $\nabla \cdot \mathbf{B} = 0$ is then an automatic consequence of the identity $\nabla \cdot \nabla \times \mathbf{A} = 0$.

Maxwell's equations are the foundation of classical electromagnetism. With $\mathbf{B}$, $\mathbf{E}$ also now functions of time t, they may be stated as follows, in units in which the speed of light is 1:

$$\begin{aligned} \nabla \cdot \mathbf{B} &= 0 & \nabla \times \mathbf{E} + \partial \mathbf{B}/\partial t &= 0 \\ \nabla \cdot \mathbf{E} &= \rho & \nabla \times \mathbf{B} - \partial \mathbf{E}/\partial t &= \mathbf{j} \end{aligned} \tag{A.2}$$

The top two equations are homogeneous, since their right-hand sides are zero: the bottom pair are inhomogeneous. The first inhomogeneous equation states Gauss's law—a reformulation of Coulomb's law for the charge density ρ, while the first homogeneous equation expresses the absence of isolated magnetic charges (magnetic monopoles). The second homogeneous equation restates Faraday's law of electromagnetic induction, whereby a changing magnetic field produces an electric field. The second inhomogeneous equation encompasses Ampere's law concerning the magnetic field produced by an electric current density $\mathbf{j}$, and also includes a term corresponding to a changing electric field that represents Maxwell's famous displacement current.

The electric field can no longer be derived solely from an electrostatic potential in the presence of changing magnetic fields, but one can still derive the magnetic field from a magnetic vector potential. The generalization is now

$$\mathbf{E} = -\nabla\varphi - \partial\mathbf{A}/\partial t \tag{A.3}$$
$$\mathbf{B} = \nabla\times\mathbf{A} \tag{A.4}$$

where $\varphi(x), \mathbf{A}(x)$ are now functions of time as well as position in space. With this generalization, the homogeneous Maxwell equations are automatically satisfied. $\mathbf{E}, \mathbf{B}$ are unchanged under the simultaneous variable potential transformations

$$\mathbf{A}(\mathbf{r}) \to \mathbf{A}(\mathbf{r}) - \nabla\Lambda(\mathbf{r}) \tag{A.5}$$
$$\varphi \to \varphi + \partial\Lambda/\partial t \tag{A.6}$$

Hence Maxwell's equations are also invariant under these transformations.

If we regard $\varphi, \mathbf{A}$ as components of a Lorentz four-vector potential $A_\mu = (\varphi, -\mathbf{A})$, this transforms under a variable potential transformation as follows:

$$A_\mu(x) \to A_\mu(x) + \partial_\mu\Lambda(x) \tag{A.7}$$

Now define a Lorentz tensor (the Maxwell–Faraday tensor) by the relation 1.7

$$F_{\mu\nu} = \partial_\mu A_\nu - \partial_\nu A_\mu \tag{A.8}$$

whose components are

$$F_{\mu\nu} = \begin{pmatrix} 0 & E_x & E_y & E_z \\ -E_x & 0 & -B_z & B_y \\ -E_y & B_z & 0 & -B_x \\ -E_z & -B_y & B_x & 0 \end{pmatrix} \tag{A.9}$$

The definition of $F_{\mu\nu}$ in terms of A_μ ensures that the homogeneous Maxwell equations are automatically satisfied. The inhomogeneous equations become

$$\partial^\mu F_{\mu\nu} = J_\nu \qquad (A.10)$$

where the four-current $J_\nu = (\rho, -\mathbf{j})$. Expressing Maxwell's equations in tensor notation makes their Lorentz covariance manifest. They are also gauge invariant, since $F_{\mu\nu}$ remains unchanged under the variable potential transformation A.7 in A_μ. The source-free Maxwell's equations (A.10 with $J_\nu \equiv 0$) may be derived as Euler–Lagrange equations by application of Hamilton's principle to variations of the action $S = \int \mathcal{L}_{EM} d^4 x$ associated with the Lagrangian density

$$\mathcal{L}_{EM} = -\frac{1}{4} F_{\mu\nu} F^{\mu\nu} \qquad (A.11)$$

The Lorentz force law 1.5

$$\mathbf{F} = e(\mathbf{E} + \mathbf{v} \times \mathbf{B}) \qquad (A.12)$$

may now be expressed in covariant form as

$$f_\mu = -e F_{\mu\nu} dx^\nu / d\tau \qquad (A.13)$$

where τ represents proper time and the force on a charge e is given by the spatial components of f_μ in its rest frame.

In ordinary non-relativistic quantum mechanics, electromagnetism is treated classically, but its action on particles is not specified by the Lorentz force law but by including interaction terms in the Hamiltonian operator that enters the fundamental dynamical equation, the Schrödinger equation 1.9

$$\hat{H}\Psi = i\hbar \partial \Psi / \partial t \qquad (A.14)$$

For particles of electric charge e and mass m subject only to an electromagnetic interaction whose potential $A_\mu = (\varphi, -\mathbf{A})$, $\hat{H}$ may be expressed by the equation

$$\hat{H} = \frac{(\hat{\mathbf{p}} - e\mathbf{A})^2}{2m} + e\varphi \qquad (A.15)$$

in which the momentum operator $\hat{\mathbf{p}} = -i\hbar \nabla$. The immediate effect of electromagnetism is therefore to modify the wave-function Ψ of a quantum system. This can affect the expected motion of the system in a way that loosely corresponds to the action of a classical Lorentz force, introduce or modify interference phenomena, or both.

Many details of atomic structure may be accounted for by representing the n electrons in an atom by a multi-particle wave-function $\Psi(\mathbf{x}_1, \ldots, \mathbf{x}_n, t)$ and writing a Hamiltonian for the atom including a term representing the electromagnetic potential due to the massive, central nucleus, as well as terms representing electromagnetic interactions between electrons. In this case, the Hamiltonian will not be an explicit function of time, permitting one to write down solutions to the Schrödinger equation in the form of a product

$$\Psi(\mathbf{x}_1, \ldots, \mathbf{x}_n, t) = \theta(t)\psi(\mathbf{x}_1, \ldots, \mathbf{x}_n) \tag{A.16}$$

where

$$\theta(t) = \exp(E/i\hbar)t \tag{A.17}$$

$$\hat{H}\psi = E\psi \tag{A.18}$$

The second of these equations is known as the time-independent Schrödinger equation. It is an eigenvalue equation, with non-trivial solutions only for certain values of E—the energy of the corresponding solution ψ_E. The resulting solution to the original Schrödinger equation is called stationary, since expectation values of electron configurations do not change with time. The existence of such stationary solutions, and in particular a solution with minimum energy, helps to explain the stability of atoms.

In a non-Abelian generalization of the inhomogeneous Maxwell equations A.10 the ordinary derivative ∂_μ is replaced by the covariant derivative D_μ, so called because $D_\mu \Psi$ transforms the same way as does the wave-function Ψ under gauge transformations of the second kind ("local" gauge transformations). The resulting generalization is

$$D^\mu \mathbf{F}_{\mu\nu} = \mathbf{J}_\nu \tag{A.19}$$

where an expression is written in boldface to indicate that it does not transform as a scalar under gauge transformations. An explicit form for D_μ includes a generalization $\mathbf{A}_\mu$ of the electromagnetic four-vector potential

$$D_\mu = \partial_\mu + ig\mathbf{A}_\mu \tag{A.20}$$

where g represents a coupling constant for the interaction that generalizes electric charge. The field is generated from its potential in a way that generalizes 1.7:

$$\mathbf{F}_{\mu\nu} = (\partial_\mu \mathbf{A}_\nu - \partial_\nu \mathbf{A}_\mu) - (ig/\hbar)\left[\mathbf{A}_\mu, \mathbf{A}_\nu\right] \tag{A.21}$$

It is therefore possible to arrive at solutions to A.19 with the same field but different sources, by suitable choice of different $\mathbf{A}_\mu$ that yield the same $\mathbf{F}_{\mu\nu}$.

APPENDIX B
Fiber Bundles

A (differentiable) *fiber bundle* is a triple $\langle E, M, \pi \rangle$, where E, M are differentiable manifolds, and $\pi : E \to M$ is a differentiable *projection* mapping of the *total space* E onto the *base space* M.[1] The inverse image $\pi^{-1}(m)$ of a point $m \in M$ is called the *fiber above* m, and the fibers above every point $m \in M$ are required to be diffeomorphic to one another. A *section* of E is a smooth map $\sigma : U \subseteq M \to E$ such that $\pi \circ \sigma$ is the identity map; a section is *global* if $U = M$, otherwise it is *local*.

The fiber bundles that are of interest in formulating gauge theories are all defined in terms of a *structure group* G, which physicists often call the gauge group of the theory, such as U(1) or SU(2). These are all Lie groups, and hence differentiable manifolds. Elements $g \in G$ act on elements $u \in \pi^{-1}(m)$ of the fiber above each point, and this action is conventionally written on the right as $R_g(u) = ug$. The right action is required to be *free*—i.e. for all $u \in E$, if $ug = u$ then $g = e$, the group identity. The fibers are then all isomorphic to one another and to the *typical fiber* F of the bundle. In a principal fiber bundle, F is the just the structure group G itself, whereas in a vector bundle, F is a vector space, so G acts on the fibers above M via a representation. Yang–Mills gauge fields may be represented on principal fiber bundles, in which case the matter fields on which they act are represented on vector bundles associated to such a principal fiber bundle. To explain how this representation works, it is easiest to begin with a matter field.

Prior to the introduction of the fiber bundle formalism, physicists usually represented a matter field by specifying its value (or values) at each point of a manifold M representing space, or (in a relativistic theory) space-time. Whereas in an analogous classical field theory each value would consist of a (n-tuple of) number(s), in a quantum theory, these field value(s) are operators on a suitable state space. Consider, for example, a classical Klein–Gordon field ϕ, whose quantized counterpart may be used to represent the behavior of charged spinless particles. Its value at space-time point x is a complex number $\phi(x)$—an element of the vector space $\mathbb{C}$ of complex numbers. The free field

---

[1] A differentiable manifold is just a geometric space with enough structure that it makes sense to speak of smooth functions on it. A formal definition appears in Nakahara (1990), and many other texts.

ϕ satisfies the Klein–Gordon equations

$$\partial_\mu \partial^\mu \phi + m^2 \phi = 0 \tag{B.1}$$
$$\partial_\mu \partial^\mu \phi^* + m^2 \phi^* = 0 \tag{B.2}$$

relating its values at distinct space-time points. These are differential equations governing a vector field—a field whose values lie in a vector space ($\mathbb{C}$). They are invariant under a constant phase transformation $\phi \to \exp i\Lambda \phi$ corresponding to the action on ϕ of an element of the group U(1).

There is an alternative way of representing a matter field that illuminates the transition, from the free matter field to the matter field interacting with a gauge field representing a fundamental force (strong, electromagnetic or electroweak). This is to represent the matter field by a section of a vector bundle, and differentiation of the field by the covariant derivative corresponding to a bundle connection. The typical fiber V is the vector space in which the matter field was previously thought of as taking values at each point of M, and the structure group G is a continuous symmetry group of the equations of motion of the free field that generalizes the group U(1) for the free Klein–Gordon field.

$\langle E, M, \pi, G, V \rangle$ constitutes a vector bundle in which E is *locally trivial*, that is, every point $m \in M$ has a neighborhood U_m such that $\pi^{-1}(U_m)$ is isomorphic to $U_m \times V$; a canonical isomorphism is given by a diffeomorphism $\chi : \pi^{-1}(U_m) \to U_m \times V$ with $\chi(u) = (\pi(u), \varphi(u))$, where $\varphi : \pi^{-1}(U_m) \to V$ satisfies $\varphi(ug) = g\varphi(u)$ for all $u \in \pi^{-1}(U_m)$ and $g \in G$. A local trivialization specifies the right action of g on $u \in V_m$ by $ug = \chi^{-1}(m, g\varphi(u))$, where $m \in U_m$ and χ is the local trivialization, with $\chi(u) = (m, \varphi(u))$: this specification is independent of χ, as figure B.1 shows.

Differentiation of a section of a vector bundle involves comparing elements in the fibers above neighboring points in the base manifold M to see how rapidly these elements are changing as one moves from point to point of M. The elements themselves are mapped onto vectors by a local trivialization. Any comparison requires extra structure in the bundle, which is provided by a connection on M. A connection defines a particular basis for comparison and thereby specifies a particular way of differentiating sections called a covariant derivative. A *connection* D on M assigns to each vector field X on M a smooth map $D_X : \Sigma \to \Sigma$ from the set of (global) sections of a vector bundle E into itself called the *covariant derivative* D_X that satisfies

$$\begin{aligned} D_X(\sigma + \tau) &= D_X(\sigma) + D_X(\tau) \\ D_{fX+gY} &= fD_x + gD_y \\ D_X(f\sigma) &= (Xf)\sigma + fD_X\sigma \end{aligned} \tag{B.3}$$

APPENDIX B FIBER BUNDLES 235

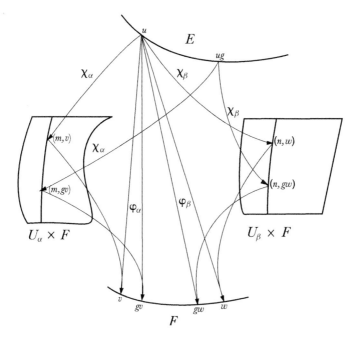

Figure B.1. Right-action of G in a vector bundle

for all sections $\sigma, \tau \in \Sigma$, and all smooth vector fields X, Y and functions f, g on M. Because of its linearity, a covariant derivative is fully specified by its action on every basis of sections e_a, in terms of which an arbitrary section may be uniquely expanded as $\sigma = \sum_a s^a e_a$, where the s^a are smooth functions on M.[2] Write X in terms of a coordinate basis $\partial_\mu \equiv \frac{\partial}{\partial x^\mu}$ for a coordinate patch $U \subseteq M$ as $X = \sum_\mu X^\mu \partial_\mu$, where μ ranges over the space-time indices 0, 1, 2, 3 and the X^μ are smooth real-valued functions on U, and abbreviate D_{∂_μ} as D_μ. Then, using the Einstein convention requiring summation over repeated indices, we have

$$D_\mu e_b = A^a_{b\mu} e_a \qquad (B.4)$$

for some smooth real-valued functions $A^a_{b\mu}$. It follows that

$$D_X \sigma = X^\mu D_\mu (s^b e_b)$$
$$= X^\mu \left[(\partial_\mu s^b) e_b + s^b D_\mu e_b \right] \qquad (B.5)$$

---

[2] A vector bundle has a basis of global sections if and only if it is trivial. But even a non-trivial vector bundle has a basis of local sections on each open set $U \subseteq M$. Each local trivialization $\chi : \pi^{-1}(U) \to U \times F$ associates a basis of local sections with a basis of the vector space F.

$$= X^\mu \left[\left(\partial_\mu s^b\right) e_b + A^a_{b\mu} s^b e_a \right]$$
$$= X^\mu \left[\partial_\mu s^a + A^a_{b\mu} s^b \right] e_a$$

The covariant derivative may be rewritten in a more familiar fashion, first by expressing D_X in terms of its coordinate components D_μ, and then by writing the components of σ as a column vector ψ,

$$D_\mu \psi = \left(\partial_\mu + \mathbf{A}_\mu\right) \psi \qquad (B.6)$$

where $\mathbf{A}_\mu$ is a square matrix with components $\left[\mathbf{A}_\mu\right]_{ab} = A^a_{b\mu}$.

These functions $A^a_{b\mu}$ are sometimes called the *components* of the connection corresponding to this covariant derivative. If they are all zero, then the covariant derivative reduces to the ordinary derivative. This is the case for a free matter field over each open set $U \subseteq M$ if one chooses an inertial coordinate system on U and a basis of local sections e_a that stay constant from point to point in this coordinate system. In such a case one can cover M by coordinate patches homeomorphic to a Euclidean space, in each of which there exists a coordinate system on which all the components of the connection are everywhere zero, and the connection is then said to be flat. But when the matter field interacts with a Yang–Mills gauge field, the connection is no longer flat. In either case, the components of the connection represent the gauge potential associated with the gauge field. Only when the connection is curved rather than flat is this gauge potential accompanied by a non-zero gauge field. The curvature of the connection then represents the gauge field strength.

The curvature of a vector bundle may be defined in terms of the covariant derivative on M as follows. Given two vector fields X, Y on M, the *curvature* is a smooth map $F(X, Y) : \Sigma \to \Sigma$ of the set of sections of E onto itself given by

$$F(X, Y)\sigma = D_X D_Y \sigma - D_Y D_X \sigma - D_{[X,Y]} \sigma \qquad (B.7)$$

Here $[X, Y]$ is the Lie bracket: a vector field defined by its action $[X, Y]f = X(Y(f)) - Y(X(f))$ on an arbitrary smooth function f on M. A bundle E is *flat* just in case $F(X, Y) = 0$ for all X, Y. The curvature is antisymmetric ($F(Y, X) = -F(X, Y)$) and linear over functions in both arguments:

$$F(fX, Y) = fF(X, Y) \qquad (B.8)$$
$$F(X, gY) = gF(X, Y)$$

It may be expressed in terms of its action on a coordinate basis in U as

$$F_{\mu\nu} = F(\partial_\mu, \partial_\nu) \qquad (B.9)$$
$$= F(D_\mu, D_\nu)$$

APPENDIX B FIBER BUNDLES   237

where $F_{\nu\mu} = -F_{\mu\nu}$. In terms of a basis of sections e_a this gives

$$F_{\mu\nu}e_a = \left[\left(\partial_\mu A^b_{a\nu}\right) - \left(\partial_\nu A^b_{a\mu}\right) + A^b_{k\mu}A^k_{a\nu} - A^b_{k\nu}A^k_{a\mu}\right]e_b \qquad (B.10)$$

which may be rewritten more elegantly as a matrix equation:

$$\mathbf{F}_{\mu\nu} = \partial_\mu \mathbf{A}_\nu - \partial_\nu \mathbf{A}_\mu + [\mathbf{A}_\mu, \mathbf{A}_\nu] \qquad (B.11)$$

There is a more geometric way of thinking of the connection and curvature of a vector bundle. On this way of thinking, a connection is defined in terms of a one-form $\bar{d}^A$ on U, a smooth map of vector fields in M into the set of smooth isomorphisms from the set Σ of sections on U onto itself that is linear over smooth functions on U:

$$\bar{d}^A X : \Sigma \to \Sigma \qquad (B.12)$$
$$\bar{d}^A (fX) = f \bar{d}^A X$$

Given a vector bundle E, one can construct a *dual bundle* E^* over M whose fibers V^*_m are isomorphic to the dual space V^* of the typical fiber V of E via a local trivialization, and whose projection map projects V^*_m onto $m \in M$. If e_a is a basis of sections of E on U then e^a forms a unique basis of sections of E^* on U where, for each $m \in U$, $e^a(m)$ is the basis of V^*_m dual to the basis $e_a(m)$ of V_m. One can now express $\bar{d}^A$ in terms of the previously defined components as

$$\bar{d}^A = A^a_{b\mu} e_a e^b dx^\mu \qquad (B.13)$$

where the dx^μ form a basis of one-forms on U that satisfy

$$dx^\mu(\partial_\nu) = 1 \text{ if } \mu = \nu \qquad (B.14)$$
$$= 0 \text{ if } \mu \neq \nu$$

We can now re-express the covariant derivative as

$$(D_X \sigma)^a = X s^a + \bar{d}^A X(s)^a \qquad (B.15)$$
$$\equiv d^0 X(s)^a + \bar{d}^A X(s)^a \qquad (B.16)$$
$$= d^A X(s)^a, \text{ where } d^A \equiv d^0 + \bar{d}^A \qquad (B.17)$$

Thinking of the connection and covariant derivative this way has the significant advantage that it applies even when E is non-trivial. In that case, these objects may still be defined over the whole of E while the expansion B.13 holds only locally, so the $A^a_{b\mu}$ cannot be smoothly defined over the whole of E. Note that since it is not $\tilde{d}^A$ but rather the one-form d^A that represents the connection here, it is not strictly correct to call the $A^a_{b\mu}$ its components.

Similarly, one can regard the curvature $F(X, Y)$ as a two-form defined on the whole of E whose local expression in a coordinate patch $U \subseteq M$ is

$$F = \frac{1}{2} F_{\mu\nu} dx^\mu \wedge dx^\nu \tag{B.18}$$

F may be simply expressed in terms of the connection one-form d^A as

$$F = d^A d^A \tag{B.19}$$

This shows that the curvature is a geometric object on E, definable independently of any coordinate system.

The $A^a_{b\mu}$ are generalizations of the components A_μ of the electromagnetic potential: they represent the components of an arbitrary Yang–Mills gauge potential. Similarly, the components of the curvature $F^b_{a\mu\nu}$ represent the components of an arbitrary Yang–Mills gauge field strength, generalizing the electromagnetic field tensor $F_{\mu\nu}$. For a free Yang–Mills gauge field without sources, all the components of the potential are zero everywhere in some coordinate system associated with a local trivialization of the vector bundle E. Otherwise, the gauge field is interacting with the matter field represented by sections of E. But this way of proceeding makes the gauge field appear parasitic on a prior matter field, such as a Klein–Gordon or Dirac field, represented on a vector bundle whose typical fiber is fixed by the specific properties of that field ($\mathbb{C}$ for a charged Klein–Gordon field, $\mathbb{R}$ for a neutral Klein–Gordon field, etc.).

There is a way of using fiber bundles to represent a gauge field independently of any assumed interaction with a matter field, and only later to consider its possible interactions with various kinds of matter field, possibly represented on vector bundles with different typical fibers. It is to represent a gauge field not on a vector bundle, but on what is called a principal fiber bundle. Whereas the typical fiber of a vector bundle is a vector space, the typical fiber of a principal fiber bundle is the structure group itself—a continuous group of transformations. These transformations may (or may not) be taken to act on a vector space via a group representation. Any interaction between a gauge field and a matter field is handled by associating some vector bundle for the matter field with the principal bundle on which acts the structure group of the gauge field, and arranging for the structure group of the associated vector bundle to be a vector representation of that structure group.

A *principal fiber bundle* $P(M, G)$ over M with structure group G consists of a differentiable manifold P and an action of G on P satisfying the following conditions:

(1) G acts freely on P on the right.

(2) M is the quotient space of P by the equivalence relation induced by G, and the resulting projection map $\pi : P \to M = P/G$ is differentiable.

(3) P is *locally trivial*; that is, every point $m \in M$ has a neighborhood U_m such that $\pi^{-1}(U_m)$ is isomorphic to $U_m \times G$, i.e. there is a diffeomorphism $\chi : \pi^{-1}(U_m) \to U_m \times G$ with $\chi(u) = (\pi(u), \varphi(u))$, where $\varphi : U_m \to G$ satisfies $\varphi(ug) = \varphi(u)g$ for all $u \in \pi^{-1}(U_m)$ and $g \in G$.

A *principal automorphism* of a principal fiber bundle P is a smooth map $h : P \to P$ from the total space into itself satisfying

$$h(ug) = h(u)g \tag{B.20}$$

for all $u \in P$ and all $g \in G$. A principal automorphism is called a *vertical automorphism* if, in addition,

$$\pi(h(u) = \pi(u) \tag{B.21}$$

where π is the projection map on P. The group of all vertical automorphisms of a principal fiber bundle P is a subgroup of $AutP$, the group of all automorphisms of P.

There is an association between a principal fiber bundle and a class of vector bundles.

An *associated vector bundle* $\langle E, M, \pi_E, G, V, P\rangle$ associated with the principal fiber bundle $P(M, G)$ is a fiber bundle with the vector space V as typical fiber, base space M, and projection map π_E onto M from a total space E constructed in the following way. G acts on V on the left via a representation ρ, written as $\rho(g) \circ v = gv$ for all $g \in G$, $v \in V$. This action may be extended to a right action on $P \times V$ as follows: $(u, v)g \to (ug, g^{-1}v)$ (the need to introduce the group inverse here is simply a consequence of choosing G to act on elements of P on the right rather than the left). If points of $P \times V$ connected in this way by some element of G are regarded as equivalent, then E is defined as the quotient space $P \times V/G$ under this equivalence relation, whose typical element is the equivalence class $[(u, v)] = [(ug, g^{-1}v)]$. So the fiber $\pi_E^{-1}(m)$ of the associated vector bundle above $m \in M$ does not consist of elements of the vector space V itself even though it is an isomorphic copy of V: any association of a particular element of V with a point $p \in \pi_E^{-1}(m)$ is relative to an arbitrary choice of $u \in \pi^{-1}(m) \subset P$.

Since the total space P of a principal fiber bundle is a differentiable manifold, at each point $u \in P$ there is a space T_uP of tangent vectors to curves in P. The projection map π maps smooth curves in P onto smooth curves in M, and thereby induces a linear map $\pi^* : T_uP \to T_{\pi(u)}P$ from the space of tangent vectors to curves at $u \in P$ onto the space of tangent vectors to curves at $\pi(u) \in M$. Since the dimension of P is greater than that of M, π^* maps a subspace $V_uP \subset T_uP$ onto the null subspace of $T_{\pi(u)}P$; V_uP is called the *vertical*

240 APPENDIX B FIBER BUNDLES

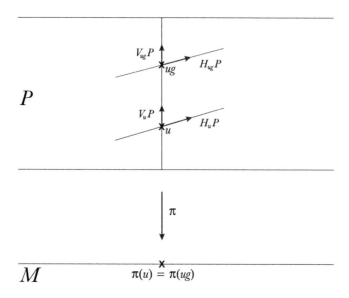

Figure B.2. Connection on a principal fiber bundle

subspace of $T_u P$. Vectors in $V_u P$ are tangent to curves in the fiber above $\pi(u)$. A vertical automorphism of P preserves the vertical subspace above each point of M.

A principal fiber bundle may also have a connection that defines neighboring points in the fibers above neighboring points of M. But this is not so closely related to an idea of covariant differentiation as is the connection on a vector bundle: a section of a principal fiber bundle is nothing like a vector field.

A *connection* on principal fiber bundle $P(M, G)$ is a smooth assignment to each point $u \in P$ of a subspace $H_u P$ of $T_u P$ (the *horizontal subspace*) such that

(1) $T_u P = H_u P \oplus V_u P$ (every vector in $T_u P$ may be uniquely decomposed into a sum of a horizontal and a vertical vector), and

(2) $R_g^*(H_u P) = H_{ug} P$ (the mapping $R_g^* : T_u P \to T_{ug} P$ induced by the right action R_g of G on P preserves horizontal subspaces).

This geometric definition is illustrated in figure B.2.

An algebraic definition may be shown to be equivalent. To any connection on a principal fiber bundle $P(M, G)$ there uniquely corresponds a connection one-form ω on P that takes values in the Lie algebra $\mathfrak{g}$ of G and which defines the horizontal subspace $H_u P$ of the tangent space at each point $u \in P$. Its definition requires some explanation.

A *one-form* on P is a smooth assignment of a vector from the dual space $T_u^* P$ to the tangent space $T_u P$ at each point $u \in P$. At each point u it is a linear map from $T_u P$ into a vector space, which in this case is not $\mathbb{R}$ but the Lie algebra $\mathfrak{g}$

of G, a vector space of the same dimension as $V_u P$, which may be defined as follows.

The left action L_g of a Lie group element $g \in G$ on G is defined by $L_g(h) = gh$, for all $h \in G$. Since G is a differentiable manifold, it admits vector fields, where a vector field is a smooth assignment of a vector from the tangent space at each point $g \in G$. L_g induces a smooth map $L_g^* : T_h G \to T_{gh} G$ from the tangent space at h to that at gh. A vector field X on G is *left invariant* if and only if it satisfies $L_g^* X|h = X|gh$ for all $g, h \in G$, i.e. it is generated by moving around the manifold in a way that is dictated by the group elements that form it.

The left invariant vector fields form the *Lie algebra* $\mathfrak{g}$ *of* G where the Lie bracket $[X, Y]$ is defined by the commutator $[X, Y]f \equiv XYf - YXf$ for all smooth functions f on G.[3] Since it is left-invariant, an element X of $\mathfrak{g}$ is determined by its value X_e at the identity $e \in G$. This sets up a vector space isomorphism between the Lie algebra and $T_e G$, in terms of which one can characterize the local structure of G as follows. Let $\{T_a\}$ be a basis for $T_e G$, and extend the Lie bracket operation to $T_e G$ via the isomorphism $\mathfrak{g} \simeq T_e G$. Then $\mathfrak{g}$ is determined by its *structure constants* $f_{\alpha\beta}^\gamma$ by $[T_\alpha, T_\beta] = f_{\alpha\beta}^\gamma T_\gamma$.

To each element $A \in \mathfrak{g}$ there uniquely corresponds a vector field $A^\#$ on P called the *fundamental vector field* generated by A. At each point $u \in P$ this defines a vector space isomorphism $\# : \mathfrak{g} \to V_u P$ between $\mathfrak{g}$ and the vertical subspace $V_u P$. The *connection one-form* ω on P effects a projection of $T_u P$ onto $V_u P$ and so indirectly defines the horizontal subspace $H_u P$ by imposing the conditions

(1) $\omega(A^\#) = A$ for all $A \in \mathfrak{g}$
(2) $R_g^* \omega_{ug}(X) = g^{-1} \omega_u(X) g$

Now define the horizontal subspace at u as $H_u P = \{X \in T_u P \mid \omega(X) = 0\}$. Condition (2) on ω ensures that R_g^* preserves horizontal subspaces.

For any smooth curve γ in M, the connection one-form defines a class of curves in P, each called a *horizontal lift* $\tilde{\gamma}$ of γ. For any point m on γ, and any point $u \in \pi^{-1}(m)$, the tangent vector $T_{\tilde{\gamma}}$ to the horizontal lift $\tilde{\gamma}$ at u lies in $H_u(P)$, and so satisfies

$$\omega(T_{\tilde{\gamma}}) = 0 \tag{B.22}$$

When expressed with respect to a "baseline curve" defined by a section of P, this becomes an ordinary differential equation for $\tilde{\gamma}$, which has a unique solution, establishing the uniqueness of the horizontal lift of γ through u. The horizontal lift $\tilde{\gamma}_C$ of a closed curve C beginning and ending at $m \in M$ need

---

[3] An abstract Lie algebra is a vector space V that is closed under a bracket operation $[,] : V \times V \to V$ which is linear in both arguments, antisymmetric, and satisfies the Jacobi identity $[u, [v, w]] + [v, [w, u]] + [w, [u, v]] = 0$ for all $u, v, w \in V$.

not close: if $\tilde{\gamma}_C$ begins at $u \in \pi^{-1}(m)$ and ends at $v \in \pi^{-1}(m)$, then $v = ug$ for some $g \in G$ (see figure 1.4). This defines a map $\tau_m : \pi^{-1}(m) \to \pi^{-1}(m)$ called the *holonomy map* satisfying

$$\tau_m(ug) = \tau_m(u)g \tag{B.23}$$

The holonomy map is an intrinsic geometric feature of the bundle that does not depend on a choice of bundle section. It will generally depend on the bundle connection ω in a way that will soon be specified (see equation B.32).

If G is an Abelian group (such as $U(1)$), then the holonomy map of closed curve C is generated by an element $g_m(C) \in G$ as follows:

$$\tau_m(u)(C) = ug_m(C) \tag{B.24}$$

But B.24 is inconsistent with B.23 if G is non-Abelian. The holonomy map defined by the horizontal lift $\tilde{C}_u$ through $u \in \pi^{-1}(m)$ of closed curve C does define the *holonomy* $\mathcal{H}(\tilde{C}_u)$, but any association of a holonomy with C itself now depends on u. If σ is a section through $u \in \pi^{-1}(m)$, then the following definition is consistent with B.23:

$$\tau_m(u)(C) = u\mathcal{H}(C) \tag{B.25}$$

provided that $\mathcal{H}(C)$ depends appropriately on u. Specifically, for all closed curves C on M we have

$$\mathcal{H}_{\sigma_2(m)}(C) = g^{-1}\mathcal{H}_{\sigma_1(m)}(C)g, \text{ for some } g \in G \tag{B.26}$$

where the dependency of the holonomy of C on σ is made clear by the subscripts. This is a consequence of the explicit form of the *holonomy of C* in a given section with $\sigma(m) = u$, where it is given by the action of a group element on $\pi^{-1}(m)$

$$\mathcal{H}_m(C) = \wp \exp\left\{-\oint_C \mathcal{A}_\mu(x)dx^\mu\right\} \tag{B.27}$$

Here $\mathcal{A}$ is a Lie-algebra-valued one-form that represents ω on M—the pull-back $\sigma^*\omega$ of ω onto M.

If ω is a connection on $P(M, G)$, and h is a vertical automorphism of P, then $h^*\omega$ is also a connection on P. If σ_1 is a section then h defines a related section σ_2 by

$$\sigma_2(x) = h[\sigma_1(x)] \tag{B.28}$$

APPENDIX B FIBER BUNDLES

Let $\mathcal{A}_1$ represent ω in σ_1. Then $\mathcal{A}_2$ represents $h^*\omega$ in σ_1, where $\mathcal{A}_1$ and $\mathcal{A}_2$ are related by 3.49. Hence

$$\mathcal{H}_{m,h^*\omega}(C) = \wp \exp\left\{-\oint_C \mathcal{A}_{2\mu}(x)dx^\mu\right\} \tag{B.29}$$

$$= g^{-1}(m)\wp \exp\left\{-\oint_C \mathcal{A}_{1\mu}(x)dx^\mu\right\} g(m) \tag{B.30}$$

$$= g^{-1}(m)\mathcal{H}_{m,\omega}(C)g(m) \tag{B.31}$$

These holonomies are defined relative to the *same* section σ_1, in which the relation between the holonomy maps $\tau_\omega, \tau_{h^*\omega}$ corresponding to $\omega, h^*\omega$ may be expressed by the following holonomy transformation:

$$\mathcal{H}_{m,h^*\omega} = g^{-1}(m)\mathcal{H}_{m,\omega}g(m) \tag{B.32}$$

where $g(m)$ is defined by $\sigma_2(m) = h[\sigma_1(m)] = \sigma_1(m)g(m)$. This is the transformation rule for a holonomy under a vertical bundle automorphism, corresponding to what Trautman (1980) called a gauge transformation of the second kind. Holonomies related by B.32 form an equivalence class. It follows that, for an Abelian structure group, holonomies are invariant under such conjugacy transformations.

Two curves $C_{m'}, C_m$ with the same image on M but different base points m, m' have related holonomies:

$$\mathcal{H}_{m',\omega} = g^{-1}(\gamma_{mm'})\mathcal{H}_{m,\omega}\, g(\gamma_{mm'}) \tag{B.33}$$

where $\gamma_{mm'}$ is a curve from m to m', and $g(\gamma)$ is defined by

$$g(\gamma) = \wp \exp\left\{-\int_\gamma \mathcal{A}_\mu(x)dx^\mu\right\} \tag{B.34}$$

It follows that, for an Abelian structure group, holonomy maps as well as holonomies are gauge independent and independent of choice of base point m.

A Yang–Mills gauge potential may be represented by a connection one-form on a principal fiber bundle, in which case the corresponding Yang–Mills field strength is represented by the associated curvature two-form—the covariant derivative of the connection one-form.

A *two-form* Ω on a differentiable manifold P is an antisymmetric tensor field of type (0,2), i.e. $\Omega : \mathcal{X}P \times \mathcal{X}P \to V$ is a bilinear map from the direct product of the set of vector fields on P with itself into a vector space V, which satisfies $\Omega(Y, X) = -\Omega(X, Y)$ for all $X, Y \in \mathcal{X}P$. The curvature two-form Ω

on a principal fiber bundle P takes values in the Lie algebra $\mathfrak{g}$ of the bundle's structure group G. It is related to the connection one-form ω on P as follows.

Given any one-form η, its *exterior derivative* $d\eta$ is a two-form whose action on vector fields X, Y is defined by

$$d\eta(X, Y) = X\eta(Y) - Y\eta(X) - \eta([X, Y]) \tag{B.35}$$

The connection one-form ω on P defines the horizontal component X^H of each vector $X \in T_u P$. One can then define the *covariant derivative* $D\eta$ of any one-form η by the equation

$$D\eta(X, Y) = d\eta(X^H, Y^H) \tag{B.36}$$

The *curvature two-form* Ω on principal fiber bundle $P(M, G)$ is then defined as the covariant derivative of ω:

$$\Omega = D\omega \tag{B.37}$$

It then follows that

$$\Omega(X, Y) = d\omega(X, Y) + [\omega(X), \omega(Y)], \text{ for all vector fields } X, Y \text{ on } P \tag{B.38}$$

Now the following identities have straightforward proofs:

$$\omega \wedge \omega = [\omega, \omega] \tag{B.39}$$

$$[\omega(X), \omega(Y)] = \frac{1}{2}[\omega, \omega](X, Y) \tag{B.40}$$

Hence B.38 may be rewritten more elegantly as

$$\Omega = d\omega + \omega \wedge \omega \tag{B.41}$$

The curvature is related to the failure of the horizontal lifts of closed curves in M to close, and is therefore also related to the holonomies of closed curves in M (see figure 1.4).[4]

Consider an infinitesimal parallelogram γ in a coordinate system $\{x^\mu\}$ on an open set $U \subseteq M$, whose corners are $O = \{0, 0, ...\}$, $P = \{\varepsilon, 0, ...\}$, $Q = \{\varepsilon, \delta, ...\}$, $R = \{0, \delta, ...\}$. Let X, Y be vector fields on $\pi^{-1}U$ whose flows yield the sides of the horizontal lift $\tilde{\gamma}$ of γ through $u \in \pi^{-1}m$, and set

---

[4] It may also be represented on an open set $U \subseteq M$ by the pull-back of Ω onto M defined by a local section σ: $\mathcal{F} \equiv \sigma^*\Omega$, which leads to the relations 3.56, 3.58, 3.61 between the local field strength and local gauge potential.

APPENDIX B FIBER BUNDLES

$\pi^*X = \epsilon V$, $\pi^*Y = \epsilon W$, where $V = \partial/\partial x^1$, $W = \partial/\partial x^2$. The Lie derivative $[X, Y]$ provides a measure of the failure of $\tilde{\gamma}$ to close, and it is vertical, for

$$\pi^*([X, Y]) = \varepsilon\delta[V, W] = 0 \tag{B.42}$$

Since X, Y are horizontal, $\Omega(X, Y) = d\omega(X, Y) = -\omega([X, Y]) = -A$, where $[X, Y] = A^{\#}$.

Now $\tilde{\gamma}$ begins at u and ends at $v \in \pi^{-1}(m)$, where $\tau(u)(\gamma) = ug(\gamma)$ for some $g \in G$ where τ is the holonomy map and g is the corresponding holonomy. But v is also related to u by lying on the flow generated by A, an element of the Lie algebra $\mathfrak{g}$. Indeed we have

$$g(\gamma) = \exp \varepsilon\delta A \tag{B.43}$$
$$= \exp\{-\varepsilon\delta\Omega(X, Y)\} \tag{B.44}$$

Hence specifying the bundle curvature at each point is equivalent to specifying the holonomy of every infinitesimal closed parallelogram in the bundle.

For an Abelian structure group, this relation between holonomies and curvature is easier to see, since in this case there is a single holonomy associated with each closed curve C in M, independent of choice of C's base point, and of a point in the fiber above this. Here we have simply

$$H(C_{\varepsilon\delta}) = \exp\{-\varepsilon\delta\Omega(X, Y)\} \tag{B.45}$$

for every infinitesimal closed parallelogram $C_{\varepsilon\delta}$. Since the structure group U(1) for electromagnetism is Abelian, it follows that the holonomies of the principal bundle $P(M,(U(1)))$ around all closed curves fix the curvature of the bundle Ω, and hence the electromagnetic field strength $F_{\mu\nu}$ which derives from this by pulling Ω back onto M.

The connection ω on $P(M, G)$ defines the covariant derivative D in an associated vector bundle $\langle E, M, G, \pi_E, V, P \rangle$; for details, see, for example, Nakahara (1990). Since the wave-function or matter field is defined as a section on an associated vector bundle, this permits one to form its derivative. The "components" $A^a_{b\mu}$ of the connection that enter into the coordinate form of the covariant derivative in E (cf. equation B.4) are then uniquely defined relative to a section σ of P by the pull-back $\sigma^*\omega$ of ω onto M. The coordinate form of the covariant derivative (B.6) transforms covariantly (i.e. in the same way as the vector field with values in V that represents the wave-function or matter field in section σ) under gauge transformations corresponding either to a change of section σ or to a vertical bundle automorphism of P that changes ω. A vector $v \in V$ is said to be *parallel-transported* along a curve $\gamma(t)$ in M just in case there is a section s of E with $s(\gamma(t)) = [(\tilde{\gamma}(t), v)]$ everywhere along

γ, where $\tilde{\gamma}$ is a horizontal lift of γ in P; such a section has zero covariant derivative along γ. Just as horizontal lifts of a closed curve C in M define a holonomy map in $P(M, G)$, so also parallel transport around C of vectors defines a holonomy map in E. The holonomy map in E may be expressed in a given section σ of P by the action on V of an element of (a representation of) G called the holonomy of C in E. This is linear, and is given explicitly by

$$\mathbf{H}_m(C) = \wp \exp\left\{-\oint_C \mathbf{A}_\mu(x) dx^\mu\right\} \tag{B.46}$$

where m is the base point of C and $\mathbf{A}_\mu$ is a square matrix with elements $[\mathbf{A}_\mu]_{ab} = A^a_{b\mu}$, the "components" of the connection in σ.

This expression for the holonomy map depends on the connection $\mathbf{A}_\mu$, which depends in turn both on the connection ω and on the choice of section σ on $P(M, G)$. Either a change of section or a vertical principal bundle automorphism may be thought of as corresponding to a gauge transformation, and in either case $\mathbf{A}_\mu$ will transform in accordance with equation 3.38:

$$\mathbf{A}'_\mu = \mathbf{U}\mathbf{A}_\mu\mathbf{U}^\dagger - \left(\partial_\mu\mathbf{U}\right)\mathbf{U}^\dagger \tag{B.47}$$

This will induce the following transformation in the holonomy map on E:

$$\mathbf{H}_m(C) \to \mathbf{U}\mathbf{H}_m(C)\mathbf{U}^\dagger \tag{B.48}$$

So holonomies transform in the same way as gauge field strengths under such gauge transformations (cf. equation 3.27). Under a change of base point $m \to m'$ for C, the holonomy map transforms as follows (cf. equation B.33):

$$\mathbf{H}_{m'}(C) = \mathbf{U}_{mm'}^{-1}\mathbf{H}_m(C)\mathbf{U}_{mm'} \tag{B.49}$$

where, if S_C is a curve that traces the part of C from m to m',

$$\mathbf{U}_{mm'} = \wp \exp\left\{-\int_{S_C} \mathbf{A}_\mu(x) dx^\mu\right\} \tag{B.50}$$

If m' does not lie on C, a slight generalization of equation B.49 applies. Suppose now that S is some smooth curve from m to m', and let C' be a curve formed by composing C with S and its "inverse" $\overline{S}$—a curve from m' to m that retraces S—as follows

$$C' = S \circ C \circ \overline{S} \tag{B.51}$$

APPENDIX B FIBER BUNDLES

Then part of the image of C' is traced out by the "tree" $S \circ \overline{S}$—a closed curve with base point m' that encloses no area. There is another curve that traces out the image of C' starting from m, namely

$$C'' = \overline{S} \circ S \circ C \tag{B.52}$$

Since the holonomy of curves that differ only by a finite number of "trees" are the same, $\mathbf{H}(C) = \mathbf{H}(C'')$. But C'', C' differ only because they have different base points, m, m' respectively. Equation B.49 implies that their holonomies are related by a similarity transformation of the form

$$\mathbf{H}_{m'}(C') = \mathbf{U}_{mm'}^{-1} \mathbf{H}_m(C'') \mathbf{U}_{mm'} \tag{B.53}$$

Consequently,

$$\mathbf{H}_{m'}(C') = \mathbf{U}_{mm'}^{-1} \mathbf{H}_m(C) \mathbf{U}_{mm'} \tag{B.54}$$

Suppose we decide to refer the holonomies of every closed curve C to a common base point o, whether or not o lies on C. For any smooth curve S connecting o to a point m on C, one can relate $\mathbf{H}_m(C)$ to the holonomy of a curve with base point o as follows

$$\mathbf{H}_o(S \circ C \circ \overline{S}) = \mathbf{U}_{mo}^{-1} \mathbf{H}_m(C) \mathbf{U}_{mo} \tag{B.55}$$

In this way, the holonomies of all curves with a common base point determine the holonomies of all curves. But note that these holonomies are thereby fixed only relative to choices of point m and curve S made independently for each closed curve C. By themselves, the holonomies of all curves with a common base point determine the holonomies of all curves only up to a gauge transformation.

APPENDIX C

The constrained Hamiltonian formalism

The constrained Hamiltonian formalism initially developed by Dirac offers a perspective on gauge theories that is sufficiently general to encompass all the gauge field theories considered in this book as well as many other dynamical theories, of particles as well as fields. It is based on a Hamiltonian formulation of a theory, which derives in turn from a Lagrangian formulation. Since most physics texts present Yang–Mills theories in a Lagrangian formulation, it is best to begin with that.

In a Lagrangian formulation of a theory, the basic dynamical equations are derived from an underlying Lagrangian for the theory as the Euler-Lagrange equations that result from application of Hamilton's principle to the action corresponding to that Lagrangian. Consider the example of a complex scalar classical field whose Lagrangian is defined by integrating the Lagrangian density 6.4 over all space

$$\mathcal{L} = (\partial_\mu \phi)(\partial^\mu \phi^*) - m^2 \phi^* \phi \tag{C.1}$$

One can regard this as a way of re-expressing the content of a field with two real-valued components (ϕ_1, ϕ_2) by setting

$$\phi = 1/\sqrt{2}\,(\phi_1 + i\phi_2) \tag{C.2}$$
$$\phi^* = 1/\sqrt{2}\,(\phi_1 - i\phi_2) \tag{C.3}$$

The action corresponding to this Lagrangian density is the time integral of the Lagrangian, i.e.

$$S = \int_{t_1}^{t_2} L\,dt = \int_R \mathcal{L} d^4 x \tag{C.4}$$

where the integral is taken over a space-time region R with temporal boundaries at t_1, t_2. The equations of motion for the field are derived by considering how S varies under independent infinitesimal variations of $\phi, \partial_\mu \phi, \phi^*, \partial_\mu \phi^*$ that vanish on the boundary of R, and requiring that S be stationary under such

APPENDIX C THE CONSTRAINED HAMILTONIAN FORMALISM 249

variations. It is a standard exercise in the calculus of variations to show that this condition is met just in case $\mathcal{L}$ satisfies the Euler–Lagrange equations

$$\frac{\partial \mathcal{L}}{\partial \phi} - \frac{\partial}{\partial_\mu}\left(\frac{\partial \mathcal{L}}{\partial (\partial_\mu \phi)}\right) = 0 \tag{C.5}$$

$$\frac{\partial \mathcal{L}}{\partial \phi^*} - \frac{\partial}{\partial_\mu}\left(\frac{\partial \mathcal{L}}{\partial (\partial_\mu \phi^*)}\right) = 0 \tag{C.6}$$

Evaluating these equations with $\mathcal{L}$ given by C.1 gives the Klein–Gordon equation

$$\partial_\mu \partial^\mu \phi + m^2 \phi = 0 \tag{C.7}$$

(The second Euler–Lagrange equation gives just the complex conjugate of this.)

Noether's first theorem implies that if the Lagrangian for a field is invariant under an infinitesimal continuous symmetry transformation of its constituent fields generated by the operation of a finite-parameter Lie group G_n, then there exists a corresponding conserved current J_μ, i.e.

$$\partial^\mu J_\mu = 0 \tag{C.8}$$

and a corresponding conserved Noether charge

$$N = \int_\Sigma J_0 d^3x \tag{C.9}$$

defined on a space-like hypersurface Σ (representing space at a time). In the example, $\mathcal{L}$, and therefore L, is invariant under infinitesimal transformations of the form

$$\phi \to \phi - i\varepsilon\phi \tag{C.10}$$
$$\phi^* \to \phi^* + i\varepsilon\phi^*$$

and the resulting conserved Noether charge is

$$N = i\int_V \left(\phi^* \frac{\partial \phi}{\partial t} - \phi \frac{\partial \phi^*}{\partial t}\right) d^3x \tag{C.11}$$

where the integral is taken over a volume corresponding to a space-like hypersurface Σ at any time. This implies conservation of electric charge $Q = qN$, where q is the charge associated with the field ϕ.

If the symmetry transformation C.10 is generalized to a transformation in which the infinitesimal parameter ε is allowed to vary with space-time position x, then the resulting transformation is no longer a symmetry of C.1. But it *is* a symmetry of the Lagrangian 6.23 for the Klein–Gordon field interacting with electromagnetism

$$\mathcal{L}_{\text{tot}} = (D_\mu \phi)(D^\mu \phi^*) - m^2 \phi^* \phi - \frac{1}{4} F^{\mu\nu} F_{\mu\nu} \qquad (\text{C.12})$$

One might expect a generalization of Noether's first theorem to such a continuously variable symmetry transformation to yield additional conservation laws. But this is not so. Instead, it is a consequence of Noether's second theorem that if the Lagrangian density is invariant under continuous symmetry transformations generated by a group $G_{r\infty}$ parametrized by a finite set of r arbitrary *functions* of the coordinates x rather than by a finite set of arbitrary parameters, then the Euler–Lagrange equations are not independent of one another: instead, the constraint functions—the expressions appearing on their left-hand sides—satisfy certain *identities*, in conjunction with the derivatives of those functions. These relations among the Euler–Lagrange equations show that the solutions to these equations will contain arbitrary functions of the independent variables x. It therefore appears that any theory with such a symmetry will be radically indeterministic! To avoid such radical indeterminism it is necessary to reinterpret the theory in some way. Such reinterpretation is facilitated by a switch to a Hamiltonian formulation of the theory, and the development within that formulation of the constrained Hamiltonian formalism.

Begin by considering a system of n Newtonian point particles, with position coordinates $q_i (i = 1, \ldots, 3n)$ and Lagrangian $L(q, \dot{q})$, ($q = (q_i, \ldots, q_{3n})$, where $\dot{q} = dq/dt$). The Euler–Lagrange equations for this system are

$$\frac{\partial L}{\partial q_i} - \frac{d}{dt}\left(\frac{\partial L}{\partial \dot{q}_i}\right) = 0 \text{ for } i = 1, \ldots, 3n \qquad (\text{C.13})$$

If we define $p^i \equiv \frac{\partial L}{\partial \dot{q}_i}$, these become

$$\frac{\partial L}{\partial q_i} = \dot{p}^i \qquad (\text{C.14})$$

Now define the *Hamiltonian* of the system by

$$H(p, q) = \sum_i p^i \dot{q}_i - L(q, \dot{q}) \qquad (\text{C.15})$$

and treat q_i, p^i as the basic independent variables instead of $q_i, \dot{q}_i$. Then the Euler–Lagrange equations yield *Hamilton's equations*

$$\dot{q}_i = \frac{\partial H}{\partial p^i} \quad ; \quad \dot{p}^i = -\frac{\partial H}{\partial q_i} \qquad (\text{C.16})$$

APPENDIX C THE CONSTRAINED HAMILTONIAN FORMALISM

The transformation from $q, \dot{q}$ to q, p is called a *Legendre transformation*. If Q is some dynamical variable on the system, we can express its changing values as follows:

$$\frac{dQ}{dt} = \frac{\partial Q}{\partial t} + \sum_i \left(\frac{\partial Q}{\partial q_i} \frac{\partial q_i}{\partial t} + \frac{\partial Q}{\partial p^i} \frac{\partial p^i}{\partial t} \right) \quad (C.17)$$

$$= \frac{\partial Q}{\partial t} + \sum_i \left(\frac{\partial Q}{\partial q_i} \frac{\partial H}{\partial p^i} - \frac{\partial Q}{\partial p^i} \frac{\partial H}{\partial q_i} \right) \quad (C.18)$$

The *Poisson bracket* $\{Q, H\}$ is defined as

$$\{Q, H\} \equiv \sum_i \left(\frac{\partial Q}{\partial q_i} \frac{\partial H}{\partial p^i} - \frac{\partial Q}{\partial p^i} \frac{\partial H}{\partial q_i} \right) \quad (C.19)$$

so that

$$\frac{dQ}{dt} = \frac{\partial Q}{\partial t} + \{Q, H\} \quad (C.20)$$

In many cases, the Hamiltonian of a system turns out to be its energy: for a system whose Hamiltonian does not depend explicitly on time, conservation of energy follows from

$$\frac{dH}{dt} = 0 + \{H, H\} = 0 \quad (C.21)$$

In quantum mechanics, the Hamiltonian becomes an operator $\hat{H}$ acting on the wave-function or state vector Ψ. The fundamental dynamical equation of non-relativistic quantum mechanics is then the Schrödinger equation 1.9

$$\hat{H}\Psi = i\hbar \partial \Psi / \partial t \quad (C.22)$$

Moreover, the basic commutation relations between operators representing dynamical variables like (components of) position and momentum can be thought to arise from the substitution

$$\{Q, R\} \rightarrow \frac{1}{i\hbar} [\hat{Q}, \hat{R}] \quad (C.23)$$

where the square bracket indicates the commutator $\hat{Q}\hat{R} - \hat{R}\hat{Q}$. For example, for the dynamical variables q_j, p^k we have

$$\{q_j, p^k\} = \sum_l \left(\frac{\partial q_j}{\partial q_l} \frac{\partial p^k}{\partial p_l} - \frac{\partial q_j}{\partial p_l} \frac{\partial p^k}{\partial q_l} \right) = \delta_{jk} \quad (C.24)$$

and so

$$[\hat{q}_j, \hat{p}^k] = i\hbar \delta_{jk} \quad (C.25)$$

which gives, for example,

$$[\hat{x}, \hat{p}_x] = [\hat{y}, \hat{p}_y] = i\hbar \qquad (C.26)$$
$$[\hat{x}, \hat{y}] = [\hat{x}, \hat{p}_y] = [\hat{p}_x, \hat{p}_y] = 0 \qquad (C.27)$$

For a typical Lagrangian, Hamilton's equations yield $6n$ first-order differential equations as the equations of motion for a system of n Newtonian point particles, which may be solved to uniquely determine the dynamical evolution of the state of the system. This may be represented by the motion of a point (p, q) in a $6n$-dimensional space called the *phase space* of the system.

The Hamiltonian formalism may be applied also to field systems, in which each point $\mathbf{x}$ in space is associated with one or more configuration variables $\phi_i(\mathbf{x})$, and corresponding momentum variables $\pi^i(\mathbf{x})$. This means that there is now a non-denumerably infinite set of independent dynamical variables, and each of the Lagrangian and Hamiltonian is formed by integrating its density over all points $\mathbf{x}$ rather than summing over a finite set of independent configuration variables. The evolution of the system may again be determined by solving Hamilton's equations, which now involve functional derivatives rather than simple partial derivatives:

$$\dot{\phi}_i = \frac{\delta H}{\delta \pi^i} \qquad \dot{\pi}^i = -\frac{\delta H}{\delta \phi_i} \qquad (C.28)$$

and the resulting development of the fields $\phi_i(\mathbf{x})$ may be pictured in an infinite-dimensional phase space.

But as previously noted, if the Lagrangian density is unchanged under a transformation from a group $G_{r\infty}$—an infinite-parameter Lie group parameterized by functions of time—then the Euler–Lagrange equations are not independent of one another. In this situation, Noether's second theorem implies that there are r independent identities satisfied by the constraint functions that appear on the left-hand side of the Euler–Lagrange equations and their derivatives. This means that the solutions to the Euler–Lagrange equations contain arbitrary functions of time. Solutions to the corresponding Hamilton's equations also contain arbitrary functions of time, so that the evolution of the point in phase space representing the state of the system is not deterministic: starting from a point at one time, it may evolve into any point in a particular region of phase space at a later time.

The constrained Hamiltonian formalism handles the apparent indeterminism in such a case by providing a principled decomposition of the evolution of the phase point into a physical part and an unphysical part. All points in the region of phase space to which the phase point may evolve from a given initial point, according to Hamilton's equations, are considered physically equivalent: they are said to lie on the same gauge orbit. While evolution from point to point

APPENDIX C THE CONSTRAINED HAMILTONIAN FORMALISM

is indeterministic, evolution from gauge orbit to gauge orbit is deterministic. "Motion" of a phase point within a gauge orbit therefore represents no physical evolution of the system, and motion from any point in one gauge orbit to any point in another gauge orbit represents the same evolution of the physical state of the system.

The key to implementing this interpretative strategy is to distinguish the two kinds of motion of a phase point. The distinction is made in terms of a classification of the identities satisfied by the constraint functions. A constraint sets a constraint function or its derivative identically equal to zero. A *primary constraint* appears in the form of an Euler–Lagrange equation, but it differs from other Euler–Lagrange equations because it expresses a relation among the canonical variables (ϕ_i, π^i) and their derivatives that follows from the definitions of these variables, and so must hold at every time (on every space-like hypersurface) independently of how the state of the system evolves. Requiring that a primary constraint continue to hold as the system evolves may lead to one or more *secondary constraints*. This process may be iterated, so that a total set of constraints is generated after a finite number of steps. The condition that all these constraints be satisfied at once defines a region of the phase space called the *constraint surface*. The motion of the phase point representing the state of the system is confined to this surface. Independent of its classification as primary or secondary, a constraint may be classified as either first class or second class. A constraint is *first class* just in case the Poisson bracket of the corresponding constraint function with every other constraint function is zero when evaluated on the constraint surface; otherwise it is *second class*.

The evolution of any dynamical field variable Q is specified in terms of the Hamiltonian density $\mathcal{H}$ by the equation

$$\frac{dQ}{dt} = \frac{\partial Q}{\partial t} + \{Q, \mathcal{H}\} \tag{C.29}$$

But since all first-class constraints are satisfied everywhere on the constraint surface, the addition to the initial Hamiltonian density $\mathcal{H}_0$ of any function $F[f_i(\phi_i, \pi^i)]$ of the first-class constraint functions $f_i(\phi_i, \pi^i)$ will leave the value of the Hamiltonian density on that surface everywhere unchanged. Motion generated by $\mathcal{H}_0$ and $\mathcal{H}' = \mathcal{H}_0 + F[f_i(\phi_i, \pi^i)]$ is therefore considered physically equivalent, even though it takes an initial phase point into distinct phase points at later (or earlier) times. Now motion generated by $\mathcal{H}'$ will produce the following evolution in Q:

$$\frac{dQ}{dt} = \frac{\partial Q}{\partial t} + \{Q, \mathcal{H}'\} \tag{C.30}$$

$$= \frac{\partial Q}{\partial t} + \{Q, \mathcal{H}_0\} + \{Q, F\} \tag{C.31}$$

The contribution of F to such motion is therefore considered unphysical: it is regarded as "motion" confined to a gauge orbit. In this sense, the first-class constraints $f_i(\phi_i, \pi^i)$ are said to generate motion within a gauge orbit, while the Hamiltonian generates motion from gauge orbit to gauge orbit. Suppose one picks an initial function $\mathcal{H}$ from the class of functions that are equivalent to $\mathcal{H}_0$ in the sense that they take on the same values everywhere on the constraint surface. Pick an arbitrary time t_0 and an arbitrary phase point p_0 on the constraint surface at t_0. Then the set of points in the *gauge orbit at* $t \neq t_0$ will be defined by evolving the phase point (ϕ_i, π^i) from t_0 to t in accordance with

$$\frac{d\phi_i}{dt} = \frac{\partial \phi_i}{\partial t} + \{\phi_i, \mathcal{H}\} + \{\phi_i, F\} \tag{C.32}$$

$$\frac{d\pi^i}{dt} = \frac{\partial \pi^i}{\partial t} + \{\pi^i, \mathcal{H}\} + \{\pi^i, F\} \tag{C.33}$$

By choosing all functions $F\left[f_i(\phi_i, \pi^i)\right]$ and evolving backward and forward from arbitrarily chosen t_0, p_0 this defines the set of gauge orbits of the theory. If $f_i(\phi_i, \pi^i)$ is a first-class constraint, then ∇f_r is a vector field X_r on the constraint surface. The gauge orbits consist of the integral curves of vector fields formed by summing all the constraint functions, each "smeared" by an appropriate, but otherwise arbitrary, smearing field. The first-class constraints *generate* "motion" within a gauge orbit in a way analogous to C.30: the Poisson bracket of a dynamical field variable Q with a sum of constraint functions, each smeared by an infinitesimal smearing field, equals the corresponding infinitesimal change in Q within the gauge orbit.

As an important example, consider the application of the constrained Hamiltonian formalism to classical electromagnetism, considered as a theory formulated in terms of electric and magnetic potentials $\varphi, \mathbf{A}$ (see appendix A). Here the electric and magnetic fields are taken to be derived from these potentials in accordance with

$$\mathbf{B} = \nabla \times \mathbf{A} \tag{C.34}$$
$$\mathbf{E} = -\nabla\varphi - \dot{\mathbf{A}} \tag{C.35}$$

The generalized configuration variables are $\varphi, \mathbf{A}$ and the Lagrangian density is

$$\mathcal{L} = \frac{1}{2}(\mathbf{E}^2 - \mathbf{B}^2) \tag{C.36}$$
$$= \frac{1}{2}\left((\nabla\varphi + \dot{\mathbf{A}})^2 - (\nabla\times\mathbf{A})^2\right) \tag{C.37}$$

APPENDIX C THE CONSTRAINED HAMILTONIAN FORMALISM

Hence
$$\frac{\partial \mathcal{L}}{\partial \dot{\varphi}} = \pi^0 = 0 \tag{C.38}$$

$$\frac{\partial \mathcal{L}}{\partial \dot{A}_i} = \pi^i = \nabla \varphi + \dot{\mathbf{A}} = -\mathbf{E} \tag{C.39}$$

Equation C.38 expresses one first-class constraint, while requiring that C.38 continue to hold as the system evolves implies another first-class constraint

$$\nabla \cdot \mathbf{E} = 0 \tag{C.40}$$

These are the only first-class constraints. Note that the Poisson bracket of the two first-class constraints vanishes everywhere, not just on the constraint surface. This implies that the Poisson algebra generated by these constraints closes. It follows that the vector fields that define the integral curves corresponding to "motion" within a gauge orbit constitute a Lie algebra. That is what makes this theory a Yang–Mills gauge theory.

$$\mathcal{L} = \frac{1}{2} \left(\sum_i (\pi^i)^2 - (\nabla \times \mathbf{A})^2 \right) \tag{C.41}$$

Now the Hamiltonian density is defined as

$$\mathcal{H} \equiv \sum_i \dot{\phi}_i \pi^i - \mathcal{L} \tag{C.42}$$

$$= \dot{\varphi}\pi^0 - \dot{\mathbf{A}} \cdot \mathbf{E} - \frac{1}{2}(\mathbf{E}^2 - \mathbf{B}^2) \tag{C.43}$$

$$= \frac{1}{2}(\mathbf{E}^2 + \mathbf{B}^2) + \dot{\varphi}\pi^0 - \varphi(\nabla \cdot \mathbf{E}) + \nabla \cdot (\varphi \mathbf{E}) \tag{C.44}$$

When we integrate over space and apply Hamilton's principle, the integral of the fourth term on the right will vanish on the boundary of the volume, so we may set

$$\mathcal{H} = \frac{1}{2}(\mathbf{E}^2 + \mathbf{B}^2) + \dot{\varphi}\pi^0 - \varphi(\nabla \cdot \mathbf{E}) \tag{C.45}$$

which is 5.28.

Hamilton's equations C.28 now become

$$\dot{\phi}_i = \frac{\partial \mathcal{H}}{\partial \pi^i} - \sum_j \frac{d}{dx_j}\left(\frac{\partial \mathcal{H}}{\partial \left(\frac{\partial \pi^i}{\partial x_j}\right)}\right) \quad \dot{\pi}^i = -\frac{\partial \mathcal{H}}{\partial \phi_i} + \sum_j \frac{d}{dx_j}\left(\frac{\partial \mathcal{H}}{\partial \left(\frac{\partial \phi_i}{\partial x_j}\right)}\right)$$
$$\tag{C.46}$$

which, with $\phi_i = (\varphi, \mathbf{A})$, $\pi^i = (0, -E_i)$ in this case, yield

$$\dot{\mathbf{E}} = \mathbf{\nabla} \times (\mathbf{\nabla} \times \mathbf{A}) \qquad (C.47)$$
$$\dot{\mathbf{A}} = -\mathbf{E} - \mathbf{\nabla}\varphi \qquad (C.48)$$

Since the relation of lying on the same gauge orbit is an equivalence relation R on the constraint surface S, one can take the quotient of S by R to form the *reduced phase space* S/R, and represent the evolution of a system by the motion of a point in S/R. In the example, while the configuration space of the original phase space S of electromagnetism was coordinatized by pairs of potential functions $\varphi, \mathbf{A}$ on M, a point in the configuration space of the reduced phase space S/R of electromagnetism is coordinatized by the holonomies of all curves in M.

General relativity may also be formulated using the constrained Hamiltonian formalism by starting with the Einstein–Hilbert action and then switching to a Hamiltonian formulation via a Legendre transformation. This procedure has been followed as a prelude to the attempted canonical quantization of the theory. The first-class constraints are of two kinds, known as the diffeomorphism constraint(s) and the Hamiltonian constraint(s); the plural is preferable, since each constraint applies independently at every point on any spatial hypersurface Σ in a foliation of the space-time manifold M. But although the Poisson brackets of the corresponding constraint functions all vanish on the constraint surface, there are pairs of constraint functions whose Poisson bracket cannot be expressed as a linear combination of constraint functions: in that sense, the Poisson algebra does not close. It follows that the vector fields that define the integral curves corresponding to "motion" within a gauge orbit do not constitute a Lie algebra. That is why general relativity is not a Yang–Mills gauge theory.

APPENDIX D

Alternative quantum representations

This appendix explains what is meant by a Hilbert space representation of the states and observables of a quantum theory. It describes several representations of the states and observables of a quantum theory of particles, and motivates and explains an important theorem due to Stone and von Neumann that relates these to each other and to a wide class of alternative representations.

In a classical theory, a state of a system at a time may be represented by a point in a configuration space—specifying the positions of all the particles, or the values of the fields at each point in space—or by a point in an associated phase space—specifying the particles' momenta as well as positions, or the fields and conjugate field momenta at each point in space. In either case, the motion of this point represents the evolution of the system's state. A dynamical variable (such as kinetic energy, or angular momentum) is represented by a function on this space, whose value at a point gives the value of the variable in the corresponding state of the system. This limited choice of representation for states and dynamical variables is of little interpretative significance, since all dynamical variables always have precise values in all states, no matter how these are represented.

In a quantum theory, the state of a system is usually represented by a vector (or more accurately a ray—a one-dimensional subspace of vectors, of different moduli ("lengths") and phases, but all "pointing in the same direction") in a complex Hilbert space **H**—a vector space with complex-number coefficients. A dynamical variable is represented by a self-adjoint operator acting on this space: the term 'observable' is often used ambiguously to denote either the variable or the operator that represents it.[1] For all but "toy" systems, the state space is infinite dimensional. All infinite-dimensional Hilbert spaces are

---

[1] A complex vector space with an inner product (,) is a *Hilbert space* if and only if (i) the space is *separable*, i.e. there is a set of vectors $x_i (i = 1, 2, 3, \ldots)$ such that every vector v in the space is a finite or denumerably infinite sum of the form $v = \sum c_i x_i$, where the c_i are complex numbers, and no $x_i (i = 1, 2, 3, \ldots)$ can itself be expressed as such a finite or denumerably infinite sum of the other $x_j (j \neq i)$, and (ii) the space is *complete*, in the sense that every Cauchy sequence converges to a limit in **H**. (A sequence of vectors $\{v_i\}$ is a Cauchy sequence if and only if, for every $\epsilon > 0$ there is an N_ϵ such

isomorphic, but one may choose among alternative concrete realizations of the state space by distinct mathematical objects, and each choice will be accompanied by a concrete representation of the operator representing each dynamical variable. However, every legitimate choice of representation for operators representing dynamical variables must respect certain fundamental algebraic relations among these operators.

Perhaps the simplest example is the representation by a complex-valued wave-function $\psi(x)$ of the state of a single spinless, non-relativistic particle that is free to move along one dimension of Euclidean space. The Hilbert space of states consists of those functions that are square-integrable, i.e.

$$\int_{-\infty}^{\infty} |\psi(x)|^2 \, dx < \infty \tag{D.1}$$

where $|\psi(x)|^2 = \psi(x)\psi^*(x)$, and the inner product is defined by $(\chi, \psi) = \int_{-\infty}^{\infty} \chi^*(x)\psi(x) dx$. The dynamical variable *position* is represented by the self-adjoint operator $\hat{x}$ whose action is given by

$$\hat{x}\psi(x) = x\psi(x) \tag{D.2}$$

and the dynamical variable *momentum* is represented by a self-adjoint operator $\hat{p}$, where

$$\hat{p}\psi(x) = -i\hbar \frac{d\psi}{dx} \tag{D.3}$$

These satisfy the Heisenberg commutation relation

$$[\hat{x}, \hat{p}] = i\hbar \hat{I} \tag{D.4}$$

which is short for

$$[\hat{x}, \hat{p}]\psi(x) \equiv \left(\hat{x}\hat{p} - \hat{p}\hat{x}\right)\psi(x) = i\hbar\psi(x) \tag{D.5}$$

Neither $\hat{x}$ nor $\hat{p}$ is defined on every vector in this Hilbert space, and so a precise statement of the Heisenberg commutation relation would require careful attention to the set of functions $\psi(x)$ implicit in its statement.[2] Some

---

that, for every $m, n > N_\epsilon$, $(v_m - v_n, v_m - v_n) < \epsilon$.) A *linear operator* on a Hilbert space H is a linear map from a subset D of H onto R ⊆ H; (roughly speaking) its *adjoint* is a linear operator $A^\dagger : R \to D$ such that, for all $u \in R, v \in D$, $(A^\dagger u, v) = (u, Av)$. An operator is *self-adjoint* if and only if it equals its adjoint.

[2] $\hat{x}, \hat{p}$ are unbounded operators, and no unbounded operator is defined on every vector in a Hilbert space. (An operator $\hat{O}$ is *bounded* if and only if there is a positive number N such that $(\hat{O}\psi, \hat{O}\psi) \leq N(\psi, \psi)$ for every ψ in the space.)

APPENDIX D ALTERNATIVE QUANTUM REPRESENTATIONS   259

other dynamical variables may be defined in terms of $\hat{x}, \hat{p}$. For example, the *kinetic energy* is represented here by the operator $\frac{\hat{p}^2}{2m} = -\frac{\hbar^2}{2m}\frac{d^2}{dx^2}$, and in certain cases the *potential energy* may be represented by an operator of the form $V(\hat{x}) = V(x)$.[3]

This is called the *Schrödinger*, or position, representation. It extends naturally to a single particle free to move in three spatial dimensions, where the actions of the position operator $\hat{\mathbf{x}}$ and the momentum operator $\hat{\mathbf{p}}$ are given by

$$\hat{\mathbf{x}}\psi(x) = \mathbf{x}\psi(x) \tag{D.6}$$

$$\hat{\mathbf{p}}\psi(x) = -i\hbar\boldsymbol{\nabla}\psi \tag{D.7}$$

In the Schrödinger representation these and other self-adjoint operators representing dynamical variables act on the space $L^2(\mathbb{R}^3)$ of square-integrable complex-valued functions on a three-dimensional space of real number triples. The basic algebraic relations are now the Heisenberg commutation relations for the (components of) position and momentum of the single particle:

$$[\hat{x}_j, \hat{p}_k] = i\hbar\delta_{jk}\hat{1} \tag{D.8}$$
$$[\hat{x}_j, \hat{x}_k] = [\hat{p}_j, \hat{p}_k] = 0$$

where e.g. x_j ($j = 1, 2, 3$) are the three components of position in a rectangular Cartesian coordinate system. These are called canonical commutation relations (CCRs) because they relate operators representing the canonical positions and momenta in a Hamiltonian formulation of the classical mechanics of a single particle.

There is a further generalization to the Schrödinger, or position, representation for a system of N spinless, non-relativistic particles. The Heisenberg commutation relations generalize in an obvious way to commutation relations for a system of N particles, given that any pair of operators representing canonical variables pertaining to distinct particles commute. The actions of the position operators $\hat{\mathbf{x}}_n$ and momentum operators $\hat{\mathbf{p}}_n$ ($n = 1, 2, \ldots, N$) are given by

$$\hat{\mathbf{x}}_n\psi(\mathbf{x}_1, \mathbf{x}_2, \ldots, \mathbf{x}_N) = \mathbf{x}_n\psi(x) \tag{D.9}$$

$$\hat{\mathbf{p}}_n\psi(\mathbf{x}_1, \mathbf{x}_2, \ldots, \mathbf{x}_N) = -i\hbar\boldsymbol{\nabla}_n\psi \tag{D.10}$$

and these now act on a Hilbert space that is the *tensor product* $L_1^2(\mathbb{R}^3) \otimes L_2^2(\mathbb{R}^3) \otimes \cdots \otimes L_N^2(\mathbb{R}^3)$ of N one-particle Hilbert spaces.[4]

---

[3] Note that this represents the potential energy of a charged particle subject to a classical electrostatic field only in a gauge in which $\mathbf{A} = 0$!

[4] Since each $L_N^2(\mathbb{R}^3)$ is separable, any vector in $L_N^2(\mathbb{R}^3)$ may be written as a linear sum of vectors of the form $\{\psi_n^i\}$. Then any vector in the tensor product Hilbert space $L_1^2(\mathbb{R}^3) \otimes L_2^2(\mathbb{R}^3) \otimes \cdots \otimes L_N^2(\mathbb{R}^3)$

If the particles are bosons, then their wave-function is required to be symmetric under all permutations of the particles' positions

$$\psi(\mathbf{x}_1, \mathbf{x}_2, \ldots \mathbf{x}_i, \ldots \mathbf{x}_j, \ldots, \mathbf{x}_N) = \psi(\mathbf{x}_1, \mathbf{x}_2, \ldots \mathbf{x}_j, \ldots \mathbf{x}_i \ldots, \mathbf{x}_N) \quad (D.11)$$

in which case the Hilbert space is restricted to the symmetrized subspace S_N of the N particle tensor product space $L_1^2(\mathbb{R}^3) \otimes L_2^2(\mathbb{R}^3) \otimes \cdots \otimes L_N^2(\mathbb{R}^3)$. By taking the infinite direct sum $F = S_1 \oplus S_2 \oplus \cdots \oplus S_n \oplus \ldots$ one can form a Hilbert space that permits the representation of the state of a system of an arbitrary number of particles: this is called a *Fock space* for the bosons. A self-adjoint *number operator* $\hat{N} = \sum n\hat{P}_n$ may now be defined on F, where each $\hat{P}_n$ is the self-adjoint operator that projects onto the n-particle subspace S_n of F. But this construction is not very interesting for non-relativistic particles, since $\hat{N}$ is then a superselection operator—it commutes with every self-adjoint operator on F representing a dynamical variable.

Returning to the basic Heisenberg relations D.8, one can form a *momentum representation* by defining position and momentum operators as follows:

$$\hat{x}_j \psi(\mathbf{p}) = i\hbar \frac{d\psi(\mathbf{p})}{dx_j} \quad \hat{p}_j \psi(\mathbf{p}) = p_j \psi(\mathbf{p}) \quad (D.12)$$

These operators also satisfy D.8 when their commutator is taken to act on appropriate functions $\varphi(\mathbf{p})$ from the space $L^2(\mathbb{R}^3)$ of square-integrable complex-valued functions on a three-dimensional space of real number triples, representing not positions but momenta. The momentum representation can be extended to a many-particle system in the same way as the position representation.

As a final example, consider the *occupation number representation* for the states of a one-dimensional simple harmonic oscillator—a particle of mass m moving in one-dimension of Euclidean space under the influence of a potential of the form $V(x) = \frac{1}{2} m\omega^2 x^2$. The Hamiltonian operator is given by

$$\hat{H} = \frac{\hat{p}^2}{2m} + \frac{1}{2} m\omega^2 \hat{x}^2 \quad (D.13)$$

If we define operators $a, a^\dagger$ by

$$a = \sqrt{\frac{m\omega}{2\hbar}} \hat{x} + \frac{i}{\sqrt{2\hbar m\omega}} \hat{p} \quad (D.14)$$

$$a^\dagger = \sqrt{\frac{m\omega}{2\hbar}} \hat{x} - \frac{i}{\sqrt{2\hbar m\omega}} \hat{p} \quad (D.15)$$

can be written as a linear sum of vectors of the form $\psi_1^i \psi_2^j \ldots \psi_N^n$. Not every vector in this tensor product space can *itself* be written in the form of a product of such vectors.

then the Heisenberg relations D.8 imply that they satisfy

$$\left[a, a^\dagger\right] = \hat{1} \tag{D.16}$$

The Hamiltonian may be written in terms of these operators as

$$\hat{H} = \hbar\omega\left(a^\dagger a + \frac{1}{2}\right) \equiv \hbar\omega\left(\hat{N} + \frac{1}{2}\right) \tag{D.17}$$

where the *occupation number operator* $\hat{N} \equiv a^\dagger a$. Assuming that the system has a unique lowest-energy state, one can represent the states of the oscillator in a Hilbert space which has a basis of vectors of the form $|n\rangle$ where $\hat{N}|n\rangle = n|n\rangle$. In state $|n\rangle$ the particle has a definite energy $E = n + \frac{1}{2}\hbar\omega$: this can be considered to be made up of n quanta, each of energy $n\hbar\omega$, plus the so-called *zero-point energy* $\frac{1}{2}\hbar\omega$ of the ground state. The action of the operators $\hat{x}, \hat{p}$ on basis vectors in this representation is then

$$\hat{x}|n\rangle = \sqrt{\frac{\hbar}{2m\omega}}\left(\sqrt{n-1}\,|n-1\rangle + \sqrt{n}\,|n+1\rangle\right) \tag{D.18}$$

$$\hat{p}|n\rangle = i\sqrt{2\hbar m\omega}\left(\sqrt{n}\,|n+1\rangle - \sqrt{n-1}\,|n-1\rangle\right) \tag{D.19}$$

What is the relation between these various representations of the states and "observables" of a system of quantum particles? This question was first raised and answered by Dirac in his transformation theory, where he introduced a convenient notation that has since become widespread (including in this appendix)! For example, representing the state of a single spinless particle by a so-called ket $|\psi\rangle$—thought of as a vector in an abstract space of states—its position-representation state is represented by the wave-function $\langle \mathbf{x}|\psi\rangle \equiv \psi(\mathbf{x})$, while its momentum-representation wave-function is $\langle \mathbf{p}|\psi\rangle \equiv \psi(\mathbf{p})$. Such states are related by a Fourier transform

$$\langle \mathbf{p}|\psi\rangle = \int \langle \mathbf{p}|\mathbf{x}\rangle\langle \mathbf{x}|\psi\rangle \mathrm{d}^3 x \tag{D.20}$$

where $\langle \mathbf{p}|\mathbf{x}\rangle = \exp -\frac{i}{\hbar}\mathbf{p}\cdot\mathbf{x}$. States of systems of particles in different representations are thought of as related to one another by analogous transformations. Operators representing dynamical variables transform in a related way. For example, an operator $\hat{O}_\mathbf{x}$ representing a dynamical variable in position representation transforms into an operator $\hat{O}_\mathbf{p}$ in the momentum representation as follows:

$$\hat{O}_\mathbf{p}\psi(\mathbf{p}) = \int \langle \mathbf{p}|\mathbf{x}\rangle \hat{O}_\mathbf{x}\psi(\mathbf{x})\mathrm{d}^3 x \tag{D.21}$$

APPENDIX D ALTERNATIVE QUANTUM REPRESENTATIONS

After von Neumann set Dirac's transformation theory in the rigorous framework of Hilbert space theory, it became possible precisely to formulate and to prove a theorem relating the various alternative representations of the states and observables of a system of quantum particles. States $|\psi\rangle_1, |\psi\rangle_2$ in distinct Hilbert space representations of the states and observables of a system of quantum particles are related by a unitary transformation if and only if

$$|\psi\rangle_2 = U_{12}|\psi\rangle_1 \qquad (D.22)$$

where $U_{12} : \mathsf{H}_1 \to \mathsf{H}_2$ is a 1–1 map from the Hilbert space H_1 of the first representation onto the Hilbert space H_2 of the second representation that is both linear and norm preserving (and hence also preserves inner products):

$$U_{12}(a_1|\varphi\rangle + a_2|\chi\rangle) = a_1 U_{12}|\varphi\rangle + a_2 U_{12}|\chi\rangle \qquad (D.23)$$

$$\langle \varphi | U_{12}^{\dagger} U_{12} | \varphi \rangle = \langle \varphi | \varphi \rangle \qquad (D.24)$$

Two Hilbert space representations are *unitarily equivalent* if and only if there is a unitary transformation U_{12} that both relates all the states of one representation to corresponding states of the other and also transforms representations of observables so that if O is represented by operator $\hat{O}_1$ on H_1, then it is represented by operator $\hat{O}_2$ on H_2, where $U_{12}^{-1}\hat{O}_2 U_{12} = \hat{O}_1$.

Dirac's transformation theory makes it plausible that at least the position and momentum representations of the states and observables are unitarily equivalent for a single, spinless non-relativistic particle, free to move in three-dimensional Euclidean space. Stone formulated and von Neumann proved a theorem which entails that they are, when appropriately reformulated. But the Stone–von Neumann theorem also shows that this class of unitarily equivalent representations extends much more widely. This theorem applies to representations of operators that satisfy what is called the Weyl form of the Heisenberg commutation relations D.8.

What is the Weyl form, and why is it relevant here? The Heisenberg commutation relations D.8 apply to unbounded operators. But not every vector in the Hilbert space can lie in the domain of an unbounded operator. The relations D.8 are therefore not well defined unless and until one specifies a domain of definition for all unbounded operators they involve. Doing this involves technical complications that may be avoided by moving to an alternative form of commutation relations proposed by Weyl, namely

$$\hat{U}(\mathbf{a})\hat{V}(\mathbf{b}) = \exp(-i\mathbf{a}.\mathbf{b})\,\hat{V}(\mathbf{b})\hat{U}(\mathbf{a}) \qquad (D.25)$$

where $\mathbf{a}, \mathbf{b}$ are vectors in the $3n$-dimensional configuration space of an n-particle system, and one thinks of $\hat{U}(\mathbf{a}), \hat{V}(\mathbf{b})$ as related to $\hat{x}_j, \hat{p}_k (j, k = 1, \ldots, 3n)$

by $\hat{U}(\mathbf{a}) = \exp(i\mathbf{a}.\hat{\mathbf{x}})$, $\hat{V}(\mathbf{b}) = \exp(i\mathbf{b}.\hat{\mathbf{p}})$. $\hat{U}(\mathbf{a})$, $\hat{V}(\mathbf{b})$ are unitary operators, and are therefore bounded and everywhere defined, so it is not necessary to attend to their domains of definition. Although the Weyl commutation relations are not equivalent to the Heisenberg CCRs, the latter are essentially the infinitesimal form of the former.[5]

The relations may be restated in a more convenient form as follows. Define new *Weyl operators* by

$$\hat{W}(\mathbf{a},\mathbf{b}) \equiv \exp(i(\mathbf{a}.\mathbf{b})/2)\,\hat{U}(\mathbf{a})\hat{V}(\mathbf{b}) \sim \exp i(\mathbf{a}.\hat{\mathbf{x}} + \mathbf{b}.\hat{\mathbf{p}}) \qquad (D.26)$$

which therefore obey the multiplication rule

$$\hat{W}(\mathbf{a},\mathbf{b})\hat{W}(\mathbf{c},\mathbf{d}) = \hat{W}(\mathbf{a}+\mathbf{c}, \mathbf{b}+\mathbf{d})\exp(-i(\mathbf{a}.\mathbf{d} - \mathbf{b}.\mathbf{c})/2) \qquad (D.27)$$

Equation D.27 is equivalent to the Weyl relations D.25. A pair of vectors $\mathbf{a}, \mathbf{b}$ defines a point in a $6n$-dimensional vector space: the quantity $(\mathbf{a}.\mathbf{d} - \mathbf{b}.\mathbf{c})$ is a symplectic form on this space, making it into a symplectic vector space.[6]

The problem now becomes that of characterizing the representations of D.27 subject to the further requirement

$$\hat{W}^\dagger(\mathbf{a},\mathbf{b}) = \hat{W}(-\mathbf{a},-\mathbf{b}) \qquad (D.28)$$

The Stone–von Neumann theorem contains the solution to this problem, as follows:

Let (V, ω) be a finite-dimensional symplectic vector space. Let $(\mathsf{H}, W(\mathbf{a},\mathbf{b}))$ and $(\mathsf{H}', W'(\mathbf{a},\mathbf{b}))$ be strongly continuous, irreducible, unitary representations of the Weyl relations over that vector space. Then $(\mathsf{H}, W(\mathbf{a},\mathbf{b}))$ and $(\mathsf{H}', W'(\mathbf{a},\mathbf{b}))$ are unitarily equivalent. Here $W(\mathbf{a},\mathbf{b})$ and $W'(\mathbf{a},\mathbf{b})$ are the images of maps, into unitary operators on the Hilbert spaces H, H' respectively, of Weyl operators that obey the Weyl relations D.27 and D.28 over V.[7]

The Stone–von Neumann theorem shows that all of the representations of the states and observables of a system of non-relativistic, spinless particles free to move in Euclidean space are formally equivalent, in the sense that any pair of

---

[5] If the representation of D.25 is in terms of unitary groups of operators strongly continuous in $\mathbf{a}, \mathbf{b}$, then (by Stone's theorem) the generators of the groups are self-adjoint operators, and on a common dense domain these operators satisfy D.8.

[6] A symplectic form on a vector space V is a bilinear form $\omega(u,v)$ that is skew-symmetric: $\omega(u,v) = -\omega(v,u)$ for all $u, v \in V$, and non-degenerate: if $\omega(u,v) = 0$ for all $v \in V$ then $u = 0$.

[7] A representation of the Weyl relations is *irreducible* if and only if the only subspaces of the Hilbert space H left invariant by the operators $\hat{U}(\mathbf{a})$, $\hat{V}(\mathbf{b})$ are H and the null subspace $\{0\}$. It is *strongly continuous* if and only if it is continuous in the operator norm $\|\hat{W}\| = \sup_{v \in \mathsf{H}} \frac{\|\hat{W}v\|}{\|v\|}$, where $\|\hat{W}v\|^2 = (\hat{W}v, \hat{W}v)$, $\|v\|^2 = (v,v)$.

"well-behaved" (i.e. unitary, irreducible, strongly continuous) representations are related by a unitary mapping. This formal equivalence makes the task of interpreting the quantum mechanics of particles easier than that of interpreting a quantum field theory, to which the Stone–von Neumann theorem does not apply.

APPENDIX E
Algebraic quantum field theory

This appendix shows how algebraic quantum field theory provides a clear mathematical framework within which it is possible to raise and answer questions about the relations among various representations of the states and observables of a quantum field theory. It motivates and explains the idea of an abstract Weyl algebra of field observables and points out the interpretative significance of the fact that the Stone–von Neumann theorem does not extend to its representations. It says what is meant by a Fock representation, and explains how this is related to the occupation number representation of a system of quantum particles. Much of it relies on the paper by Ruetsche (2002) and the appendix to that by Earman and Fraser (2006).

The Heisenberg relations

$$[\hat{x}_j, \hat{p}_k] = i\hbar\delta_{jk}\hat{I} \qquad (E.1)$$
$$[\hat{x}_j, \hat{x}_k] = [\hat{p}_j, \hat{p}_k] = 0$$

generalize formally to equal-time commutation relations (ETCRs) for field systems such as the following for operators corresponding to a real classical scalar field $\varphi(\mathbf{x}, t)$:

$$[\hat{\varphi}(\mathbf{x}, t), \hat{\pi}(\mathbf{x}', t)] = i\hbar\delta^3(\mathbf{x} - \mathbf{x}')\hat{I} \qquad (E.2)$$
$$[\hat{\varphi}(\mathbf{x}, t), \hat{\varphi}(\mathbf{x}', t)] = [\hat{\pi}(\mathbf{x}, t), \hat{\pi}(\mathbf{x}', t)] = 0$$

as well as anticommutation relations for field operators acting on states of fermionic systems such as electrons and quarks. But the presence of the delta function $\delta^3(\mathbf{x} - \mathbf{x}')$ means that these field commutators are not really well defined. To arrive at a well-defined algebraic generalization of the Heisenberg relations it is necessary to introduce "smeared" field operators—field operators parametrized by a family of "test" functions peaked around points like $(\mathbf{x}, t)$ that fall off sufficiently fast away from there (perhaps restricted even to functions of compact support). This gives rise to a basic algebra of operators of the form $\hat{\varphi}(f(\mathbf{x}, t))$, $\hat{\pi}(f(\mathbf{x}, t))$ for a real scalar field, with analogous generalizations for fields of other kinds (complex, vector, etc.). As appendix D explained, it is

266 APPENDIX E ALGEBRAIC QUANTUM FIELD THEORY

also necessary to replace the Heisenberg form of the canonical commutation relations by a Weyl form in which all operators are bounded and can therefore be defined on all vectors in a Hilbert space on which they act. Just as the pair of vectors $(\mathbf{a}, \mathbf{b})$ defining the Weyl operator $\hat{W}(\mathbf{a}, \mathbf{b})$ for a particle system serves to pick out a point in the finite-dimensional phase space of that particle system, so also a pair of test functions (g, f) picks out a point in the infinite-dimensional phase space of a field system. Particle Weyl operators $\hat{W}(\mathbf{a}, \mathbf{b})$ therefore generalize to *field Weyl operators* $\hat{W}(g, f)$.

Now on the classical phase space for a field theory like that of the Klein–Gordon field there is a so-called symplectic form $\sigma(f, g)$ that generalizes the form $(\mathbf{a}.\mathbf{d} - \mathbf{b}.\mathbf{c})$ on the phase space of a classical particle system. The multiplication rule 7.8 accordingly generalizes to

$$\hat{W}(g_1, f_1)\hat{W}(g_2, f_2) = \hat{W}(g_1 + g_2, f_1 + f_2) \exp\left(-i\sigma(f, g)/2\right) \quad \text{(E.3)}$$

which specifies a so-called abstract Weyl algebra for the Klein–Gordon field and provides the required rigorous form of the ETCR's E.2.[1] The explicit expression for the symplectic form in this case is given by the following integral over a space-like "equal-time" hyperplane Σ

$$\sigma(f, g) = \int_\Sigma d^3x (g_1 f_2 - g_2 f_1) \quad \text{(E.4)}$$

We now face the problem of characterizing the representations of the Weyl algebra specified by equation E.3. This is the analogous problem for a quantum field theory to that considered in appendix D for a quantum particle theory. The problem is now set in the context of an algebraic approach to quantum field theory, so before we continue it is appropriate to reflect on just what that amounts to.

In the algebraic approach to quantum field theory, observables are represented by an abstract algebra $\mathcal{A}$ of operators, and states are represented by linear functionals s on this algebra. So if $\hat{A}_1, \hat{A}_2$ are elements of $\mathcal{A}$, then

$$s(a\hat{A}_1 + b\hat{A}_2) = as(\hat{A}_1) + bs(\hat{A}_2) \quad \text{(E.5)}$$

Such a state is intended to yield the expectation value for a measurement of an arbitrary observable in $\mathcal{A}$. $\mathcal{A}$ itself is taken to be a C^* algebra: a complete, normed

---

[1] The *Weyl algebra* itself is constituted by a set of abstract operators $\{\hat{A}\}$ generated from the $\hat{W}(g, f)$ satisfying E.3 as well as $\hat{W}^*(g, f) = \hat{W}(-g, -f)$. It is closed under complex linear combinations. The * operation satisfies $(c\hat{A})^* = \bar{c}\hat{A}^*$, where $\bar{c}$ is the complex conjugate of c. The algebra possesses a unique norm $||\hat{A}||$ satisfying $||\hat{A}^*\hat{A}|| = ||\hat{A}||^2$. The Weyl algebra is also closed under this norm, making it a C^* algebra.

vector space over the complex numbers whose elements may be multiplied in such a way that $\forall \hat{A}_1, \hat{A}_2 \in \mathcal{A}$, $\|\hat{A}_1\hat{A}_2\| \leq \|\hat{A}_1\| \|\hat{A}_2\|$, with an *involution* operation $*$ satisfying conditions modeled on those of the Hilbert space adjoint operation, plus $\forall \hat{A} \in \mathcal{A}$, $\|\hat{A}^*\hat{A}\| = \|\hat{A}\|^2$.[2] Abstract states s on $\mathcal{A}$ are linear functionals satisfying

$$s(\hat{A}^*\hat{A}) \geq 0 \tag{E.6}$$
$$s(\hat{I}) = 1 \tag{E.7}$$

The bounded operators of a Hilbert space $\mathcal{B}(\mathsf{H})$ constitute one concrete realization of a C^* algebra. In the context of the algebraic approach to quantum field theory, we seek a representation in some Hilbert space of an abstract C^* algebra of smeared field operators with states on them. Every representation of the Weyl algebra specified by the Weyl relations E.2 will give rise to such a representation, since the Weyl algebra constitutes a C^* algebra. A *representation* of an abstract C^* algebra $\mathcal{A}$ on a Hilbert space H is a $*$-homomorphism $\pi : \mathcal{A} \to \mathcal{B}(\mathsf{H})$ of that algebra into the algebra of bounded linear operators on H, i.e. a structure-preserving map of elements of $\mathcal{A}$ onto a C^* algebra constituted by elements of that algebra which satisfies the condition

$$\left(\varphi, \pi(\hat{A})\psi\right) = \left(\pi(\hat{A}^*)\varphi, \psi\right) \text{ for all } \varphi, \psi \in \mathsf{H} \tag{E.8}$$

Such a representation is *faithful* if and only if $\pi(\hat{A}) = 0 \to \hat{A} = 0$, and *irreducible* if and only if the only subspaces of the Hilbert space H left invariant by the operators $\{\pi(\hat{A}) : \hat{A} \in \mathcal{A}\}$ are H and the null subspace $\{0\}$. Every representation of a Weyl C^* algebra is faithful. Two representations π, π' of an abstract C^* algebra $\mathcal{A}$ are *unitarily equivalent* if and only if there is a unitary map $U : \mathcal{B}(\mathsf{H}_\pi) \to \mathcal{B}(\mathsf{H}_{\pi'})$ such that $\pi'(\hat{A}) = U\pi(\hat{A})U^{-1}$ for all $\hat{A} \in \mathcal{A}$.

The Stone–von Neumann theorem does not generalize to representations of field Weyl algebras like those specified by E.3. While such an algebra does possess Hilbert space representations, these are not all unitarily equivalent to one another. Indeed, there is a continuous infinity of inequivalent representations of equation E.3's algebra.

One important kind of representation is called a *Fock representation*. This is related to the occupation number representation for the quantum harmonic oscillator considered in appendix D. To get the idea of a Fock representation, recall the discussion of the real Klein–Gordon field in chapter 5, section 5.1.

---

[2] Specifically, we have the following conditions:

$$\forall \hat{A}_1, \hat{A}_2 \in \mathcal{A} : (\hat{A}_1 + \hat{A}_2)^* = \hat{A}_1^* + \hat{A}_2^*, \; (\hat{A}_1\hat{A}_2)^* = \hat{A}_2^*\hat{A}_1^*$$
$$\forall \lambda \in \mathbb{C}, \forall \hat{A} \in \mathcal{A} : (\lambda\hat{A})^* = \bar{\lambda}\hat{A}^*, \; (\hat{A}^*)^* = \hat{A}.$$

The general solution to the classical Klein–Gordon equation (5.1)

$$\partial_\mu \partial^\mu \phi + m^2 \phi = 0 \tag{E.9}$$

may be expressed as

$$\phi(\mathbf{x}, t) = \int \{a(\mathbf{k}) e^{i(\mathbf{k} \cdot \mathbf{x} - \omega_\mathbf{k} t)} + a^*(\mathbf{k}) e^{-i(\mathbf{k} \cdot \mathbf{x} - \omega_\mathbf{k} t)}\} d^3 x \tag{E.10}$$

where $\omega_\mathbf{k}^2 = \mathbf{k}^2 + m^2$ corresponds to the relativistic energy–momentum relation $E^2 = \mathbf{p}^2 c^2 + m^2 c^4$ with $E = \hbar \omega_\mathbf{k}$, $p = \hbar k$ and here and in the rest of this appendix we have chosen units so that $c = \hbar = 1$. The canonical conjugate field $\pi(x^\mu)$ is defined by

$$\pi(x^\mu) = \frac{\partial \mathcal{L}}{\partial \dot{\varphi}(x^\mu)} = \dot{\varphi}(x^\mu) \tag{E.11}$$

where $\mathcal{L}$ is the Klein–Gordon Lagrangian density $\mathcal{L} = \frac{1}{2}[(\partial_\mu \varphi)(\partial^\mu \varphi) - m^2 \varphi^2]$. On quantization, ϕ, π become operators $\hat{\phi}, \hat{\pi}$, and the solution to the quantized Klein–Gordon equation is

$$\hat{\phi}(\mathbf{x}, t) = \int \{\hat{a}(\mathbf{k}) e^{i(\mathbf{k} \cdot \mathbf{x} - \omega_\mathbf{k} t)} + \hat{a}^\dagger(\mathbf{k}) e^{-i(\mathbf{k} \cdot \mathbf{x} - \omega_\mathbf{k} t)}\} d^3 x \tag{E.12}$$

where the commutation relations for the operators $\hat{a}(\mathbf{k})$ and its adjoint $\hat{a}^\dagger(\mathbf{k})$ that follow from this and equations E.2 are

$$[\hat{a}(\mathbf{k}), \hat{a}^\dagger(\mathbf{k}')] = \delta^3(\mathbf{k} - \mathbf{k}') \hat{I} \tag{E.13}$$

$$[\hat{a}(\mathbf{k}), \hat{a}(\mathbf{k}')] = [\hat{a}^\dagger(\mathbf{k}), \hat{a}^\dagger(\mathbf{k}')] = 0 \tag{E.14}$$

If we define a so-called number operator $\hat{N}(\mathbf{k}) \equiv \hat{a}^\dagger(\mathbf{k}) \hat{a}(\mathbf{k})$, then these give

$$[\hat{a}(\mathbf{k}), \hat{N}(\mathbf{k}')] = \delta^3(\mathbf{k} - \mathbf{k}') \hat{a}(\mathbf{k}) \tag{E.15}$$

$$[\hat{a}^\dagger(\mathbf{k}), \hat{N}(\mathbf{k}')] = -\delta^3(\mathbf{k} - \mathbf{k}') \hat{a}^\dagger(\mathbf{k}) \tag{E.16}$$

It follows that $\hat{a}(\mathbf{k}), \hat{a}^\dagger(\mathbf{k})$ act respectively as raising and lowering operators on eigenstates $|n_\mathbf{k}\rangle$ of the number operator with $\hat{N}(\mathbf{k})|n_\mathbf{k}\rangle = \delta^3(0) n_\mathbf{k} |n_\mathbf{k}\rangle$:

$$\hat{N}(\mathbf{k}) \hat{a}^\dagger(\mathbf{k}) |n_\mathbf{k}\rangle = \hat{a}^\dagger(\mathbf{k})[\hat{N}(\mathbf{k}) + \delta^3(0)]|n_\mathbf{k}\rangle \tag{E.17}$$

$$= \delta^3(0)(n_\mathbf{k} + 1) \hat{a}^\dagger(\mathbf{k}) |n_\mathbf{k}\rangle \tag{E.18}$$

Hence $\hat{a}^\dagger(\mathbf{k})|n_\mathbf{k}\rangle$ is an eigenstate of $\hat{N}(\mathbf{k})$ corresponding to eigenvalue $n_\mathbf{k}+1$. Similarly, $\hat{a}(\mathbf{k})|n_\mathbf{k}\rangle$ is an eigenstate of $\hat{N}(\mathbf{k})$ corresponding to eigenvalue $n_\mathbf{k}-1$. Provided the system has a unique state of lowest energy, by repeatedly applying the lowering operators one arrives at that ground state—the so-called *vacuum state* $|0\rangle$—a simultaneous eigenstate of every number operator $\hat{N}(\mathbf{k})$ with eigenvalue $n_\mathbf{k}=0$. The Hamiltonian operator $\hat{H}$ for the Klein–Gordon field has the form

$$\hat{H} = \int \frac{1}{2}\left[\hat{\pi}^2 + \left(\boldsymbol{\nabla}\hat{\phi}\right)^2 + m^2\hat{\phi}^2\right] d^3x \qquad (E.19)$$

which becomes

$$\hat{H} = \int \frac{1}{2}\omega_\mathbf{k}\left[\hat{a}(\mathbf{k})\hat{a}^\dagger(\mathbf{k}) + \hat{a}^\dagger(\mathbf{k})\hat{a}(\mathbf{k})\right] d^3k \qquad (E.20)$$

The commutation relations for the raising and lowering operators then give

$$\hat{H} = \int \omega_\mathbf{k}\left[\hat{a}^\dagger(\mathbf{k})\hat{a}(\mathbf{k}) + \frac{1}{2}\delta^3(0)\right] d^3k = \int \omega_\mathbf{k}\left[\hat{N}(\mathbf{k}) + \frac{1}{2}\delta^3(0)\right] d^3k \quad (E.21)$$

If one follows custom in ignoring as unmeasurable the infinite zero-point energy associated with the delta function, one can therefore try to interpret the total energy of a Klein–Gordon field as consisting of the sum of the energies $\omega_\mathbf{k}$ of all its constituent quanta of momentum $\mathbf{k}$. Similarly, the total momentum represented by the operator

$$\hat{\mathbf{p}} = \int \frac{1}{2}\mathbf{k}\left[\hat{a}(\mathbf{k})\hat{a}^\dagger(\mathbf{k}) + \hat{a}^\dagger(\mathbf{k})\hat{a}(\mathbf{k})\right] d^3k \qquad (E.22)$$

might be interpreted as consisting of the sum of the momenta of all its constituent quanta. A total number operator $\hat{N}$ may also be defined as

$$\hat{N} \equiv \int \hat{N}(\mathbf{k}) d^3k \qquad (E.23)$$

whose eigenvalues might indicate the total number of quanta present in the field. The vacuum state satisfies $\hat{N}|0\rangle = 0|0\rangle$, in accordance with its interpretation as a state in which no quanta are present. Other states of the quantized Klein–Gordon field may then be built up from the vacuum state by successive applications of linear combinations of raising and lowering operators; indeed every state in the representation may be approximated to arbitrary precision in this way. A state $|n_\mathbf{k}\rangle$ that can be formally "created" from the vacuum state $|0\rangle$ by application of the raising (or "creation") operator $\hat{a}^\dagger(\mathbf{k})$

$$|n_\mathbf{k}\rangle = \hat{a}^\dagger(\mathbf{k})|0\rangle \qquad (E.24)$$

is a simultaneous eigenstate of $\hat{\mathbf{p}}$ and $\hat{H}$ with eigenvalues $\mathbf{k}$, $\omega_\mathbf{k}$ respectively. It is naturally thought to contain one quantum whose energy and momentum values obey the usual relativistic relation. Repeated action with this and other "creation" operators is naturally thought to result in a state containing multiple quanta of various energies and momenta, always obeying this relation. But a typical state will be a superposition of such states, with no determinate number of quanta, and no determinate energy or momentum.

The algebraic approach makes it possible to place this heuristic treatment of a Fock representation on a sounder mathematical footing, and to state precisely what counts as a Fock representation of an abstract Weyl algebra. Instead of focusing on field operators defined at each space-time point, one considers a corresponding abstract algebra of operators which have Hilbert space representations as "smeared" fields. In a Fock representation of a Weyl algebra, a creation or annihilation operator is parametrized not by momentum, but by an element of a complex Hilbert space H_1 (called, suggestively, the *one-particle Hilbert space*). For all $f, g \in H_1$, their commutation relations are

$$[a(f), a(g)] = \left[a^\dagger(f), a^\dagger(g)\right] = 0 \quad \left[a(f), a^\dagger(g)\right] = (f,g)\hat{I} \qquad (E.25)$$

permitting the definition of a number operator $\hat{N}(f) = a^\dagger(f)a(f)$ with

$$[\hat{a}(f), \hat{N}(g)] = (f,g)\hat{a}(f) \qquad (E.26)$$
$$[\hat{a}^\dagger(f), \hat{N}(g))] = -(f,g)\hat{a}^\dagger(f) \qquad (E.27)$$

and a total number operator $\hat{N} = \sum_i a^\dagger(f_i)a(f_i)$ over an orthonormal basis $\{f_i\}$ for H_1. A symmetric Fock space $F(H_1)$ is built up from H_1 as the infinite direct sum of symmetrized tensor products of H_1 with itself: $F(H_1) = \mathbb{C} \oplus s(H_1) \oplus s(H_1 \otimes H_1) \oplus \ldots$. The creation and annihilation operators are defined over a common dense domain D of $F(H_1)$.[3] A representation of the Weyl algebra specified by E.3 is a *Fock representation* in $F(H_1)$ if and only if there is a unique vacuum state $|0\rangle$ in D with $a(f)|0\rangle = 0$ for all $f \in H_1$, and D is the span of $\{a^\dagger(f_1)a^\dagger(f_2) \ldots a^\dagger(f_n)|0\rangle\}$. In a Fock representation, the total number operator $\hat{N}$ is a densely defined self-adjoint operator independent of the basis used to define it with spectrum $\{0, 1, 2, \ldots\}$. Any representation of

---

[3] A set of vectors in a Hilbert space is *dense* just in case every vector in the space is arbitrarily close in the Hilbert space norm to a member of that set.

APPENDIX E ALGEBRAIC QUANTUM FIELD THEORY

the Weyl algebra defined by E.3 with such a number operator is either a Fock representation or a direct sum of Fock representations.

But the Fock representation of a free quantum field like the Klein–Gordon field is only one among an infinite number of unitarily inequivalent representations of the Weyl form of the basic ECTRs. One way to get a handle on this multiplicity is to associate representations of a Weyl algebra with states defined on that algebra. An abstract state s on an abstract Weyl algebra $\mathcal{W}$ (with identity $\hat{I}$) is a map from $\mathcal{W}$ into real numbers satisfying

$$s(\hat{A}^*\hat{A}) \geq 0 \tag{E.28}$$
$$s(\hat{I}) = 1 \tag{E.29}$$
$$s(a\hat{A} + b\hat{B}) = as(\hat{A}) + bs(\hat{B}) \tag{E.30}$$

A state s is *pure* just in case it cannot be expressed as a linear sum of other states. A representation of $\mathcal{W}$ in a Hilbert space H is a map $\pi : \mathcal{W} \to \mathcal{B}(\mathsf{H})$ from $\mathcal{W}$ into the set $\mathcal{B}(\mathsf{H})$ of bounded self-adjoint operators on H such that the images of elements of $\mathcal{W}$ themselves constitute a concrete Weyl algebra under the corresponding algebraic operations on $\mathcal{B}(\mathsf{H})$. Since $\mathcal{W}$ is a C^* algebra, each state s on $\mathcal{W}$ defines a representation π_s of the operators in $\mathcal{W}$ by self-adjoint operators on a Hilbert space H_s, in accordance with the Gelfand–Naimark–Segal theorem:

Any abstract state s on a C^* algebra $\mathcal{A}$ gives rise to a unique (up to unitary equivalence) faithful representation (π_s, H_s) of $\mathcal{A}$ and vector $\Omega_s \in \mathsf{H}_s$ such that

$$s(\hat{A}) = <\Omega_s|\pi_s(\hat{A})|\Omega_s> \text{ for } \hat{A} \in \mathcal{A}$$

and such that the set $\{\pi_s(\hat{A})\Omega_s : \hat{A} \in \mathcal{A}\}$ is dense in H_s. This representation is irreducible if s is pure.[4]

Each vector $|\psi\rangle$ in the space H of a representation of $\mathcal{W}$ defines an abstract state s by $s(\hat{A}) = <\psi|\pi(\hat{A})|\psi>$, and so to any vector that represents a state in a representation of $\mathcal{W}$ there corresponds a unique abstract state on $\mathcal{W}$. But if the GNS representations of abstract states s, s' are not unitarily equivalent, then s cannot be represented as a vector or density operator on $\mathsf{H}_{s'}$. Since a representation π will map the elements of $\mathcal{W}$ into a proper subset of the set of bounded self-adjoint operators on H, a concrete Hilbert space representation of $\mathcal{W}$ will contain additional candidates for physical magnitudes represented by operators in $\mathcal{B}(\mathsf{H})$, over and above those represented by elements of $\mathcal{W}$.

---

[4] Recall that an irreducible representation is one in which the only subspaces of H_s that are invariant under the operators $\pi_s(\hat{A})$ are H_s and the null subspace.

APPENDIX F

Interpretations of quantum mechanics

An interpretation of quantum mechanics is an account of what the world is like if that theory is true. To be convincing, the interpretation must explain how the observations we take to support quantum mechanics in fact do so, given that the world is the way that interpretation says it is. There is general agreement on the formal framework of the non-relativistic quantum mechanics of particles, but disagreement on its interpretation has persisted since that formal framework was first established in the 1920s and 1930s. Appendix D sketched part of the formal framework of quantum mechanics. After heated initial debates among the founders, something resembling a consensus emerged on its interpretation which persisted until the 1950s and 1960s, since when it has faced persistent and varied challenges from both physicists and philosophers. What passed as an initial consensus has come to be known as the Copenhagen interpretation.

F.1 The Copenhagen interpretation

A central tenet of this interpretation is that the quantum mechanical description provided by the state vector is both predictively and descriptively complete. The most complete description of a system at a given time permits only probabilistic predictions of its future behavior, and so the indeterminism of the theory reflects the underlying indeterminism of the processes to which it applies. Moreover, this description, though complete, fails to assign precise values to all dynamical variables.

There are many versions of "the" Copenhagen interpretation. One version, going back to Dirac and von Neumann, understands the completeness claim as follows. The quantum state vector yields a complete description of the real properties of an individual system. This view endorses the following interpretative principle connecting a system's quantum state to its dynamical properties, which I shall call the *state–property link*:

A system has a dynamical property locating the value of dynamical variable A in a (measurable) set Δ in state ψ just in case a measurement of A would certainly reveal that property:[1]

$$prob_\psi(A \in \Delta) = 1 \qquad (F.1)$$

The state–property link severely restricts the limits of precision within which the values of a system's dynamical variables are well defined. It implies that a pair of dynamical variables represented by non-commuting operators do not both have precise values in a typical state, and that there are pairs of dynamical variables that *never* both have precise values in *any* state.[2] But the interpretative principle does not favor any particular representation of the fundamental commutation relations of the theory. Since it is based on the probabilistic predictions of the theory, which are invariant under a unitary transformation corresponding to a change of representation, it will deliver the same results when applied in any of the unitarily equivalent interpretations of the theory. On this version of the Copenhagen interpretation, it is the system's instantaneous quantum state alone that determines what dynamical properties it then has.

Bohr and other proponents of the Copenhagen interpretation often speak of the complementarity of different descriptions of quantum phenomena, where each of two complementary descriptions is required to completely characterize a phenomenon, but the two descriptions cannot both be given together by combining them into a single overerarching description. One way to try to make sense of such talk in the Copenhagen interpretation is to associate complementary descriptions with certain pairs of unitarily equivalent representations of the basic Heisenberg commutation relations of the theory (for example, the Schrödinger or position representation, and the momentum representation). As appendix D shows, the position operator $\hat{\mathbf{x}}$ acts by multiplication in the Schrödinger representation, while the momentum operator $\hat{\mathbf{p}}$ acts by multiplication in the momentum representation. These two representations may be considered to offer complementary descriptions if one assumes that the Schrödinger representation is suited to describe phenomena taken to involve particles with precise (though possibly unknown) positions, while the momentum representation is suited to describe phenomena taken to involve particles with precise (though possibly unknown) momenta. These

---

[1] If A is represented by an operator $\hat{A}$ whose eigenvectors $\{\psi_i\}$ span the Hilbert space H, then this is equivalent to the *eigenvalue–eigenstate link*: A has value a in state ψ if and only if ψ is an eigenstate of $\hat{A}$ with eigenvalue a: $\hat{A}\psi = a\psi$. If the state of the system is represented not by a vector but by a density operator W, F.1 generalizes to $A \in \Delta$ if and only if $\mathrm{Tr}WP^{\hat{A}}(\Delta) = 1$, where $P^{\hat{A}}(\Delta)$ is the element of the spectral measure of $\hat{A}$ corresponding to the property $A \in \Delta$.

[2] The classic example involves a component x of a particle's position and the corresponding component p_x of its momentum. But a rigorous application of the interpretative principle implies that there is no state in which *either* of these dynamical variables has a precise value. A more illuminating example is provided by any pair of distinct Cartesian components of the angular momentum of a spin-1/2 particle. Each component will have a precise value in some state, but there is no state in which both have precise values.

descriptions could not be taken at face value while accepting the state–property link, but one can read Bohr himself as admitting, or even insisting, that no single complementary description can be precisely correct.

In understanding complementarity, it is important to ask what makes one rather than the other of a pair of complementary descriptions appropriate. A strict application of the state–property link suggests that a description is appropriate in a quantum state if and only if it is in terms of a dynamical variable to which that principle assigns a precise value in that state. But that would imply that neither a description in terms of position nor a description in terms of momentum is ever appropriate! Bohr's stress on the importance of the entire experimental arrangement involving a quantum phenomenon suggests a different answer: that it is the experimental arrangement rather than the quantum state that determines the appropriateness of describing a system by one rather than another of a pair of complementary descriptions. But that raises the problem of saying just what constitutes an experimental arrangement, and what it is about one such arrangement that makes a particular complementary description appropriate.

A general problem for all versions of the Copenhagen interpretation is to say exactly what constitutes a measurement, and how such processes are compatible with the dynamics of the theory. This is especially problematic for those versions that (unlike Bohr's) take measurement to induce "collapse"—a discontinuous, stochastic physical transition of a system's state vector to a vector in which the state–property link implies that the measured observable definitely has the measured value. For such collapse is incompatible with the continuous, unitary evolution of the state vector prescribed by the Schrödinger equation.

F.2 Bohmian mechanics

Bohmian interpretations privilege the Schrödinger, or position, representation. They maintain that this is the only representation that is a reliable guide to the underlying ontology of the theory, namely an ontology of precise particle positions, continuously evolving under the influence of the many-particle wave-function, or at least the velocity field derived from the gradient of its phase. Indeed Goldstein (1996) complains about the so-called naive realism about operators that leads people to believe that not only all operators, but also all unitarily equivalent representations, are on a par when it comes to understanding quantum ontology.

On a Bohmian interpretation, a quantum particle theory describes the motion of a system of n point particles. This motion is most conveniently represented by the motion of a point in a configuration space of dimension $3n$ whose coordinates consist of the (always precisely defined) $3n$ position coordinates of the n particles. The point's motion is continuous and smooth (differentiable), as are the motions of the particles themselves in physical space. For a non-relativistic system of spinless particles, the motion of the kth particle of mass m_k is governed by a guidance equation involving the position-representation wave-function $\Psi(\mathbf{x}_1, \mathbf{x}_2, \ldots, \mathbf{x}_n, t)$ of the system:

$$\frac{d\mathbf{x}_k}{dt} = \frac{1}{m_k} \nabla_k S \tag{F.2}$$

where

$$\Psi = |\Psi| \exp(iS/\hbar) \tag{F.3}$$

It follows that each particle always has a precisely defined velocity and momentum, as well as a precisely defined position. This does not conflict with the indeterminacy relations between position and momentum derived from the basic Heisenberg commutation relations D.8. The indeterminacy relations put constraints on the relation between the probability distribution for position measurements and that for momentum measurements: they say nothing about simultaneous joint measurements of position and momentum on a single system. Moreover, self-adjoint operators do not represent dynamical variables of a particle other than (functions of) position,[3] so what is called a momentum measurement does not in fact reveal the momentum of a particle, but rather a relation between the particle and a specific experimental arrangement (typically falsely!) believed to reveal its momentum. Like all measurements, its result is recorded in particle positions; and while these may depend on how the particle is moving, they do not record its actual momentum.

The guidance equation specifies how the motions of the particles depend on $\Psi(\mathbf{x}_1, \mathbf{x}_2, \ldots, \mathbf{x}_n, t)$: this wave-function obeys the (deterministic) Schrödinger equation, from which it follows that the particles' motion is also deterministic. Probability enters epistemically. If this approach is to constitute an interpretation of quantum mechanics rather than a rival theory, the particle positions must be assumed initially to be distributed with probabilities (strictly, probability densities) given by the squared modulus of the wave-function: it then follows that they will continue so to be distributed as the system evolves.

A measurement is a process described by the Schrödinger equation that correlates the particles' wave-function with that of an apparatus. Its result is recorded in the positions of the particles making up the apparatus. Immediately after an ideal measurement, the different components of the system's wave-function no longer overlap, so that the system's effective wave-function (the only part that influences their subsequent motion through the guidance

---

[3] This is true in the "pure" Bohmian interpretation: there is also a variant that treats spin as an independently possessed dynamical variable.

equation) is just the component that is an eigenvector of the measured observable. So the results of repeated ideal measurements are just what they would have been if measurement had caused the system's wave-function to "collapse" onto that eigenvector. The system's effective wave-function "collapses," even though its total wave-function simply evolves according to the Schrödinger equation.

While the point in configuration space representing all the particle positions moves only under the influence of magnitudes (including the wave-function) defined at that point, the way the wave-function affects each particle through the guidance equation ensures that a particle's motion at each instant depends on the interactions experienced by all the other particles at that instant, no matter how far away they are. So the motions of particles in a multi-particle system are influenced non-locally.[4]

F.3 Everettian interpretations

The dissertation by Everett (1957) founded an alternative interpretative tradition of interpretations of quantum mechanics often associated with the idea of many worlds.[5] Everett's relative-state formulation of quantum mechanics is an attempt to solve the measurement problem faced by a version of the Copenhagen interpretation that incorporates the state–property link, by dropping the collapse dynamics from the standard von Neumann/Dirac theory of quantum mechanics.

According to contemporary developments of Everett's view, the state vector of the universe evolves linearly and deterministically; but, as a result of quantum interactions between a subsystem and its environment, the state of that subsystem is almost always representable as a mixture of pure states that is approximately diagonal in some fixed basis called a *decoherence basis*

$$\hat{W}_S = \sum_i w_i \hat{P}_i \qquad (F.4)$$

where $\hat{W}_S = \text{Tr}_{E_S} \hat{P}_\Psi$ is the mixed state that results when the universal state vector Ψ is traced over S's environment E_S, and each $\hat{P}_i$ projects onto a vector ψ_i in the decoherence basis.[6] It is as if a system's environment is constantly "measuring" observables that are diagonal in the decoherence basis, keeping S

---

[4] For further details of Bohmian mechanics, see the entry in the *Stanford Electronic Encyclopedia of Philosophy* at http://plato.stanford.edu/entries/qm-bohm/.

[5] See, for example, the entries in the *Stanford Electronic Encyclopedia of Philosophy* at plato.stanford.edu/entries/qm-everett/ and plato.stanford.edu/entries/qm-manyworlds/.

[6] Even when the state of a compound quantum system is represented by a vector Ψ, the state of a component S cannot generally be so represented. The compound state Ψ suffices to predict probabilities

in not just one state ψ_i but *every* state ψ_i in the decoherence basis. Linearity then ensures that each *relative state* ψ_i evolves essentially independently, unless and until not only the states ψ_i of S, but also the state $\psi_i^{E_S}$ of the environment E_S correlated to each ψ_i, come to overlap and interfere. A relative state ψ_i together with its correlated environment state $\psi_i^{E_S}$ define an Everett branch. Applying the state–property link (or perhaps something weaker) to a branch determines what properties a system and its environment have on that branch, and so what properties each has relative to the properties of the other. The goal of the interpretation is to show that these properties suffice to recover the approximately classical world that we experience as a *quasi-classical domain* on each of a class of branches. The changing quantum properties of sufficiently complex quantum systems on such a branch would account for the familiar, mostly classical, behavior of the large-scale objects these systems constitute.

An Everettian interpretation models measurement as a quantum process involving an interaction between two (or more) physical subsystems of the universal system—one of which is designated an *observer system*, while another is the system it measures. While there is no physical "collapse," it is a consequence of this model that an observer system on a branch following an ideal measurement should then predict the statistics for future measurement outcomes on the measured system by assigning it a "collapsed" state vector. A result of a measurement is recorded in the relative state of the observer system at the conclusion of the interaction: there will typically be many such states, one for each branch constituting a quasi-classical domain. It is supposedly because the records of many such measurements on "our" branch accord with the statistical predictions of quantum mechanics that we are justified in believing the theory. I say supposedly, because such an interpretation faces problems concerning the specification of the decoherence basis, the identity conditions for an observer, the significance of probability when every outcome of a measurement occurs on some branch, and the associated epistemology.[7]

The state–property link itself does not privilege any particular representation of the fundamental commutation relations of the theory. In so far as it relies on (at least a modified form of) the state–property link in assigning properties

---

of measurement outcomes on S. But, as an alternative, one can calculate these by assigning a state to S alone. The state of S is then represented by a self-adjoint operator $\hat{W}_S$ on S's Hilbert space H_S known as a *density* operator. The sum of the eigenvalues w_i of a density operator is 1, and so it may be written in diagonal form as $\hat{W}_S = \sum_i w_i \hat{P}_i$, where the $\{\hat{P}_i\}$ project onto orthogonal one-dimensional subspaces that span H_S. $\hat{W}_S^2 = \hat{W}_S$ if and only if some w_j is 1, and the rest are 0. In that case, the state of S is said to be *pure*, and may be represented alternatively by a vector ψ_j in the subspace projected onto by $\hat{P}_j$; otherwise, the state of S is said to be *mixed*. If the state (pure or mixed) of a compound system $S + S'$ is represented by the density operator $\hat{W}$, it follows that the state of S is represented by $\hat{W}_S = \text{Tr}_{S'} \hat{W}$, where $\text{Tr}_{S'} \equiv \sum_k (\psi_k, \hat{W}\psi_k)$, the $\{\psi_k\}$ are a basis of orthogonal, unit-normed vectors for $\mathsf{H}_{S'}$, and the action of $\hat{W}$ on $\mathsf{H}_{S'}$ is defined in a natural way.

[7] For interesting contemporary discussions of these problems, see Wallace (2005, 2006), Greaves (2004, 2007), Lewis (2007, 2007a), Price (2006), and references therein.

to quantum systems on a branch, an Everettian interpretation does not itself privilege any representation. But contemporary Everettian interpretations do appeal to the existence of a decoherence basis in their solution to the measurement problem and the wider problem of accounting for our experience of an approximately classical world, and this may be thought to privilege observables whose associated operators are diagonal in such a basis. Since decoherence is (almost) never more than approximate, there is no exact decoherence basis, but many equally good approximations to such a basis. So still no single representation is privileged within such an Everettian interpretation.

F.4 Modal interpretations

Like Bohmian mechanics, modal interpretations are no-collapse interpretations that reject the state–property link. But unlike Bohmian mechanics, a modal interpretation still allows a system's quantum state at a time some role in specifying what dynamical properties it then has. The term 'modal' is particularly appropriate as a name for an interpretation according to which a system's quantum state determines what dynamical properties it *must* have (using the state–property link as a sufficient condition for possessing a property), while prescribing a further range of *possible* dynamical properties, at least some of which it also has (thereby denying that the state–property link yields a necessary condition for possessing a property).[8]

A modal interpretation seeks to use these additional dynamical properties to solve the measurement problem, and to account for our experience of an approximately classical world. Some modal interpretations further hold out the promise of an account of correlations involved in violation of Bell-type inequalities that involves no violation of Local Action, and can be squared with fundamental Lorentz invariance.[9]

Unlike Bohmian mechanics, most modal interpretations do not take any representation of the fundamental commutation relations of the theory to be privileged in all situations. But a variety of modal interpretations do take a particular basis of states to be privileged when it comes to assigning the possible properties of a subsystem of a compound system. These are interpretations that appeal to the *biorthogonal decomposition theorem*:

Given a vector Ψ in a tensor-product Hilbert space, $H_1 \otimes H_2$, there exist bases $\{\varphi_i\}$ and $\{\psi_j\}$ for H_1 and H_2 respectively such that Ψ can be written as a linear combination

---

[8] For further details of modal interpretations, see the entry in the *Stanford Electronic Encyclopedia of Philosophy* at http://plato.stanford.edu/entries/qm-modal/.

[9] A principle of Local Action was stated in chapter 2. Recent work has shown how hard it would be for a modal interpretation to deliver on this promise, or to solve the measurement problem: see the previous footnote.

of terms of the form $\varphi_i \otimes \psi_i$. If the absolute value (modulus) of the coefficients in this linear combination are all unequal then the bases are unique.

The basic idea behind this kind of modal interpretation is to assign correlated properties to subsystems S_1, S_2 when $S_1 + S_2$ is in state Ψ by applying the state–property link to each as if the pair were in correlated states $\{\varphi_k, \psi_k\}$, for some k, where $\{\varphi_i\}$ and $\{\psi_j\}$ are bases for the Hilbert spaces $\mathsf{H}_1, \mathsf{H}_2$ of S_1, S_2 respectively. Such an interpretation thereby privileges these bases, but only relative to state Ψ: as the state of $S_1 + S_2$ evolves, the bases in its biorthogonal decomposition will evolve with it.

Bub (1997) proposed a modal interpretation that privileges a specific observable R by taking it always to have a precise value on a quantum system: one possibility is that R is a discretized position variable, in which case the interpretation has Bohmian mechanics as a kind of limit, but Bub (1997) left the identity of R open, while arguing that it could be settled by the physical process of decoherence. A system will have dynamical properties in this modal interpretation in addition to those determined by R's precise value: what these are is jointly specified by R and the system's quantum state.[10] Bub's modal interpretation does not appeal to the biorthogonal decomposition theorem. By privileging observable R, it privileges a representation in which $\hat{R}$ is diagonal.

Berkowitz and Hemmo (2006) propose a modal interpretation of quantum mechanics in which no observable and no basis is privileged: every observable on a subsystem S of a system $S + S'$ of particles has a precise value. For every orthonormal basis $\{\alpha_i\}$ for H_S consisting of eigenvectors of $\hat{A}$, observable A has exactly one value, a_j say, where $\hat{W}_S \hat{P}_j \neq 0$: here $\hat{W}_S$ represents the (mixed) state of S, obtained by tracing the quantum state of $S + S'$ over $\mathsf{H}_{S'}$, and $\hat{P}_k$ projects onto the ray spanned by an eigenvector α_k corresponding to the eigenvalue a_k. There are well-known no-go theorems by Bell, Gleason, Kochen and Specker, and others that may appear to rule out such an interpretation (see, for example, the discussion in Bub (1997)). The modal interpretation of Berkowitz and Hemmo (2006) evades their conclusion by allowing properties to be assigned *contextually*, so that S may have property P relative to one orthonormal basis $\{\alpha_i\}$ for H_S, while failing to have P relative to another orthonormal basis $\{\beta_j\}$, where $\hat{P}\alpha_k = \alpha_k$, $\hat{P}\beta_l = \beta_l$ for some $\alpha_k \in \{\alpha_i\}$, $\beta_l \in \{\beta_j\}$. Further, these properties are assigned to S only *relative to* S'. In a different partition of $S + S'$ into $T + T'$, where S is a subsystem of T, S may fail to have a property P relative to T' that it does have relative to S', or vice versa. Berkowitz and Hemmo (2006) argue that such relativized properties suffice to account for our experience as of a classical world, while the resulting modal interpretation holds out the prospect of a genuinely relativistic modal interpretation. Nothing in this modal interpretation privileges any particular representation.

---

[10] The dynamical properties of a system are represented by the non-zero projections of the quantum state onto the eigenspaces of $\hat{R}$.

Bibliography

Aharonov, Y. A. (1984) "Aharonov–Bohm Effect and Non Local Phenomena" in *Proceedings of the International Symposium on the Foundations of Quantum Mechanics in the Light of New Technology*. Singapore: World Scientific.

Aharonov, Y. and Bohm, D. (1959) "Significance of Electromagnetic Potentials in the Quantum Theory," *Physical Review* 115, 485–91.

Aitchison, I. J. R. and Hey, A. J. G. (2003) *Gauge Theories in Particle Physics, Volume II: Non-Abelian Gauge Theories*. Bristol: Institute of Physics Publishing.

Albert, D. (1992) *Quantum Mechanics and Experience*. Cambridge, Mass.: Harvard University Press.

Anandan, J. (1983) "Holonomy Groups in Gravity and Gauge Fields." In: G. Denardo and H. D. Doebner (eds.) *Proceedings of the Conference on Differential Geometric Methods in Physics, Trieste 1981*. Singapore: World Scientific.

Anandan, J. (1993) "Remarks Concerning the Geometries of Gravity and Gauge Fields." In: B. Hu, M. Ryan, and C. Vishveshwara (eds.) *Directions in General Relativity*. Cambridge: Cambridge University Press.

Ashtekar, A. and Isham, C. (1992) "Inequivalent Observable Algebras: Another Ambiguity in Field Quantisation," *Physics Letters B* 274, 393–8.

Ashtekar, A. and Rovelli, C. (1992) "A Loop Representation for the Quantum Maxwell Field," *Classical and Quantum Gravity* 9, 1121–50.

Auyang, S. Y. (1995) *How is Quantum Field Theory Possible?* Oxford: Oxford University Press.

Barrett, J. W. (1991) "Holonomy and Path Structures in General Relativity and Yang–Mills Theory," *International Journal of Theoretical Physics* 30, 1171–1215.

Batterman, R. W. (2003) "Falling Cats, Parallel Parking and Polarized Light," *Studies in History and Philosophy of Modern Physics* 34, 527–57.

Bell, J. S. (1987) *Speakable and Unspeakable in Quantum Mechanics*. Cambridge: Cambridge University Press.

Belot, G. (1998) "Understanding Electromagnetism," *British Journal for the Philosophy of Science* 49, 531–55.

Berkowitz, J. and Hemmo, M. (2006) "A Modal Interpretation of Quantum Mechanics in Terms of Relational Properties." In: W. Demopoulos and I. Pitowsky (eds.) *Physical Theory and Its Interpretation*. New York: Springer.

Bernstein, H. and Phillips, T. (1981) "Fibre Bundles and Quantum Theory," *Scientific American* 245, pp.122–37.

Bertlmann, R. A. (1996) *Anomalies in Quantum Field Theory*. Oxford: Oxford University Press.

Bohm, D. (1952) "A Suggested Interpretation of Quantum Theory in Terms of 'Hidden Variables,' I and II." *Physical Review* 85, 166–183.

Brown, H. (1999) "Aspects of Objectivity in Quantum Mechanics." In: J. Butterfield and C. Pagonis (eds.) *From Physics to Philosophy* pp. 45–70, Cambridge: Cambridge University Press.

Brown, H. and Brading, C. (2004) "Are Gauge Symmetry Transformations Observable?" *British Journal for the Philosophy of Science* 55, 645–665.

Brown, H. and Harré, R. (eds.) (1988) *Philosophical Foundations of Quantum Field Theory*. Oxford: Oxford University Press.

Bub, J. (1997) *Interpreting the Quantum World*. Cambridge: Cambridge University Press.

Butterfield, J. (1989) "The Hole Truth," *British Journal for the Philosophy of Science* 40, 1–28.

Butterfield, J. (2006) "Against Pointillisme about Mechanics," *British Journal for the Philosophy of Science* 57, 709–53.

Cartwright, N. (1983) *How the Laws of Physics Lie*. Oxford: Oxford University Press.

Chambers, R. G. (1960) "Shift of an Electron Interference Pattern by Enclosed Magnetic Flux," *Physical Review Letters* 5, 3–5.

Chan, Hong-Mo and Tsou, Sheung Tsun (1993) *Some Elementary Gauge Theory Concepts*. Singapore: World Scientific.

Clifton, R. (2000) "The Modal Interpretation of Algebraic Quantum Field Theory," *Physics Letters A* 271, 167–77 (reprinted in Clifton (2004)).

Clifton, R. (2004) *Quantum Entanglements: Selected Papers of Rob Clifton*, Butterfield, J. and Halvorson, H. (eds.). Oxford: Oxford University Press.

Clifton, R. and Halvorson, H. (2001) "Are Rindler Quanta Real? Inequivalent Particle Concepts in Quantum Field Theory," *British Journal for the Philosophy of Science* 52, 417–70.

Dieks, D. (2000) "Consistent Histories and Relativistic Invariance in the Modal Interpretation of Quantum Mechanics," *Physics Letters A* 265, 317–25.

Dirac, P. A. M. (1931) "Quantised Singularities in the Electromagnetic Field," *Proceedings of the Royal Society of London*, A133, 60–72.

Dirac, P. A. M. (1964) *Lectures on Quantum Mechanics*. New York: Yeshiva University Press.

Earman, J. (2003) "Rough Guide to Spontaneous Symmetry Breaking." In: Brading, K. and Castellani, E. (eds.) *Symmetries in Physics*. Cambridge: Cambridge University Press.

Earman, J. (2004) "Curie's Principle and Spontaneous Symmetry Breaking," *International Studies in the Philosophy of Science* 18, 173–198.

Earman, J. and Fraser, D. (2006) "Haag's Theorem and Its Implications for the Foundations of Quantum Field Theory," *Erkenntnis* 64, 305–44.

Earman, J. and Norton, J. (1987) "What Price Spacetime Substantivalism? The Hole Story," *British Journal for the Philosophy of Science* 38, 515–25.

Earman, J. and Ruetsche, L. (2005) "Relativistic Invariance and Modal Interpretations," *Philosophy of Science* 72, 557–83.

Einstein, A. (1948) "Quanten-Mechanik und Wirklichkeit," *Dialectica* 2, 321–4.

Elitzur, S. (1975) "Impossibility of Spontaneously Breaking Local Symmetries," *Physical Review D*, 3978–82.

Essén, H. (1996) "The Darwin Magnetic Interaction Energy and its Macroscopic Consequences," *Physical Review E*, 5228–39.

Everett, H., III (1957) " 'Relative State' Formulation of Quantum Mechanics," *Reviews of Modern Physics* 29, 454–62.

Feynman, R. P. (1965) in *Nobel Lectures, Physics 1963–1970*. Amsterdam: Elsevier, 1972.

Feynman, R. P. and Hibbs, A. R. (1965) *Quantum Mechanics and Path Integrals*. New York: McGraw-Hill.

Feynman, R. P., Leighton, R. B., and Sands, M. L. (1965a) *The Feynman Lectures on Physics*. Reading, Mass.: Addison-Wesley.

Fort, H. and Gambini, R. (1991) "Lattice QED with Light Fermions in the P Representation," *Physical Review D* 44, 1257–62.

Fort, H. and Gambini, R. (2000) "U(1) Puzzle and the Strong CP Problem from a Holonomy Perspective," *International Journal of Theoretical Physics* 39, 341–49.

Friedman, M. (1983) *Foundations of Space-time Theories*. Princeton: Princeton University Press.

Galileo Galilei (1632/1967) *Dialogue Concerning the Two Chief World Systems*, trans. Stillman Drake, 2nd revised edition. Berkeley: University of California Press, 1967.

Gambini, R. and Pullin, J. (1996) *Loops, Knots, Gauge Theories and Quantum Gravity*. Cambridge: Cambridge University Press.

Gambini, R. and Setaro, L. (1995) "SU(2) QCD in the Path Representation: General Formalism and Mandelstam Identities," *Nuclear Physics B* 448, 67–92.

Gambini, R. and Trias, A. (1981) "Geometrical Origin of Gauge Theories," *Physical Review D* 23, 553–5.

Gambini, R. and Trias, A. (1983) "Loop-Space Quantum Formulation of Free Electromagnetism," *Lettere al Nuovo Cimento* 38, 497–502.

Gell-Mann, M. and Hartle, J. (1993) "Classical Equations for Quantum Systems," *Physical Review D* 47, 3345–82.

Georgi, H. (1993) "Effective Field Theory," *Annual Review of Nuclear and Particle Science* 43, 209–52.

Ghirardi, G. (2004) *Sneaking a Look at God's Cards: Unraveling the Mysteries of Quantum Mechanics*. Princeton: Princeton University Press.

Ghirardi, G. C., Rimini, A., and Weber, T. (1986) "Unified Dynamics for Microscopic and Macroscopic Systems," *Physical Review D* 34, 470–91.

Giles, R. (1981) "Reconstruction of Gauge Potentials from Wilson loops," *Physical Review D* 24, 2160–68.

Giulini, D. (1995) "Asymptotic Symmetry Groups of Long-Ranged Gauge Configurations," *Modern Physics Letters* 10, 2059–70.

Giulini, D. (2003) "Superselection Rules and Symmetries." In: Joos, Zeh, Kieffer, Giulini, Kupsch and Stamatescu, *Decoherence and the Appearance of a Classical World in Quantum Theory*, 2nd edition. New York: Springer.

Goldstein, S. (1996) M. Daumer, D. Durr, S. Goldstein, and N. Zanghi, "Naive Realism about Operators," *Erkenntnis* 45, 379–97.

Greaves, H. (2004) "Understanding Deutsch's Probability in a Deterministic Multiverse," *Studies in History and Philosophy of Modern Physics* 35, 423–56.

Greaves, H. (2007) "On the Everettian Epistemic Problem," *Studies in History and Philosophy of Modern Physics* 38, 120–52.

Gribov, V. N. (1977) "Instability of Non-Abelian Gauge Theories and Impossibility of Choice of Coulomb Gauge," translated from a lecture at the 12th Winter School of the Leningrad Nuclear Physics Institute, *SLAC-TRANS-0176*.

Hartmann, S. (2001) "Effective Field Theories, Reductionism and Scientific Explanation," *Studies in History and Philosophy of Modern Physics* 32, 267–304.

Healey, R. A. (1989) *The Philosophy of Quantum Mechanics: an Interactive Interpretation*. Cambridge: Cambridge University Press.

Healey, R. A. (1991) "Holism and Nonseparability," *Journal of Philosophy* 88, 393–421.

Healey, R. A. (1994) "Nonseparable Processes and Causal Explanation," *Studies in History and Philosophy of Science* 25, 337–74.

Healey, R. A. (1995) "Substance, Modality and Spacetime," *Erkenntnis* 42, 287–316.

Healey, R. A. (1997) "Nonlocality and the Aharonov–Bohm Effect," *Philosophy of Science* 64, 18–41.

Healey, R. (1998) "The Meaning of Quantum Theory." In: *The Great Ideas Today*, 74–105. Chicago: Encyclopedia Britannica, Inc.

Healey, R. A. (2004) "Holism and Nonseparability in Physics," in the *Stanford Electronic Encyclopedia of Philosophy*. <http://Plato.Stanford.edu/entries/physics-holism>.

Healey, R. A. (2004a) "Gauge Theories and Holisms," *Studies in History and Philosophy of Modern Physics* 35, 643–66.

Healey, R. A. (2006) "Symmetry and the Scope of Scientific Realism." In: W. Demopoulos and I. Pitowsky (eds.) *Physical Theory and Its Interpretation*. New York: Springer.

Holland, P. R. (1993) *The Quantum Theory of Motion*. Cambridge: Cambridge University Press.

Howard, D. and Stachel, J. (eds.) (1989) *Einstein and the History of General Relativity: Einstein Studies*, Volume 1 Boston: Birkhäuser.

Huggett, N. (2000) "Philosophical Foundations of Quantum Field Theory," *British Journal for the Philosophy of Science* 51, 617–37.

Jackson, J. D. (1999) *Classical Electrodynamics* 3rd edition. New York: Wiley.

Kay, B. and Wald, R. M. (1991) "Theorems on the Uniqueness and Thermal Properties of Stationary, Nonsingular, Quasifree States on Spacetimes with a Bifurcate Killing Horizon," *Physics Reports* 207, 49–136.

Kuhlmann, M., Lyre, H. and Wayne, A. (eds.) (2002) *Ontological Aspects of Quantum Field Theory*. Singapore: World Scientific.

Kuhn, T. S. (1970) *The Structure of Scientific Revolutions*, 2nd edition. Chicago: University of Chicago Press.

Lange, M. (2002) *An Introduction to the Philosophy of Physics: Locality, Fields, Energy and Mass*. Oxford: Blackwell.

Leeds, S. (1999) "Gauges: Aharonov, Bohm, Yang, Healey," *Philosophy of Science* 66, 606–627.

Lewis, D. K. (1970) "How to Define Theoretical Terms," *Journal of Philosophy* 67, 427–446.

Lewis, D. K. (1983) *Philosophical Papers*, Volume I. Oxford: Oxford University Press.

Lewis, D. K. (1986) *Philosophical Papers*, Volume II. Oxford: Oxford University Press.

Lewis, D. K. (2001/2007) "Ramseyan Humility," to appear in D. Braddon-Mitchell and R. Nola (eds.) *Naturalistic Analysis*. Cambridge, Mass.: MIT Press.

Lewis, P. (2007) "Uncertainty and Probability for Branching Selves," *Studies in History and Philosophy of Modern Physics* 38, 1–14.

Lewis, P. (2007a) "Quantum Sleeping Beauty", *Analysis* 67, 59–65.

Lyre, H. (2004) "Holism and Structuralism in U(1) Gauge Theory," *Studies in History and Philosophy of Modern Physics* 35, 643–670.

Mandelstam, S. (1962) "Quantum Electrodynamics without Potentials," *Annals of Physics* 19, 1–24.

Manton, N. S. (1983) "Topology in the Weinberg–Salam Theory," *Physical Review D* 28, 2019–2026.

Marder, L. (1959) "Flat Space-Times with Gravitational Fields," *Proceedings of the Royal Society of London*, A252, 45–50.

Martin, C. A. (2002) "Gauge Principles, Gauge Arguments and the Logic of Nature," *Philosophy of Science* 69, S221–S234.

Mattingly, J. (2006) "Which Gauge Matters," *Studies in History and Philosophy of Modern Physics* 37, 243–262.

Mattingly, J. (in press) "Classical Fields and Quantum Time-Evolution in the Aharonov-Bohm Effect," forthcoming in *Studies in History and Philosophy of Modern Physics*.

Maudlin, T. (1989) "The Essence of Spacetime." In: A. Fine and J. Leplin (eds.) *PSA 1988* Volume 2, 82–91.

Maudlin, T. (1994) *Quantum Non-Locality and Relativity*. Oxford: Blackwell.

Maudlin, T. (2007) *The Metaphysics Within Physics*. Oxford: Oxford University Press.

Maxwell, J. C. (1881) *An Elementary Treatise on Electricity*, ed. W. Garnett. Oxford: Clarendon Press.

Nakahara, M. (1990) *Geometry, Topology and Physics*. Bristol: Institute of Physics Publishing.

O'Raifeartaigh, L. (1979) "Hidden Gauge Symmetry," *Reports on Progress in Physics* 42, 159–223.

O'Raifeartaigh, L. (1997) *The Dawning of Gauge Theory*. Princeton: Princeton University Press.

Peshkin, M. and Tonomura, A. (1989) *The Aharonov–Bohm Effect*. New York: Springer.
Polyakov, A. N. (1974) "Particle Spectrum in Quantum Field Theory," *Soviet Physics JETP* 41, 988–95. (Russian original in *JETP Letters 20*, 194–5.)
Price, H. (2006) "Probability in the Everett World: Comments on Wallace and Greaves" <http://philsci-archive.pitt.edu/archive/00002719/>.
Quine, W. V. O. (1966) *The Ways of Paradox and Other Essays*. New York: Random House.
Redhead, M. L. G. (1983) "Quantum Field Theory for Philosophers." In: P. D. Asquith and T. Nickles (eds.) *PSA 1982* Volume 2, pp. 57–99. East Lansing, Mid.: Philosophy of Science Association.
Redhead, M. L. G. (1988) "A Philosopher Looks at Quantum Field Theory," in: Brown and Harré (1988), pp. 9–23.
Rovelli, C. (2004) *Quantum Gravity*. Cambridge: Cambridge University Press.
Rubakov, V. (2002) *Classical Theory of Gauge Fields*. Princeton: Princeton University Press.
Ruetsche, L. (2002) "Interpreting Quantum Field Theory," *Philosophy of Science* 69, 348–378.
Ryder, L. H. (1996) *Quantum Field Theory*, 2nd edition. (Cambridge: Cambridge University Press.
Singer, I. M. (1978) "Some Remarks on the Gribov Ambiguity," *Communications in Mathematical Physics* 60, 7–12.
Sklar, L. (1974) *Space, Time and Spacetime*. Berkeley: University of California Press.
Smolin, L. (2001) *Three Roads to Quantum Gravity*. New York: Basic Books.
Stein, H. (1967) "Newtonian Space-Time," *The Texas Quarterly* 10, 174–200.
Stein, H. (1970) "On the Notion of Field in Newton, Maxwell and Beyond." In: R. H. Stuewer (ed.) *Historical and Philosophical Perspectives of Science*, pp. 264–87. Minneapolis: University of Minnesota Press.
Struyve, W. and Westman, H. (2006) "A New Pilot-Wave Model for Quantum Field Theory," arXiv:quant-ph/0602229 v1 28 Feb 2006.
Suppes, P. (1957) *Introduction to Logic*. Princeton, N.J.: van Nostrand.
Teller, P. R. (1995) *An Interpretive Introduction to Quantum Field Theory*. Princeton: Princeton University Press.
Teller, P. R. (2000) "The Gauge Argument," *Philosophy of Science* 67, S466–S481.
t'Hooft, G. (1974) "Magnetic Monopoles in Unified Gauge Theories," *Nuclear Physics B* 79, 276–84.
t'Hooft, G. (ed.) (2005) *50 Years of Yang–Mills Theory*. Singapore: World Scientific.
Tonomura, A. *et al.* (1986) "Evidence for Aharonov–Bohm Effect with Magnetic Field Completely Shielded from Electrons," *Physical Review Letters* 56, 792–795.
Trautman, A. (1980) "Fiber Bundles, Gauge Fields and Gravitation." In: A. Held (ed.) *General Relativity and Gravitation*. New York: Plenum Press.
van Fraassen, B. C. (1980) *The Scientific Image*. Oxford: Clarendon Press.
van Fraassen, B. C. (1991) *Quantum Mechanics: an Empiricist View*. Oxford: Clarendon Press.

Wallace, D. (2002) "Worlds in the Everett Interpretation," *Studies in History and Philosophy of Modern Physics* 33, 637–661.

Wallace, D. (2005) "Quantum Probability from Subjective Likelihood: Improving on Deutsch's Proof of the Probability Rule." <http://philsci-archive.pitt.edu/archive/00002302/>.

Wallace, D. (2006) "Epistemology Quantized: Circumstances in which We Should Come to Believe in the Everett Interpretation." Studies in History and Philosophy of Modern Physics 38, 311–32.

Weinberg, S. (1992) *Dreams of a Final Theory*. New York: Pantheon.

Weingard, R. (1988) "Virtual Particles and the Interpretation of Quantum Field Theory," In: H. Brown and R. Harré (eds.) (1988).

Weyl, H. (1918) "Gravitation und Elektrizität," *Preussische Akademie der Wissenschaften Sitzungsberichte*, 465–478.

Weyl, H. (1929) "Elektron und Gravitation," *Zeitschrift für Physik* 56, 330–52.

Wightman, A. (1956) "Quantum Field Theory in Terms of Vacuum Expectation Values," *Physical Review* 101, 860–66.

Wu, T. T. and Yang, C. N. (1975) "Concept of Nonintegrable Phase Factors and Global Formulation of Gauge Fields," *Physical Review D* 12, 3845–57.

Wu, T. T. and Yang, C. N. (1975a) "Some Remarks about Unquantized non-Abelian Gauge Fields," *Physical Review D* 12, 3843–4.

Yang, C. N. and Mills, R. L. (1954) "Conservation of Isotopic Spin and Isotopic Spin Gauge Invariance," *Physical Review* 96, 191–5.

Index

action 248
　at a distance 31, 34, 44, 49, 54, 84, 220
　effective 146
　electromagnetic 19, 132, 164–5, 231
　ghost 146, 167–8
　gravitational 256
　group 11, 15, 17, 27, 29, 69, 234–5, 238–42, 246
　Klein-Gordon 161
　local, *see* local action
　quantum 183
　Yang-Mills 167, 175
　see also interaction
Aharonov 124 n.
Aharonov-Bohm effect
　action at a distance in 31, 34, 44, 49, 54, 84, 220
　attempted local accounts of 31–40
　electric analog 41, 157–9
　experimental demonstrations of 21–3, 35
　fiber bundle treatment of 26–31
　fringe shift in 22–3
　geometry and topology in 40–4
　gravitational analog 78–81
　indeterminism in 25–6, 30, 50, 56
　lessons for classical electromagnetism in 54–7
　locality in 44–54
　non-separability in 46–53, 78, 123–8
　unobservability of gauge in 25
Aitchison 160, 164
Albert 203 n.
algebra
　C^* 266–7
　Lie **241**
　Poisson 189 n.
　Weyl 266 n.
Ampère 230
Anandan 72, 79–80
anomalies 149, 182–3
Aristotle 226
Arntzenius 120 n.
Ashtekar 71, 189–91, 196, 198, 207–9, 226
automorphism 150, 181
　principal bundle 17, **239**

vector bundle 29
vertical bundle 17, 29, **239**
Weyl algebra 173
Auyang 162, 200, 203
　event ontology 200–3

Bargmann 190–1
Barrett 72
base point 70
base space, *see* space, base
Batterman 40–2
Bell 210, 278–9
Belot 30, 50, 124
Berkowitz 219, 279
Berkeley 211
Bernstein 99
Bertlmann 149, 183
Bohm 209
Bohmian
　mechanics 36, 52, 205, 209, 211, **274–6**, 278–9
　interpretation 52, 209–11, 274–5
　see also Aharonov-Bohm effect
Bohr 214, 273–4
boson 66, 146, 169, 206, 260
　gauge 207, 223 n., 225, 227
　Goldstone 170, 172
　Higgs 149
　vector 169, 175, 205, 207, 214
Brading 158–9
de Broglie wavelength 23
Brown 158–60, 164, 203 n.2
Bub 203 n.2, 219, 279
bundle, *see* fiber bundle
Butterfield 88, 124

Cartesian 123, 127
　coordinates 5, 153, 259, 273 n.
Cartwright 40
Cauchy sequence 257 n.
CCR's, *see* commutation relations, canonical
CERN 228
Chambers 21, 57

Chan 75
charge
  color 94, 166
  conjugation 167 n., 173, 197
  conservation 160, 166–7, 181–2, 203
  electric 19, 59, 75, 105–6, 110, 147, 165, 181, 217–8, 231–2, 249
  magnetic 75
  Noether 161–2, 169, 173, 217, 249
  operator 181
Chern-Simons number 178–180, 197
Clifton 224 n., 208, 211–12, 214, 219
color (of quarks) 94–5, 114, 169, 183
  see also charge, color
commutation relations 66, 132, 134, 194, 268–70, 273, 277–8
  canonical 132, 139–40, 144, 204, 251, **259**, 266
  Heisenberg 186–8, 259, 262–3, 266, 273, 275
  Lie algebra 59, 144, 148
  Weyl form 187–8, 204, 262–3
  equal time 132, 134–5, 140, 186–7, 265
connection 184
  affine 77, 85
  bundle 8, 12–13, 16–18, 20, 28, 30, 42, 79, 97, 100–4, 108, 112, 119, 234, 242
  Levi-Civita 79, 85, 87, 96
  linear 77, 80, 84, 87, 96
  one-form, see one-form, Lie-algebra valued
  and curvature 236–8, 243–5
  and covariant derivative 234–6, 245
  on a vector bundle 184, 192, 234–8, 245–6,
  on a principal bundle 96–8, 110, 112, 118, 162 n., 184, 192, 200–201, 240
constraint 136, 139–40, 140, 173, 253
  diffeomorphism 256
  function 37 n., 139, 181, **250**, 252–4
  Gauss 140, 181–2
  Hamiltonian 256
  first-class 138–41, 144, 173–4, 181, **253**, 254–6
  second-class 139–41, **253**
  primary 137, **253**
  secondary 137–8, **253**
  surface 139, 173–4, 253, 255–6
Copenhagen interpretation 53, 205, 212–15, 272–4, 276

Coriolis 169
Coulomb condition 76–7, **133**, 140, 143
Coulomb force 169
Coulomb gauge 39 n., 76, 133–5, 140, 143, 209–10
Coulomb's law 229–30
covariance
  gauge 25, 64, 108, 165, 245
  general 29
  Lorentz 6, 21, 54, 132–6, 168, 206, 231
covariant derivative, see derivative, covariant
Creary 40
curvature
  bundle 9, 13–14, 17, 28, 31, 42, 65–6, 75
  space-time 29, 78–9, 85, 88, 124
  of a vector bundle 66, 236–8
  of a principal bundle 75, 77–8, 84–5, 87–8, 96, 110, 162 n., 244–5
curve **70**

Darwin 37–9
Dehmelt 114
derivative
  covariant 15–16, 20, 64–6, 87, 99–100, 147–8, 163–5, 184
  on a vector bundle 234–7, 245–6
  on a principal bundle 243–4
  exterior 32, 43, 68, 244
  functional **140 n.**, 252
Dieks 219
Dirac
  and the Copenhagen interpretation 272, 276
  bracket 141
  constrained Hamiltonian formalism 135–9, 144, 181, 192, **248–56**
  equation 2
  field 18, 166, 202, 238
  ket 191
  monopole, see magnetic monopole, Dirac
  phase-factor 27, 49, 51, 56, 64, 73, 105–6
  spinor 196
  transformation theory 261–2

Earman 88, 96, 174, 219, 265
eigenvalue 80, 202, 269–70, 277 n., 279
  -eigenstate link 273 n.
  equation 232

eigenstate 177, 180, 191, 202, 206, 212, 268–70
Einstein 1–2, 7–8, 45, 54, 98, 126, 128, 226, 229, 235, 256
electromagnetism
  classical 3–7, 12–14, 54–7, 136–9, 229–32
electron 23–4, 35–8, 99–102, 116, 129, 157–8, 175, 203, 210, 232
Elitzur 174
empirical adequacy 35, 82, 112, 114
empiricism
  constructive 83, 114–18
EM properties view
  see new localized,
  no new **54**, 83–4
energy 36, 39, 79–80, 164, 175–9, 206, 224–6, 232, 251, 259–61, 269–70
  zero-point 206 n., 261, 269
epistemology 26, 98, 112–23, 198, 209, 277
  realist 49, 111, 117
  scientific 116
  voluntarist 83, 114
Essén 39 n.
ETCR's, *see* commutation relations, equal time
Euclidean 116, 152–3, 178, 236, 258, 260, 262–3
Euler-Lagrange equations 18–19, 132, 135–6, 147, 161, 165, 231, 248–53
Everett 53, 205, 215–18, 276–8
explanation 41–4, 114, 156, 166–7, 175
  causal 42, 44
  local 44–58, 113
  inference to the best 119–23

Fadeev 145–6
Faraday 4, 113, 129, 155–7, 226–8, 230
fermion 66, 145–6, 168, 175, 183, 187, 195–7, 210–11, 215, 265
Feynman 23, 40, 49, 51, 54–5, 73, 136, 141 n., 145, 203
fiber
  typical 9–12, 42, 66, 73, 76–7, 99, 192, **233**, 234, 237–9
  above a point 10, 12, 17, 20, 27–9, 72, 78, 100–1, 103, 107, 109, 200, **233**, 240, 245
fiber bundle 8
  associated 15–18, **239**
  frame 80–1, 84, 87

phase 99, 103
principal 9–11, **238–9**
tangent 80
trivial 10–12
  locally trivial **234**
vector 11, 239
field
  component 37–40
  current 34–9, 50, 52, 55, 88–9, 127
  electric 3–5, 21, 155, 189–91, 194, 209, 217, 230
  electromagnetic 6, 21, 28, 129
  gauge, *see* gauge field
  magnetic 4–5, 9, 13, 21, 23–6, 32–3, 49, 209, 229–30
  Maxwell 131, 139–41, 186–92, 196, 198, 206–7, 209–11
  strength 13, 27–30, 43–54, 62, 66, **84 n.**, 243, 245
  vector **241**
  Yang-Mills 143, 167–8, 185, 194, 199, 214, 243
flux, magnetic 23–4, 41, 44
Fock space 132, 134, 191, 198, 206–8, 211–13, 260, 270
  representation 132, **267–70**
force
  Coulomb 169
  electric 229
  fundamental 129, 234
  Lorentz 6, 21, 231
  lines of 226
  nuclear 94, 206
  strong 58, 94, 169
  weak 169
Fort 196, 198
Fourier expansion 132, 134, 190 n.3
Fourier transform 261
van Fraassen 83, 114, 203 n.3, 222
frame
  field 79, 87, 96–7
  inertial 6, 154, 186
Fraser 266
frequency, positive/negative **134–5**
Friedman 117 n.8
function
  constraint, *see* constraint function
  wave- 3, 7, 14–17, 24–39, 50–66, 86–8, 99–111, 127, 158–9, 177, 190 n.7, .8, 194–9, 202–5, 213, 231–2, 245, 251, 258–1, 275–6

functional
   differentiation **140 n.3**, 142, 252
   generating 142, 145
   integral 196
   wave- 140, 144, 184, 190 n.7, 196, 209

Galilean
   boost 152
   relativity 153–4
   ship 155–9
Galileo 151–2
Gambini 71, 139, 191–8
gauge
   argument 18, 66, 147, 149, 159–67, 169, 171
   axial 143, 168
   Coulomb, *see* Coulomb gauge
   Lorenz 39 n., 133, 135–6
   orbit 30 n., 76–7, 174, 252–6
   principle 159–64
gauge field 26, 66, 80, 84–7, 104, 113–14, 131, 142–9, 159–70, 183–4, 189–99, 220–3, 233–8
   strength 62, 64, 66, **84 n.**, 236–8, 246
   theory 7–8, 20, 55, 75, 141 n., 248
gauge invariance 8, 25–39, 49–56, 73–4, 105–9, 120, 133, 145, 160–171, 184, 191–7, 207–17, 231
gauge potential 41, 80, 84–5, 99, 102–4, 123, 200–1, 225
   electromagnetic 164, 181
   gravitational 77, 84–5, 96
   Yang-Mills 61, 66, 73–4, 84–5, 106, 109, 113, 143–7, 185, 196–8, 236–8, 243–4
gauge potential properties 85–7, 97, 105, 113
   localized 86–104, 112, 118–24, 209, 218–21
   non-localized 104–10, 112
gauge potential properties view
   see localized, no, non-localized
gauge theories 1–20
   classical 58–81
      non-Abelian Yang-Mills 58–77
      general relativity 77–81
   quantized Yang-Mills 129–48
gauge transformation 2–4, 6, 9, 13–15, 18–20, 25, 28–30
   "local" (of second kind) 2, **15**, 17–20,
   101, 149, 154–9, 172, 175–6, 181–5, 232
   "global" (of first kind) **15**, 18–19, 173, 176, 181–2
   "large" **176**, 175–82, 197–8
   "small" 175, **176**, 180–2
Gauss
   constraint 144, 181–2
   law 139–40, 144, 217, 230
Gelfand 172, 207 n., 208, 271
Gelfand-Naimark-Segal theorem **271**
Gell-Mann 217
general relativity 1–2, 29, 36, 54, 77–88, 96–8, 110, 125–6, 130, 150–4, 201, 207–8, 226, 256
Georgi 225
Ghirardi 53, 203
ghost 117–19
   field 145–6, 149, 167–9
Giles 74
Giulini 181–2
Gleason 279
global 15 n.7, 32, 160, 181, 186, 203, 214, 216
   gauge transformation, *see* gauge transformation, "global"
   section, *see* section, global
GNS representation 172–3, 214, **271**
GNS theorem, *see* Gelfand-Naimark-Segal theorem
God 67 n., 117 n.
Goldilocks 105
Goldstein 274
Goldstone 170–4
gravity 1–2, 36, 40, 54, 77–81, 85, 98, 114, 125–30, 150, 160, 208, 220–6
Greaves 277 n.
Gribov ambiguity 74–7, 143
Group
   Abelian **12**, 58
   compact 77
   non-Abelian 130
   structure 11, **233**
   gauge 72, 76, **233**
   hoop **72**
   Lie 59 n.2, 233
   Lorentz 79, 200
   Poincaré 77, 200
   U(1) **12**, 27
   SU(2) 59, 150

SU(3) 148
SU(5) 106
SU(N) 60, 143

Halvorson 208, 212–14
Hamilton's principle 136, 231, 248, 255
Hamiltonian 138, **250**, 251–2, 254
  constraint, *see* constraint, Hamiltonian
  Darwin 38
  density 137–8, 253–5
  equation of motion 133, 137–8, **250**
  formalism, constrained 248–256
  formulation 25, 40, 38, 77 n.1, 119, 131, 136–7, 143, 256, 259
  operator 7, 14–15, 23, 140, 144, 177–180, 206, 231–2, **251**, 260–1, 269
  system, constrained 1–2, 135–41, 173–4, 181–2
Harré 203 n.,
Hartle 217
Hartmann 223 n.
hath **194**, 196
Healey 46, 88, 91, 97, 124, 203
Heisenberg
  picture **217**
  see also commutation relations, Heisenberg
Hemmo 219, 279
Hey 160, 164
Hibbs 141 n.
Higgs
  mechanism 149, 170–5, 223 n.
  model 174
  boson 149
Hilbert space **257 n.**
Hodge decomposition theorem 32
  failure in pseudo-Riemannian spaces 33
hole argument 88, 96–8
holism 123–8
  physical property **125**
Holland 31, 52 n.
holonomy 14, 27–8, 69–74, 112, 120–1, 186, 190–4, 198, 211, 242–7
  loop, *see* hoop
  map, **242**
  non-Abelian 106–10
  of principal bundle 27, **242**
  of vector bundle 28, 73, 107–8, 192–3, **246**

property 41, 81, **110**, 111–12, 118–28, 185, 200, 205, 210–11, 218–21
holonomy interpretation 110–28, 210
  epistemological considerations 112–119
  metaphysical consequences of 123–8
  semantic considerations 122–3
  objections considered 119–22
t'Hooft 75
hoop 70–4, 112, 118, 122, 184, **186**, 190–5, 208, 210
  group *see* group, hoop
horizontal
  component 12, 244
  lift 12–13, 16, 27–9, 72, 78, 103, **241**-6
  subspace 13 n., 78, **240**-2
Howard 98
Hubble space telescope 116
Huggett 203 n.
Humean supervenience, *see* supervenience, Humean

indeterminism 25–6, 30, 50, 96, 173, 250–2, 272
instantons 178–9
instrumentalism 49
interactions
  electromagnetic 7–20, 45, 53, 80, 109–20, 130, 147, 166–9, 195, 232
  electroweak 1, 130, 160, 223
  gravitational 1, 78–81, 114, 130
  strong 1, 58, 130, 145, 160, 166–9, 197, 206, 211
  weak 58, 128–30, 169–170, 175, 211, 214, 225
intrinsic property 46–7, 85, 103–111, 124–5, 152–7, 179, 185
invariance, *see* symmetry
  Lorentz, *see* Lorentz invariance
  gauge, *see* gauge invariance
Isham 189–92, 198, 207–9

Jackson 34
Jacobi identity, 241 n.
Josephson 80
Jupiter 46–7

K meson 80
Kay 208
Kennedy 32

Klein-Gordon
    equation 19, 131–2, 234, 249, 268
    field 18, 131–2, 206, 233, 238, 266–7, 269–271
    Lagrangian density 18, 161, 131, 248
Kochen 279
Kuhlmann 203 n.
Kuhn 226–7

Lagrangian
    Darwin, see Darwin Lagrangian
    for electromagnetism 231
    formulation for interacting fields 146–8
    for a complex Klein-Gordon field 18, 161, 248
    for a real Klein-Gordon field 131
    for QCD 147
    for QED 147
    quark 148
    Yang-Mills 143
        in the presence of a source 167
Lange 36
Laplace's equation 32
Laplacian 134
Leeds 55, 119
Leeds's view 99–102
Legendre transformation 251, 256
Leibniz equivalence 97–8
Leighton 23, 49
Levi-Civita connection 79, 85, 87, 96
Lewis, D. 91–8, 124, 126
Lewis, P. 277 n.7
Lie algebra **241**
Lie group, see group, Lie
light cone 45 n.
local
    action 44–53
    interaction 23, 45, 200–2
    section, see section, local
    trivialization 10, 234
"local" 15
    see also gauge transformation, "local"
locality
    in Aharonov-Bohm effect 23, 30–1, 33, 44–54
    relativistic 36, 45–53, 127
    synchronic 124 n.
localized gauge potential properties
    view 85–104

action on quantum particles 86–7, 99, 102, 104
epistemological problems for 118–19, 122–3
semantic difficulties of 89–95, 98, 122–3
localized properties 54, 98
    see also new localized EM properties view,
    non-localized properties 52, 110–12, 127, 211, 220
loop **70–1**
    representation 184–99
    supervenience, see supervenience, loop
    transform 190–9, 208
    Wilson, see Wilson loop
    /path representation 195–8
Lorentz
    boost 154–7
    force 6, 21, 231
    invariance 143, 154, 164, 167, 210, 278
    transformation 6, 133 n.
Lorenz 133
    condition 133
    gauge 39 n., 133
        quantization 135–6
Lyre 54, 203 n.2

Mach 91
magnetic monopoles 8, 12, 16, 74–5, 230
    Dirac 75
    Polyakov-'t Hooft 75
Mandelstam 55 n., 184
    identities 74, 120, 195
manifold 9–10, 43, 71–3, 77, 105, 201
    base 78, 201, 234
    space-time 8–9, 12, 15–17, 29, 54–6, 66–7, 70, 75–7, 85–9, 96–7, 104–5, 111–12, 118, 256
    differentiable 150, **233**, 234, 238–243, 256
Manton 178–9
Marder 79
    space-time 80, 85, 88
Martin 160
matrix
    adjoint **59 n.1**
    Hermitian **59 n.1**
    trace of **59 n.1**
    unitary 59
    see also Pauli spin matrices

matter 12, 18–20, 40, 66–7, 74, 124, 146–8, 159–69, 181–2, 196, 200–2, 211–15, 233–8, 245
Mattingly 34–40, 44, 55, 88, 127
Maudlin 55, 88, 94, 109 n., 120 n., 122 n., 203 n.3
Maudlin's interpretation 102–4
Maxwell 4, 226–30
  equations 3–8, 32–4, 84, 129–39, 164–6, **229–32**
  field 131, 139–42, 147, 184–91, 196, 198, 206–11
  theory 4–7, 129, 208–9
  -Faraday tensor 6, **230**
Mercury 150–3
van der Meer 227
metaphysics 41, 54, 102–4, 123–8, 203, 210, 215
Millikan 40, 114
Mills 2, 8, 58, 130, 162
  see also Yang-Mills
Minkowski
  Fock space 212–13
  metric 77–87, 135
  quanta 208–14
  representation 213
  space-time 8, 22, 178 n., 150, 186
Möbius strip 10–12
modal interpretations 218–19, 278–9
model
  Bohmian, of spin
  Higgs 174
  of a theory 8, 70, 75, 78, 88, 96–7, 127, **150**, 152–9, 170, 175–80, 212–24
  Schwarzschild 150, 153
Model, Standard, see Standard Model
momentum
  representation, see representation, momentum

Naimark 172, 207 n., 208, 271
Nakahara 32, 233 n., 245
new localized EM properties view 55, 91, 99, 102
  epistemological problems for 55–6
new non-localized EM properties view 56–7, 102
Newton 116, 117 n.7
  theory of mechanics 7, 113–17, 121–8, 152–4, 226
  theory of gravity 36, 54

Newtonian space-time 8
no new EM properties view 54, 83–4
  and action at a distance 54–5
no gauge potential properties view 83–5, 88, 112
  Wu-Yang objection to 84, 113
  general relativity objection to 84–5
non-localized gauge potential properties view 104–110, 112
Nobel
  prize 227–8
  lecture 203
Noether
  charge 161–2, 165–6, 173, 217, 249
  current 119, 165, 217
Noether's theorems 183
  first theorem 119, 173, 181–2, 249–50
  second theorem 173–5, 250–52
non-local 23–5, 124, 127–8, 276
non-localized properties 52–7, 110, 112, 124 n., 127, 183, 200, 205–14, 220–21
  see also new non-localized EM properties view,
  non-localized gauge potential properties view
non-separability 46, 52–3, 78, **124**-8, 169, 210, 218–21, 225
  strong **124**-26
Norton 88, 96, 98
von Neumann 262, 272, 276
  see also Stone-von Neumann theorem

observation 3, 91–9, 112–19, 151–62, 215, 228
one-form
  field 28, 43, 87
  harmonic 32–5, 43
  Lie-algebra valued 12–13, 28, 54–5, 66–7, 76–8, 84, 86, 112, 184 n., **240–4**
ontology 40, 200–6, 210–11, 218, 221–8, 274
O'Raifeartaigh 160–6

parallel-transport 61, 79, **245**
particle
  charged 7, 14, 18, 20, 28, 39–56, 79–81, 90, 108, 220, 233, 259 n.3
  classical 36, 45, 86, 88, 91, 99, 118, 204, 250, 252, 266
  elementary 1, 18, 131, 172, 195, 223

294   INDEX

particle (cont.)
  interpretations 205–9
  neutral 86
  non-relativistic 205, 258–262
  point 36, 53, 94, 250, 252, 274
  quantum 14–17, 25–36, 44, 51–4, 66, 78–81, 86–8, 99–105, 110–12, 127, 261–5
  quantum ⎯⎯ theory 185, 199, 204, 266, 274
  spinless 18, 80, 233, 261–3, 275
patchwork principle 126–7
path 70, 141, 184, 194
  integral 76, 141–2, 145–6, 168, 183
path-ordered exponential 63
Pauli spin matrices 60, 175
Peshkin 22, 52, 102
phase 12, 17, 21, 27, 34–8, 50–1, 56, 61–3, 78–81, 86–8, 100, 105, 108–9, 119, 201–2, 257
  bundle 99–103, 162,
  difference 24–5, 80, 100, 108–10, 127, 158
    properties 110, 127
  factor 27, 51, 101, 105, 119
    see also Dirac phase factor
  generalized 58, 60–6, 106–7, 110, 159–66, 201
  gravitational 80
  shift in Aharonov-Bohm effect 21–5, 38–9, 78, 80, 86, 100, 157 n.
  space 30 n., 139, 173, 181, 188, 189 n., **252**-3, 257
    reduced 30 n., 174, **256**, 266
  transformation 3, 58–9, 64–6, 159, 170, 181
    constant **15**, 19, 161–2, 167, 171, 181–2, 234
    variable **15**, 17–20, 99, 158, 160, 162–3, 166
  tropes 104
Phillips 99
photon 7, 130, 146–7, 169, 189–90, 191 n.10, 199, 205, 207, 211, 214
  virtual 40
pi meson 95, 169, 183
Planck's constant 14
Poincaré group 77, 200
Poincaré's theorem 43
Poisson
  bracket 189 n., **251**

  algebra, see algebra, Poisson
Polyakov 75
Popov 145–6
positivism 49, 83, 91
potential
  electric scalar 3–5, 54–6, 157, 180, **229**
  electromagnetic four-vector **6**, 16, 18, 230
  gauge, see gauge potential
  magnetic vector 4–5, 23–5, 32, 48–52, 102, 119, 156, 186–9, **229–30**
  transformation 4–6, 157, 180
  variable 5–6, 8, 15, 17–18, 186, 230–1
Price 277 n.7
properties
  holonomy, see property, holonomy
  intrinsic, see intrinsic property
  localized, see localized properties
  non-local 124 n., see also non-localized properties
  phase-difference 110, 127
  qualitative 47
Pullin 71, 139, 191–6

quantization of classical fields
  canonical 131–3
  Dirac's method, see Dirac, constrained Hamiltonian formalism
  in Coulomb gauge 133–5
  in Lorenz gauge 135–6
  path-integral 141–2, 145–6
  loop quantization 189–95
quantization of Yang-Mills fields 129–48
  canonical quantization 143–4
  path-integral quantization 145–6
  loop quantization 192–5
quantum mechanics
  interpretations 272–9
  representations 257–4
quantum chromodynamics 1, 147–8, 166–7, 169, 196–7
quantum electrodynamics 7, 40, 99, 129–30, 147, 166, 196, 210, 221–5
quantum field theory 129–48
  algebraic 265–71
  anomalies in 182–3
  as an effective theory 223–6
  ghost fields in 167–9
  non-separability in 221
  problems of interpreting 203–19

INDEX 295

significance of loop representations in 185
spontaneous symmetry-breaking in 169–75
Quine 200

Ramsey sentence 91–3
realization 91
  formula **92**
  unique 91–6
Redhead 203 n.2
reference 88–92, 221
Reichenbach 83
relation
  composition 125
  external 103, 124, 202
  extrinsic 103–4, 156
  intrinsic 125, 154, 156–9
  order 111
  part-whole 125
relationism 111
relativistic covariance, *see* covariance, Lorentz
relativity, *see* general relativity, special relativity
  principle 151–4
renormalizability 166–7, 169–70, 183, 223–5
representation
  Bargmann 190–1
  connection 190–9, 210
  faithful
    of a relational structure 89
    of the Weyl relations 204 n.
    of a C*-algebra **267**
  fundamental 28
  fiber bundle 66, 72–3
  Fock 132–5, 189–99, 203–9, 211–14, **267–71**
  GNS 172–3, 214, **271**
  group 70–4, 238
  inequivalent 173, 189, 198, 204–12, 267, 271
  loop 184–99
  loop/path representation 195–8
  matrix 28, 106, 192
  Minkowski 213
  momentum 199, 214, **260**-2, 273
  number 260, 265–7
  position 30, 99–100, 211, 214, 259–61, 273–5
  quantum 257–64
  irreducible 187, 263, **267**, 271 n.
  Rindler 213
  Schrödinger 213, 259, 273–4
  unitary 189–90, 263–4
  well-behaved 204, 264
  of a C*-algebra 267
  of a situation 8, 76, 78, 84, 94, 98, 126, 150, 153–5, 158–9, 179
  of a Weyl algebra 172, 189–90, 207–8, 211–12, 266-**7**, 270–1
  of the CCR's 187, 204, 213
rest, state of 113–21, 151
Riemannian curvature 29, 79
Riemannian metric 3, 33–5
Rimini 53
Rindler
  Fock space 212–13
  number operator 213
  quanta 208, 212–14
  representation 213
Rothschild lecture 227
Rovelli 190–6, 207, 220 n., 226
Rubakov 178–9
Rubbia 227
Ruetsche 203 n., 219, 265
Rutherford 114
Ryder 160–2

Salam, *see* Weinberg-Salam theory
Sands 23, 49
Saturn 115
Saunders 216 n.
Schrödinger
  equation **14**
    time-independent 232
  picture 216 n.
  representation 213–4, **259**, 273–4
Schwarzschild model 150, 153
Schwinger 183
section (of bundle)
  global 10, 12, **233**
  local 10, **233**
Segal 172, 207 n., 208, 271
semantic difficulties 89–95, 98, 122–3
semi-classical 178–9
separability **46**-57
  weak **46**, 48, 51–2
Setaro 196
Simons, *see* Chern-Simons number
Singer 75–7, 143

Sirius 114
situation 8, 27, 76, 150–63, 175–80
Sklar 117 n.
Smolin 220 n.
smooth **70**
  piecewise **70–1**
solar system 47, 150
solder-form 77–81, 84, 96
space
  base 8–13, **233**
  configuration 30, 188, 256–**7**, 262, 274–6
  dual 237, 240
  Euclidean, *see* Euclidean
  Fock, *see* Fock space
  Hilbert, *see* Hilbert space
  phase, *see* phase
  pseudo-Riemannian 35
  quotient 30 n., 239, 256
  state 204, 233, 257–8
  tangent 13 n., 77–8, 240–1
  total 10–15, **233**
  vector 10–15, *see also* Hilbert space
space-like
  curve 127
  hyperplane 266
  hypersurface 36, 50–6, 186, 249, 253
  loop 127
space-time
  curvature, *see* curvature, space-time
  manifold, *see* manifold, space-time
  Marder, *see* Marder space-time
  metric 2–3, 33, 77–87, 96–7, 124, 135
  Minkowski, *see* Minkowski space-time
  Newtonian 8
  relationism, 111
  substantivalism 97, 111
special relativity 8, 44, 46, 152–7, 162, 179, 186, 222
Specker 279
spinor **60 n.**
Stachel 98
Standard Model 1–2, 7–8, 20, 58, 66, 73 n., 159–60, 172, 200–3, 220–7
state
  mixed **276–7 n.6**
  -property link **272–3**
  pure **276–7 n.6**
  quasi-free 208

relative-state formulation 276
  quantum 101, 184, 215, 218, 272–4, 278–9
  -vector 202, 251, 272–4, 276–7
Stein 117 n.8, 203
Stokes's theorem 43
Stone's theorem 263 n.5
Stone-von Neumann theorem 187–9, 204, 213, 257, 262–**3**, 265–7
strong CP problem 167 n., 185, **197–8**
structure
  constant **59**, 148, 241
  surplus 8, 21, 30, 54, 84
Struyve 210–11
Sun 47, 150, 153
supervenience
  Humean **124**-8
  loop 122
Suppes 91
surplus structure, *see* structure, surplus
symmetry 91–8, **150**, 159, 170, 182
  breaking 75, 169–75, 176
  continuous 182, 234, 249–50
  CP- 167 n., 197
  empirical **152**, 150–9, 162, 167, 174–185
    strong **152-4**
    formal 8, 119, 149, **153**-6, 160, 168, 176, 181–3, 185, 209
  gauge 8, 25–9, 34, 49, 89–91, 95–104, 119–22, 149–83, 185, 209, 221–3
  isospin 58, 109
  theoretical **153**, 152–9, 174–8
  transformation 249–50
  of a Weyl algebra 173
symplectic form **263 n.6**, 266

tangent space, *see* space, tangent
Teller 164, 202–6, 212, 222–3
theoretical terms
  Lewis's definition of 91–2
  problems defining 91–9
theory
  effective 223–6
  gauge, *see* gauge theory
  quantized 2, 18–19, 66–7, 73 n., 76, 99, 119–20
    see also quantization of classical fields, quantization of Yang-Mills fields

quantum, *see* quantum mechanics, quantum chromodynamics, quantum electrodynamics
quantum field, *see* quantum field theory
relativity, *see* general relativity, special relativity
Yang-Mills 2, 120–6, 154, 182
   classical 74, 85, 110, 119, 124–6, 185, 198
   non-Abelian 58–77, 84–6, 147, 110, 185, 207–9, 217, 223
   quantized 2, 197, 200–19
     see also quantization of Yang-Mills fields
Tonomura 21, 52, 57, 102
topology 8, 33, 73–7, 104–5, 143, 178–9
   in Aharonov-Bohm effect 40–4
trace, *see* matrix
transformation
   conjugacy 110 n. 243
   gauge, *see* gauge transformation
   Legendre, *see* Legendre transformation
   Lorentz, *see* Lorentz transformation
   phase, *see* phase transformation
   potential, *see* potential transformation
   similarity 73–4, 109, 192–3, 247
   symmetry 249–50
   unitary 59, 107–8, 173, 199, 262, 273
Trautman 8–9, 13–14, 17–18, 29, 64, 69, 77–8, 101, 136, 243
tree **71–2**, 120–1, 186, 193–4, 247
Trias 191
trope 103–4
Tsou 75
two-form 32, 238, **243**
   Lie-algebra valued 13, 28–9, 54, 66–8, **243–4**

unification 129, 160, 166, 229
unitary equivalence **262**
Unruh effect 212–15

vacuum 33, 42, 45
   classical 175–6, 179–80
   degenerate 170, 173–80
   expectation value 142, 170
   fluctuations 212
   quantum 173, 175–6

   state 136, 142, 173–5, 197, 212–14, 269–70
   2 (theta) 149, 167, 175–82, 185, 197–8
   -vacuum transition amplitude, *see* functional, generating
van Fraassen 83, 114, 203 n.3, 222
vector field
   fundamental **241**
   harmonic 32–3
vector field, *see* field, vector
vector space, *see* space, vector
vertical subspace **239–40**
vertical automorphism, *see* automorphism, vertical
von Neumann 187, 257, 262, 272, 276
   see also Stone-von Neumann theorem

Wald 208
Wallace 216, 277 n.2
wave-function, *see* function, wave-
Wayne 203 n.2
Weber 53
Weinberg 159, 227 n.
Weinberg-Salam theory 183, 225
Weingard 168
Westman 210–11
Weyl 4, 160, 162–3
   1918 gauge theory 2–3
   1929 gauge theory 3, 160, 162
   use of term 'eich' 2
Weyl algebra, *see* algebra Weyl
Weyl form of commutation relations, *see* commutation relations
Weyl operator 188
Wightman 142
Wilson loop 73, 192–3
Wu 8, 26–32, 51, 64, 74–5, 84, 105, 113

Yang 2, 8, 58, 130, 162, *see also* Wu
Yang-Mills
   action, *see* action, Yang-Mills
   field, *see* field, Yang-Mills
   field strength 84 n., 243
   gauge potential, *see* gauge potential, Yang-Mills
   theory, *see* theory, Yang-Mills Abelian
   Lagrangian, *see* Lagrangian, Yang-Mills

Printed in Great Britain
by Amazon

77561078R00181